Mitochondria

ADVANCES IN BIOCHEMISTRY IN HEALTH AND DISEASE

Series Editor: Naranjan S. Dhalla, PhD, MD (Hon), DSc (Hon) Winnipeg, Manitoba Canada

Editorial Board: A. Angel, Toronto, Canada; I. M. C. Dixon, Winnipeg, Canada; L. A. Kirshenbaum, Winnipeg, Canada; Dennis B. McNamara, New Orleans, Louisiana; M. A. Q. Siddiqui, Brooklyn, New York; A. K. Srivastava, Montreal, Canada

Volume 1: S. K. Cheema (ed.), *Biochemistry of Atherosclerosis*

Volume 2: S. W. Schaffer and M-Saadeh Suleiman (eds.), *Mitochondria: The Dynamic Organelle*

Mitochondria

The Dynamic Organelle

Edited by

Stephen W. Schaffer
University of South Alabama, Mobile, AL

and

M-Saadeh Suleiman
Bristol Royal Infirmary, Bristol, United Kingdom

Stephen W. Schaffer
Department of Pharmacology
University of South Alabama
Mobile, AL 36688-0001
USA
swschaffer@jaguar1.usothal.edu

M-Saadeh Suleiman
Bristol Royal Infirmary
Bristol BS2 8HW
UNITED KINGDOM
M.S.Suleiman@bristol.ac.uk

ISBN-13: 978-0-387-69944-8   e-ISBN-13: 978-0-387-69945-5

Library of Congress Control Number: 2007921199

© 2007 Springer Science+Business Media, LLC
All rights reserved. This work may not be translated or copied in whole or in part without the written permission of the publisher (Springer Science+Business Media, LLC, 233 Spring Street, New York, NY 10013, USA), except for brief excerpts in connection with reviews or scholarly analysis. Use in connection with any form of information storage and retrieval, electronic adaptation, computer software, or by similar or dissimilar methodology now known or hereafter developed is forbidden.
The use in this publication of trade names, trademarks, service marks, and similar terms, even if they are not identified as such, is not to be taken as an expression of opinion as to whether or not they are subject to proprietary rights.

Printed on acid-free paper.

9 8 7 6 5 4 3 2 1

springer.com

Preface

The term mitochondrion is derived from Latin, with mitos meaning thread and chondrion meaning granules. Indeed, under the light microscope, mitochondria often appear as rods or granules within the cytoplasm. For decades after initial visualization of mitochondria by light microscopy, mitochondrial function remained clouded. However, with the development of differential centrifugation and electron microscopy, it was discovered that a chief function of the mitochondria was the generation of ATP for the remainder of the cell. For many years, the energy generating function of the mitochondria was considered the primary, if not the sole function of the mitochondria. During that period, investigators attempted to obtain information on the mechanism of ATP synthesis and the regulation of electron transport. In the first chapter of the book, Dr. Hassinen summarizes those studies, providing clear pictures on the transformation of reducing equivalents into a proton gradient and the mechanism by which the F_1F_0 ATPase utilizes the proton gradient to generate ATP. He also summarizes the key regulatory steps of the citric acid cycle, which is the major source of reducing equivalents for the electron transport chain. In the heart, most of the carbon that feeds into the citric acid cycle is derived from fatty acid metabolism. Although fatty acid utilization provides most of the ATP for contraction, a proper balance must be maintained between the utilization of fatty acids and that of glucose. In the second chapter, Drs. Folmes and Lopaschuk discuss the interrelationship between carbohydrate and fatty acid metabolism and how the interrelationship is affected in various disease states. Correction of metabolic dysregulation is introduced as a novel approach toward the treatment of heart disease. In the third chapter, Drs. Sugden and Holmes describe the profound influence of the nuclear regulators, known as the peroxisome proliferator-activated receptors (PPARs), on lipid and carbohydrate metabolism. The physiological and pharmacological ligands that activate the receptors are discussed, along with pathological conditions associated with dysregulation of the receptors.

   Today, it appears that crosstalk between the mitochondria and the remainder of the cell is an extremely important function of the mitochondria. Not only do the mitochondria provide metabolites to ensure continuation of several metabolic pathways, but they also regulate the levels of key cytosolic constituents. This is made possible by the presence of transporters in the inner mitochondrial membrane that specifically regulate the movement of substrates in and out of the mitochondria. In the fourth chapter, Drs. Kaplan and Mayor describe the structure of a transport protein that shuttles citrate out of the mitochondria, providing a carbon source for fatty acid, triglyceride and cholesterol biosynthesis. The chapter also clarifies the mechanism by which the transporter insures the integrity of the membrane barriers protecting the mitochondria from the accumulation of undesirable substances. In chapter 5, Dr. Vary outlines the role of mitochondria in ammonia detoxification, gluconeogenesis, protein synthesis and acid-base balance. The complex regulation of rate-limiting mitochondrial events, such as flux through pyruvate and branched chain ketoacid dehydrogenases, by transcription factors, hormones and cellular nutrition are reviewed. In chapter 6, Dr. King discusses the important role played by the mitochondria in amino acid metabolism. She reviews evidence that amino acids are involved in mitochondrial-dependent cytoprotective reactions that are dependent on amino acid metabolism rather than protein synthesis.

Mitochondria are dynamic organelles. They can migrate, undergo morphological changes and accumulate in response to physiological and pathological biogenetic stimuli. Dr. Al-Mehdi summarizes the dynamic changes induced upon endothelial mitochondria by shear-stress, flow adaptation and flow cessation.

Recent studies suggest that the mitochondria may also generate initiators of important signaling pathways. Of particular interest is the activation of cell altering protein kinases by mitochondrial-generated free radicals. How the mitochondria maintain a balance between the generation and destruction of reactive oxygen and nitrogen species is presently unclear. Dr. Turrens argues that it is a complex process, involving multiple sources of reactive oxygen species, several antioxidant enzymes, antioxidant vitamins and an assortment of antioxidant small molecules, all which normally act to prevent the accumulation of damaging levels of oxidants. Like reactive oxygen species, mitochondrial calcium homeostasis must be tightly regulated to ensure mitochondrial and cellular viability. In chapter 9, Drs. Griffiths, Bell, Balaska and Rutter discuss the transporters involved in calcium accumulation by the mitochondria. They also discuss the coupling mechanisms that link modest mitochondrial calcium accumulation with the stimulation of energy production. Also addressed in the chapter are potential pharmacological interventions that might be expected to prevent the damaging effects of excessive mitochondrial calcium accumulation that arise in certain pathological states. Besides the transport of calcium, the mitochondria also regulate the transport of other cations. In chapter 10, Dr. O'Rourke assigns an important cytoprotective role to the potassium channels. He describes experimental evidence showing that multiple potassium channels are cytoprotective, however, he argues that the lack of structure/function relationships has limited the development of ion selective agents with therapeutic significance.

Failure to maintain proper levels of calcium and reactive species within the mitochondria alters mitochondrial function, leading to the formation of the mitochondrial permeability transition pore. The formation of the pore renders the mitochondria leaky to organic and inorganic solutes, causing it to lose the ability to generate ATP and to control matrix volume. Drs. Halestrap, Clarke and Khalilin describe the properties of the pore, including the importance of the mitochondrial permeability transition in the depletion of cellular ATP and the loss of cell viability. They point out that events that accompany ischemia-reperfusion, such as reactive oxygen species generation, calcium overload, adenine nucleotide depletion, inorganic phosphate accumulation and collapse of the membrane potential, act as initiators or facilitators of pore formation. It follows that effective therapy against ischemia-reperfusion damage requires rapid closing of the mitochondrial permeability transition pore. In chapter 12, Drs. Suleiman and Schaffer review the mechanisms underlying the onset of cellular apoptosis, an orderly type of death that is initiated either by intrinsic, mitochondrial events or by extrinsic stimuli. Because apoptosis is genetically programmed and proceeds along specified pathways, it is highly regulated. Drs. Suleiman and Schaffer describe the effects of several cardioactive agents, including angiotensin II, calcium, reactive species and β-adrenergic agonists, on the apoptotic pathway. They maintain that the inhibition of apoptosis might serve as an important therapeutic approach in the treatment of heart disease. Traditionally, apoptosis and necrotic oncosis represent distinct pathways, with only apoptosis considered a highly regulated, genetically programmed pathway. However, in chapter 13, Drs. Murphy and Steenbergen argue that the two forms of cell death share common steps in accidental insults, such as ischemia-reperfusion. They point out that in both forms of cell death, the mitochondria serve as a source of reactive oxygen species and other pro-apoptotic factors, can contribute to ATP depletion and modulate calcium homeostasis. Therefore, they maintain that a better understanding of mitochondrial regulation should provide novel therapeutic approaches for reducing both apoptosis and necrosis. One mechanism of protecting the cell against both forms of death is preconditioning. In chapter 14, Drs. Phillip, Downey and Cohen propose a signaling cascade triggered by G_i coupled receptors that lead to the opening of the K_{ATP} channel. They suggest that attenuation of the mitochondrial permeability transition may serve as the end effector of the preconditioning event. In chapter 15, Drs. Shokolenko, LeDoux and Wilson ascribe a key role for mitochondrial DNA damage in apoptosis. They point out that mitochondrial DNA is much more susceptible to oxidative damage than nuclear DNA. Moreover, the synthesis of most mitochondrially expressed peptides is regulated by a single promoter. Since the proteins encoded by the mitochondria are components of electron transport, mitochondrial DNA damage reduces flux through the electron transport and enhances generation of superoxide. Therefore, it is logical to propose that mitochondrial DNA repair, by reversing mitochondrial DNA damage, can rescue cells from excessive superoxide generation and oxidant-induced apoptosis.

Recent evidence that mitochondria are directly involved in cell signaling initiated by pro-apoptotic factors, as well as cell survival stimuli, indicates that the status of the mitochondria determines the viability of the cell. Future therapy is likely to take advantage of this central role of the mitochondria to shift the balance in favor of either cell survival or death. Moreover, with the emergence of mitochondria as an important regulator of cell signaling, new interest in the pathophysiology of the mitochondria should yield fascinating results.

Stephen W. Schaffer
M-Saadeh Suleiman

Contents

Part 1 Mitochondrial Metabolism

1. Regulation of Mitochondrial Respiration in Heart Muscle 3
   Ilmo Hassinen

2. Regulation of Fatty Acid Oxidation of the Heart 27
   Clifford D. L. Folmes and Gary D. Lopaschuk

3. Regulation of Mitochondrial Fuel Handling
   by the Peroxisome Proliferator-Activated Receptors 63
   Mary C. Sugden and Mark J. Holness

4. Molecular Structure of the Mitochondrial
   Citrate Transport Protein 97
   Ronald S. Kaplan and June A. Mayor

5. Regulation of Pyruvate and Amino Acid Metabolism 117
   Thomas C. Vary, Wiley W. Souba and Cristopher J. Lynch

6. Amino Acids and the Mitochondria 151
   Nicola King

Part 2 The Dynamic Nature of the Mitochondria

7. Mechanotransduction of Shear-Stress at the Mitochondria 169
   Abu-Bakr Al-Mehdi

Part 3 Mitochondria as Initiators of Cell Signaling

8. Formation of Reactive Oxygen Species in Mitochondria 185
   Julio F. Turrens

9. Mitochondrial Calcium: Role in the Normal
   and Ischaemic/ Reperfused Myocardium 197
   *Elinor J. Griffiths, Christopher J. Bell, Dirki Balaska
   and Guy A. Rutter*

10. Mitochondrial Ion Channels 221
    Brian O'Rourke

Part 4 Mitochondria as Initiators of Cell Death

11. The Mitochondrial Permeability Transition
    Pore – from Molecular Mechanism to Reperfusion Injury
    and Cardioprotection 241
    Andrew P. Halestrap, Samantha J. Clarke and Igor Khalilin

12. The Apoptotic Mitochondrial Pathway – Modulators,
    Interventions and Clinical Implications 271
    M-Saadeh Suleiman and Stephen W. Schaffer

13. The Role of Mitochondria in Necrosis Following Myocardial
    Ischemia-Reperfusion 291
    Elizabeth Murphy and Charles Steenbergen

Part 5 Mitochondria as Modulators of Cell Death

14. Mitochondria and Their Role in Ischemia/Reperfusion Injury 305
    Sebastian Phillip, James M. Downey and Michael V. Cohen

15. Mitochondrial DNA Damage and Repair 323
    Inna N. Shokolenko, Susan P. Ledoux and Glenn L. Wilson

Index 349

Part 1
Mitochondrial Metabolism

1
Regulation of Mitochondrial Respiration in Heart Muscle

Ilmo Hassinen

1.1. Introduction

Heart muscle with its continual and heavy use of energy is strongly dependent on the most efficient biological energy provider, the process of oxidative phosphorylation, which is responsible for the aerobic conversion and conservation of the combustion energy of fuel substrates to ATP, the universal cellular energy currency. Its key reactions are localized in mitochondria. The myocardial mitochondrion may be regarded as an archetype of its kind and therefore is also a classical experimental model in research on oxidative phosphorylation.

Mitochondria are organelles comprised of an outer membrane and inner membrane, enclosing a matrix space. The latter contains the terminal catabolic pathways such as the tricarboxylic acid cycle, fatty acid β-oxidation and pyruvate dehydrogenase all of which extract hydrogen and electrons from fuel substrates with the formation of NADH and reduced ubiquinone, a lipid-soluble isoprenoid quinone in the mitochondrial inner membrane. Heart muscle mitochondria are characterized by maximized inner membrane area, achieved by intricate folding, which in electron micrographs appears as a tight packaging of inner membrane invaginations, the cristae.

The inner membrane contains the respiratory chain, a series of redox enzymes, which reduce oxygen to water with electrons donated by dehydrogenations in intermediary metabolism. The respiratory enzyme complexes are in principle redox-driven proton pumps, which use the energy derived from oxidation-reduction reactions to translocate protons from the matrix space across the inner mitochondrial membrane. This process leads to establishing an electrochemical potential (gradient) of protons across the inner membrane. The electrochemical potential ($\Delta\mu_H+$) has two components, an 'electric' $F \cdot \Delta\psi$, produced by the membrane potential ($\Delta\psi$) and a 'chemical' $2.3RT \cdot \Delta_p H$ provided by the transmembrane gradient of hydrogen ions.

The proton electrochemical potential serves as a primary form of existence of conserved energy, which is utilized by the reversible proton-pumping F_1F_o-ATPase to synthesize ATP from ADP and inorganic phosphate. Import of ADP and export of ADP is performed by the ADP/ATP translocase. For the operation of this (chemiosmotic) principle of energy conservation, the mitochondrial inner membrane must be proton-tight, so that fuel substrates or metabolites can permeate this membrane only by means of various exchange translocators.

Like all intracellular organelles surrounded by two membranes, the mitochondria contain their own miniature chromosome, the mitochondrial DNA (mtDNA), which codes for 13 mitochondrial peptides, all having the character of insoluble intrinsic membrane protein. The mutation frequency of mtDNA is one order of magnitude higher than that of nuclear DNA (Zeviani et al. 1998), and these mutations can lead to mitochondrial diseases. MtDNA derangements include both point mutations and deletions. The latter are the cause of a heart disease phenotype, the Kearns-Sayre syndrome characterized by a cardiac conduction defect (Channer et al. 1988) in addition to the symptoms related to skeletal muscle problems.

1.2. Heart Muscle Mitochondria

In heart muscle most of the copious mitochondria are aligned in interfibrillar rows with a pitch of nearly one per sarcomere. The mitochondria appear as tight packages of inner membrane cristae within the outer membrane. Another population of mitochondria have a subsarcolemmal location and there is some evidence (Riva et al. 2005) of functional differences between these two populations (Manneschi and Federico 1995; Weinstein et al. 1986).

The amount of subsarcolemmal mitochondria increases in certain mitochondrial diseases, and the trichrome staining properties of heaps of subsarcolemmal mitochondria gives rise to the histological picture of 'ragged red fibers' (Reichmann et al. 1996) characteristic of mitochondriopathies.

Since the respiratory chain is composed of intrinsic membrane proteins, maximizing inner membrane area is supportive of a high rate of oxidative phosphorylation. In fact, the protein occupancy of the membrane is so high that there is not much room left for translational, lateral movement of the enzyme complexes in the membrane. Although 'supercomplexes' can be isolated with defined methodology (Schagger 2002), the reactions in the 'respiratory chain' are probably more dependent on collision frequency between the complexes (Gupte and Hackenbrock 1988) and intramembrane electron carriers than on a structural organization of a chain.

1.3. The Respiratory Chain

1.3.1. Complex I

Complex I catalyses the initiating reaction of the respiratory chain, the reduction of ubiquinone by NADH. The mechanism of complex I action is the least understood of the mitochondrial respiratory complexes. This is due to the extreme complexity of the enzyme, apparent redundancy of the number of redox centers and their location with respect to the location of its lipid-soluble electron acceptor, ubiquinone in the membrane.

In vertebrates, complex I is comprised of 46 dissimilar subunits with a total molecular mass of 900 kDa. Its bacterial counterpart, NDH-1, has only 14 (or 13 due to gene fusion) subunits, all of which are homologues of the subunits of the mammalian enzyme so that the molecular mass is 550 kDa. Seven of these subunits are homologues of the seven mitochondrially coded subunits of the mammalian enzyme, the remaining seven subunits being homologues of the nuclearly coded mammalian complex I subunits. It is becoming evident that these 14 subunits represent the functional core of the mammalian enzyme.

The purified enzyme is large enough to be visible in electron micrographs obtained with negative staining. Both mammalian and bacterial enzymes have an L-shaped form (Guenebaut et al. 1998), with an intramembrane arm and a peripheral arm. Resolution of the holoenzyme to subcomplexes (Sazanov et al. 2000) reveals that all of the redox-active centers are located in the peripheral arm of the enzyme, which contains the NADH-binding site, one molecule of flavin mononucleotide (FMN) and eight iron-sulfur centers. The hydrophobic subunits (ND1-ND6 and ND4L) coded in mtDNA reside in the membrane arm of the complex. The ND1 subunit, which reacts with ubiquinone, has no identified redox center. It is also the target of numerous pathogenic mutations, typically resulting in the Leber hereditary ophthalmic neuropathy (LHON).

Complex I is a redox-driven proton pump with a stoichiometry of $4H^+/2e^-$, i.e. four protons per one molecule of NADH oxidized (Honkakoski and Hassinen 1986; Wikstrom 1984). This high H^+/e^- ratio cannot be explained with a classical 'Mitchelian' redox loop, because the enzyme contains only one hydrogen carrier amongst several electron carriers. Several hypothetical models of complex I function ranging from Q-cycle type mechanisms (Ragan, 1987; Dutton et al., 1998) to reductant-induced oxidation (Brandt 1997) and conformation-driven proton pump (Ohnishi and Salerno 2005) have been proposed. Sequence homologies to Na^+/H^+ antiporters (Friedrich and Weiss 1997) or H^+ conductors (Hassinen and Vuokila 1993) suggest that the ND2, ND5, ND4 and ND4L subunits may be involved in proton translocation (Kurki et al. 2000; Garofano et al. 2003; Holt et al. 2003; Flemming et al. 2003; Kervinen et al. 2004)

1.3.2. Complex III

Complex III (ubiquinol:cytochrome c oxidoreductase) conveys the electrons from reduced ubiquinone to ferricytochrome c, which becomes reduced. The complex is composed of 11–12 dissimilar subunits, and one of these, cytochrome b, is coded in mitochondrial DNA. Cytochrome b contains two heme B groups, namely, a low potential heme (BL -30 mV) and a high potential heme (BH $+90$ mV). With respect to the mitochondrial inner membrane these hemes are located on opposite sides, one near the inner face and the other near the outer face. The enzyme has also two quinone-binding domains in the vicinity of the two heme groups of cytochrome b, and an iron-sulfur protein center reacting with cytochrome c_1, which further reacts with cytochrome c on the outer face of the inner membrane.

Complex III is a proton pump with a stoichiometry of $4H^+/2e^-$. This can be fully explained by the Q-cycle originally proposed by Peter Mitchell (1975), which is based on spectroscopic findings suggestive of the presence of two different cytochromes b. The 11-subunit complex III from bovine heart has been crystallized, and on the basis of x-ray diffraction pattern its 3-D structure is known at 3-Å resolution (Iwata et al., 1998). Its detailed spatial structure fits the Q-cycle mechanism of action.

1.3.3. Complex IV

Complex IV (cytochrome c oxidase) catalyses the reduction of oxygen to water by ferricytochrome c with concomitant pumping of protons (Wikström 1984) with a stoichiometry of $2H^+/2e^-$. Since the electron transport reaction is vectorial and oriented from cytochrome c on the outer face of the membrane to the oxygen reduction site on the inner face of the membrane, a membrane potential (negative inside) is generated. Thus complex IV contributes to the formation of the electrochemical potential by two modes, translocation of electrical charges alone and protons.

Complex IV of mammals has 13 dissimilar subunits. Three of them (I-III) are encoded by mtDNA. Bacterial cytochrome c oxidase is composed of only three subunits and these are homologues of the mitochondrially coded ones. Two mitochondrially-coded subunits contain all of the four redox-active groups (two hemes and two copper atoms) of the enzyme (heme A and the binuclear heme A_3-Cu_B center in subunit I, and Cu_A in subunit II). The low number of subunits in the bacterial enzyme indicates that these three subunits constitute the functional core of the enzyme, and the other subunits apparently contribute to its stability and regulation. The remaining 10 subunits of the mammalian enzyme are nuclear coded, and at least two of them are involved in the allosteric regulation of the enzyme.

Cytochrome c oxidase has been crystallized from a few tissues and species and its 3-D structure has been determined from x-ray diffraction data, e.g. Paracoccus denitrificans (Iwata et al. 1995) and bovine heart (Tsukihara et al. 1996).

1.4. ATP Synthesis

Complex V (ATP synthetase, F_1F_o-ATPase) is a reversible H^+-ATPase, which pumps protons across a membrane at the expense of ATP hydrolysis. Conversely, it is able to join ADP and inorganic phosphate by extracting water with concomitant formation of ATP.

Complex V has three main parts, a membrane-embedded domain F_o, a catalytic domain F_1, and a joining stalk. The F_1 domain is a heterohexamer composed of three α-subunits and three β-subunits. Pairs of α and β subunits form three catalytic sites, which adopt three different conformations for three partial reactions of ATP synthesis. In the middle of the stalk there is a rod-formed γ subunit anchored in the F_o domain that extends to the center of the F_1 domain. The b and δ subunits also join the F_1 and F_o domains and immobilize F_1 in relation to F_o.

The F_1F_o-ATPase operates in ATP synthesis like a "nanomechanical engine" (Kato-Yamada et al. 1998; Ueno et al. 2005). In the F_o domain there is a 'rotor' composed of 9 to 12 c subunits, depending on species. Subunit a of the F_o moiety channels protons into the rotor. Protons are forced through the channel by their electrochemical potential and cause the rotor to rotate. The γ subunit rotates with the rotor, and its bent end in the center of the stationary F_1 part deforms the α-β subunits and their catalytic sites in a sequential manner through three functionally different conformations (Oster and Wang 2003). Thus, the electrochemical energy of the mitochondrial inner membrane is transformed through conformational changes to the F_1 catalytic sites, which in turn drives ATP synthesis. Three ATP molecules are formed during one turn of the rotor and subunit γ. The number of protons required for one turn of the rotor is equal to the number of c subunits in it. With 12 c subunits 12 protons are needed for 3 ATP molecules, i.e. the H^+/ATP ratio is 4. The common experimental finding is that three ATP molecules are formed when one molecule of NADH is oxidized by mitochondria. This is in good agreement with the notion that four protons or electrical charges are translocated by each of complexes I, III and IV.

The structure of the F_1-ATPase has been determined by x-ray crystallography at 2.8 Å resolution (Abrahams et al. 1994).

The proton ATPase/ATP synthase reaction is reversible, which gives rise to problems under conditions in which the electron transport and proton pumping reactions of the respiratory chain are blocked as in ischemia or anoxia. Dissipation of the proton electrochemical potential through proton leakage leaves the H^+-ATPase as the only supporter of ($\Delta\mu_H+$). Thus, in ischemia F_1F_o-ATPase enhances the depletion of ATP.

However, there exists an endogenous inhibitor peptide IF_1, which binds to F_1F_o-ATPase under conditions of intracellular acidification or decreased mitochondrial membrane potential (Cabezon et al. 2000). There are big species differences in the IF_1 content of the myocardium. The concentration is high in the low-beat hearts of larger mammals, whereas in the fast-beat hearts of small rodents the IF_1 concentration is low. Ischemia-induced inhibition of F_1F_o-ATPase can be demonstrated even in the isolated perfused rat heart (Ylitalo et al. 2001). Ischemic preconditioning, i.e. a brief ischemia/reperfusion, enhances the inhibition of the ATPase during subsequent ischemia (Vuorinen et al. 1995a; Ylitalo et al., 2001; Penna et al. 2004). However, the significance of F_1F_o-ATPase inhibition during the preconditioning process, the most efficient endogenous mechanism of protection against ischemia/reperfusion damage, remains uncertain (Green et al. 1998).

1.5. Metabolite Translocation in the Inner Membrane.

As far as oxidative phosphorylation and aerobic glycolysis are concerned, the main transport proteins in the mitochondrial inner membrane are ADP/ATP translocase, phosphate translocase, glutamate/aspartate translocase and 2-oxoglutarate translocase. The outer membrane is permeable to most small-molecular weight substrates of the mitochondrion. In creatine kinase (CK)-containing tissues such as brain and muscle, a mitochondrial isomorph of the CK exists that is bound to the intermembrane surface of the inner membrane in close structural and functional linkage to the ADP/ATP translocase. In connection of its cytosolic isoenzyme, the CK form a "creatine phosphate cycle" which provides a means of facilitated diffusion of ATP between the mitochondria and the ATP-consuming reactions in myofilaments (Bessman and Geiger 1981; Kupriyanov et al. 1984). The mitochondrial inner membrane is impermeant to NADH and NAD^+, so that the glutamate/aspartate and 2-oxoglutarate translocators (malate-asparate shuttle) are indispensable for transferring reducing equivalents of cytosolic NADH to the mitochondrial matrix under aerobic conditions.

1.6. Mitochondrial Redox Enzymes of Intermediary Metabolism Having Inner Membrane Ubiquinone as Electron Acceptor

1.6.1. Complex II (Succinate:Ubiquinone Oxidoreductase)

For historical reasons, succinate dehydrogenase (succinate:ubiquinone oxidoreductase) is commonly listed as a component of the respiratory chain as complex II, although it is not capable of proton pumping. This enzyme of the tricarboxylic acid cycle contains covalently bound flavin-adenine dinucleotide (FAD) as the prosthetic group and three iron-sulfur centers (bi- tri and tetranuclear). It is anchored to the mitochondrial inner membrane by means of two b-type cytochromes, whose role in the electron transfer is unknown, and two ubiquinone-binding proteins.

A common textbook error is the classification of succinate dehydrogenase as a flavin-reducing enzyme, although the net reaction catalyzed is: succinate + ubiquinone ↔ fumarate + ubiquinol. None of the subunits of SDH is encoded in mitochondrial DNA (Anderson et al. 1981).

1.6.2. Electron Transfer Flavoprotein:Ubiquinone Oxidoreductase

In fatty acid β-oxidation the electrons are fed into the respiratory chain via two routes: In the first, fatty acyl-CoA is oxidized to enoyl-CoA by the FAD-containing acyl-CoA oxidase enzyme, which becomes reduced. This is reoxidized by the electron transfer flavoprotein (ETF) (Beckmann and Frerman 1985), which in turn is reoxidized by ETF dehydrogenase (ETF:ubiquinone oxidoreductase), a metalloflavoprotein of the mitochondrial inner membrane (Ruzicka and Beinert 1977). The electrons are channeled to ubiquinone, a substrate of complex III of the respiratory chain. The next β-oxidation reaction catalyzed by 3-hydroxyacyl-CoA dehydrogenase produces NADH, which is oxidized by complex I (for references, see Schulz 1991). Therefore electrons are fed into the respiratory chain via two routes, one of which bypasses complex I, thus lowering the ATP/O ratio.

1.6.3. Glycerol-3-Phosphate Dehydrogenase

Two glycerol-3-phosphate dehydrogenases exist. In the cytosol there is an NAD-linked glycerol phosphate dehydrogenase (Ostro, and Fondy 1977), and in the mitochondrial inner membrane there is an FAD-containing flavoenzyme (glycerol-3-phosphate:ubiquinone oxidoreductase), which is linked to the respiratory chain (Klingenberg 1970). The glycerol-3-phosphate-binding site of the enzyme faces the intermembrane compartment and is then accessible to the cytosolic glycerol-3-phosphate, which cannot penetrate the inner mitochondrial membrane. These two enzymes constitute the glycerophosphate shuttle (Houstek et al. 1975), a mechanism of reoxidation of cytosolic NADH during aerobic glycolysis. In heart muscle the mitochondrial content of the glycerol phosphate dehydrogenase flavoenzyme is low, and the glycerophosphate shuttle is not significant in the myocardium, where aerobic reoxidation of cytosolic NADH occurs by the malate-aspartate shuttle (LaNoue and Williamson 1971). The glycerophosphate shuttle is significant in some other tissues such as the β-cells of the pancreatic islets, which show a high rate of glucose oxidation (MacDonald 1981).

1.7. Levels of Regulation of Electron Transport in the Mitochondrion

The activity of the mitochondrial respiratory chain is geared to the cellular requirements of ATP, but this subtle regulation operates at several levels. The majority of the ATP-consuming reactions occur in the cytosol, such that the translocation of ATP, ADP, Pi and respiratory substrates, although the entry of reducing equivalents into the respiratory chain must also be considered. In fact, a unifying view of the mitochondrial respiratory control in intact cells has not been achieved, and several competing hypotheses have been proposed. The major in vivo regulation has been attributed to the ADP/ATP translocator, the "substrate level" dehydrogenations (Katz et al. 1989; Heineman and Balaban 1990) and the near-equilibria between the ADP/ATP translocator, F_1F_o-ATPase and redox-linked proton pumps (Wilson et al. 1977; Hassinen 1986; Hassinen and Hiltunen 1975). In skeletal muscle, Michaelis-Menten kinetics approximates the ADP-dependent respiratory control reasonably well (Chance et al. 1985). Thus, for evaluation of respiratory control an integrating discussion of metabolic control is needed.

1.8. Substrate Level Regulation of Supply of Reducing Equivalents for the Respiratory Chain

1.8.1. Carbohydrate Oxidation

Under aerobic conditions the product of glycolysis is pyruvate, which is oxidized in mitochondria by the pyruvate dehydrogenase complex (PDH) to acetyl-CoA, which enters the tricarboxylic acid cycle. The enzyme is interconvertible by means of phosphorylation to an inactive form (PDH_i) and dephosphorylation to an active dephosphorylated form (PDH_a). Free Ca^{2+} and Mg^{2+} ions activate the dephosphorylation by a phosphoprotein phosphatase (Midgley et al. 1987). Because mitochondrial ATP is found as a magnesium salt, changes in the mitochondrial ATP concentration influence mitochondrial $[Mg^{2+}]_f$ (Wieland and Portenhauser 1974; Denton et al. 1975). Thus, the mitochondrial [ATP]/[ADP] ratio becomes a determinant of the $[PDH_i]/[PDH_a]$ ratio. Because the heart rate and contraction force are linked to $[Ca^{2+}]_f$ transients that follow the action potentials, the time-averaged cytosolic $[Ca^{2+}]_f$ becomes dependent on the mechanical work output of the heart. Moreover, it has been shown (MacGowan et al. 2001) that the mitochondrial $[Ca^{2+}]_f$ follows the time-averaged cytosolic $[Ca^{2+}]_f$ although the transients become filtered out (Leisey et al. 1993). This means that the activation state of PDH is effectively regulated by the mechanical workload of the heart despite the fact that the [ATP]/[ADP] ratio is not markedly influenced by it. The phosphorylation of PDH_a to the inactive PDH_i is catalyzed by a protein kinase, which is activated by increased acetyl-CoA/CoA and $NADH/NAD^+$ ratios (Ravindran et al. 1996; Peters 2003). Fatty acid oxidation, which is determined by fatty acid availability, increases the mitochondrial acetyl-CoA concentration (Ala-Rämi et al. 2005). Thus, fatty acid oxidation converts PDH to its inactive form and is an efficient suppressor of myocardial carbohydrate oxidation. Also ketone body oxidation elevates the myocardial concentration of acetyl CoA and subdues glucose oxidation. It follows that pyruvate oxidation becomes regulated by the acetyl-CoA/CoA, ATP/ADP, $NADH/NAD^+$ ratios and cardiac mechanical workload.

1.8.2. Fatty Acid Oxidation

The regulation of fatty acid oxidation is poorly understood, largely because β-oxidation proceeds to completion and intermediate metabolites do not accumulate. Administration of a fatty acid increases cardiac oxygen consumption. This has been attributed to a decoupling effect of fatty acids on mitochondria (Kornberg 1966), a decrease in the P/O ratio related to the by-passing of complex one by a portion of the reducing equivalents (see above) or a reduction of NADH and the resulting increase in the thermodynamic driving force of cell respiration (Hassinen et al. 1990).

The regulation of fatty acid oxidation is possible at the level of fatty acid uptake into the cells, their import into the mitochondria (via the acylcarnitine/carnitine transporter) and the process of β-oxidation itself. For the latter, it is known that the semiquinone form of ETF is a feedback inhibitor of acyl-CoA dehydrogenase (Beckmann and Frerman 1985), NADH inhibits 3-hydroxyacyl-CoA dehydrogenase, and acetyl-CoA inhibits thiolase enzyme (for references, see Schulz 1991).

It has been shown in liver mitochondria that an increase in the mitochondrial matrix volume enhances β-oxidation. It is also known that matrix volume is affected by calcium. Thus, it is possible that calcium regulates β-oxidation, forming a link between ATP-consuming heart muscle contractions and ATP-yielding fatty acid oxidation. Moreover, the matrix volume is also affected by mitochondrial pyrophosphate, and mitochondrial pyrophosphatase is inhibited by calcium. However, in heart muscle the pyrophosphate concentration is not affected by metabolic perturbations (Griffiths and Halestrap 1993). When contractility of an isolated perfused heart is adjusted by changes in the extracellular $[Ca^{2+}]$ during fatty acid infusion, the alterations in the mitochondrial redox state of flavins (and $NADH/NAD^+$) are small compared with changes in oxygen consumption or the free $[ATP]\cdot[ADP]/[P_i]$ ratio (Vuorinen et al. 1995b). This suggests that the rate of fatty acid oxidation is regulated more by calcium-induced changes in ATP consumption than by direct effects of calcium on β-oxidation.

In heart muscle, the main determinant of fatty acid oxidation is their availability, and even low concentrations of fatty acids are able to almost totally suppress carbohydrate oxidation as has been recently shown by the ^{13}C NMR isotopomer analysis methodology (Ala-Rämi et al. 2005). The mitochondrial acylcarnitine/carnitine transferase I is inhibited by malonyl-CoA, an intermediate of fatty acid synthesis. In lipogenic tissues malonyl-CoA is a major regulator of fatty acid oxidation and ketogenesis (McGarry and Foster 1979). Malonyl-CoA is also synthesized by cytosolic acetyl-CoA carboxylase, which is regulated by 5'-AMP-activated protein kinase in heart muscle, where its only metabolic fate is the decarboxylation by the malonyl-CoA enzyme, whose sole role is likely the regulation of fatty acid oxidation (Dyck et al. 1998; Belke et al. 1998).

1.8.3. Tricarboxylic Acid Cycle

The mitochondrial tricarboxylic cycle is the most central metabolic pathway with input from carbohydrate, fatty acid and amino acid oxidation. It feeds the respiratory chain by reduction of NAD or ubiquinone. The regulation of tricarboxylic acid (TCA) cycle is somewhat controversial as is the mutual significance of the TCA cycle and respiratory chain in the regulation of cell respiration.

There are two modes of regulation of the TCA cycle. First, there is flux regulation, but an equally important mode of regulation involves the maintenance of the pool of TCA cycle intermediates. The TCA cycle is designed for oxidation of acetyl-CoA. The net degradation or oxidation of any of the cycle intermediates necessitates their prior export from the cycle. The intermediates act by mass action to accelerate TCA flux but are not consumed by the cycle itself. In spite of that, the total pool size of the TCA cycle in heart muscle undergoes large fluctuations and some of its intermediates function as feedback regulators of cytosolic metabolic pathways.

1.8.3.1. Flux Regulation

Three enzymes of the TCA cycle have properties suitable for flux regulation. (1) The mitochondrial concentration of oxaloacetate (OxAc) is within the range of the K_m of citrate synthase (Siess et al. 1984), such that the reaction velocity becomes dependent on the mitochondrial concentration of OxAc, which in turn depends on the status of the malate dehydrogenase reaction (Forsander 1970). It is known that the activity of malate dehydrogenase is near-equilibrium, such that the concentration of OxAc varies as a function of the mitochondrial $NADH/NAD^+$ ratio, which in turn is dependent on the activity of the respiratory chain. (2) Calcium is an activator (Nichols et al. 1994), NADH a product inhibitor and ADP an allosteric activator (Ehrlich and Colman 1982) of NAD-dependent isocitrate dehydrogenase, which accounts for the feedback regulation of the enzyme by the respiratory chain. (3) Calcium is an activator of 2-oxoacid dehydrogenase at concentrations encountered in mitochondria (McCormack and Denton 1990). As described above, the mitochondrial $[Ca^{2+}]$ follows the time averaged cytosolic calcium. Thus, the activity of 2-oxoacid dehydrogenase is dependent upon the mechanical work output of the heart.

1.8.3.2. Pool Size Regulation

An important aspect of the TCA cycle is the maintenance of its pool size. Although carbon shuffling between intermediates occurs, TCA cycle flux does not affect total pool size. However, cycle metabolites are also used in some other metabolic routes, biosynthetic for example. Thus, the pool requires continual replenishment, a process referred to 'anaplerosis' by Kornberg (1966). It should also be borne in mind that anaplerosis is important not only because it elevates the levels of TCA intermediates but also because it regulates flux through other metabolic pathways. A classical example is citrate, which regulates phosphofructokinase, a rate limiting enzyme in glycolysis. Because the pool size is dependent on both replenishment and disposal, it is also influenced by the metabolic state of the cell. This happens for example during the metabolism of odd-number carbon fatty acids, which produce propionate, which in turn feeds into the TCA cycle via propionyl-CoA carboxylase, thereby producing succinyl-CoA. These TCA cycle disposal mechanisms are sometimes called 'cataplerotic'

There are several means of anaplerosis in the heart muscle: (1) coupled alanine and aspartate transaminations which produce oxaloacetate, (2) the purine nucleotide cycle which by sequential purine deamination and amination produces ammonia and fumarate, (3) glutamate dehydrogenase which produces 2-oxoglutarate, a reaction that is dependent on the mitochondrial $NAD^+/NADH$ ratio, (4) pyruvate carboxylase which produces oxaloacetate, (5) 'malic' enzymes (decarboxylating malate dehydrogenases) that produce malate from pyruvate by their reverse reactions. Pyruvate carboxylation in the myocardium is well documented (Peuhkurinen and Hassinen 1982), but its rate during biotin deficiency and the kinetic properties of the malic enzyme suggests that the latter is the main route of pyruvate carboxylation in heart muscle (Sundqvist et al. 1989). Ala-Rämi et al. (2005) have shown that anaplerotic flux is dependent on the work output of the heart.

The opposing 'cataplerotic' disposal reactions have been less characterized. The redox state of the malic enzyme is such that the net rate is anaplerotic. Only during high rates of anaplerosis can malic enzyme function in the direction of cataperosis (Sundqvist et al. 1987).

1.9. Respiratory Chain Regulation

1.9.1. Substrate-Level or Respiratory Chain

The rate of oxidative phosphorylation and oxygen consumption is adjusted according to the energy needs of the cell, and the phenomenon of "respiratory control" in suspensions as first described by Lardy and Wellman (1952). Nonetheless, one must take into consideration the regulation exerted by the "rate-limiting" steps of mitochondrial electron transport, including (1) control by substrate supply affecting the production of NADH, adjusted by the various regulators of mitochondrial and cytosolic dehydrogenations, (2) control over mitochondrial matrix dehydrogenases, (3) control by the mitochondrial availability of ADP and P_i, (4) control of cytochrome oxidase by the $ATP/ADP \cdot P_i$ ratio through equilibrium reactions in respiratory complexes I and III, (5) allosteric or covalent regulation of cytochrome oxidase, (6) oxygen availability and (7) regulation of expression of the genes coding for the subunits of the respiratory chain complexes.

1.9.2. NADH Limitation

The production of NADH or other kinds of reducing equivalents has two important roles in the regulation of cell respiration. First, the oxidation-reduction state of the mitochondrial matrix determines the redox gap and Gibbs free energy change throughout the respiratory chain, ending with the reduction of oxygen. Irreversible thermodynamics predicts that the reaction rate is proportional to the driving force. Second, if the reaction rate of oxygen reduction is determined by the redox state of cytochrome oxidase by means of the reversible redox-driven proton pumps and H^+-ATPase, the set point of the redox state of cytochrome oxidase would be indefinite, without the control of the redox state of the NADH/NAD$^+$ couple.

A significant amount of evidence implicating an important regulatory role for NADH production has been obtained by means of in vivo phosphorus NMR of the canine heart. It has been shown that during phenylephrine-induced work output transitions the myocardial concentrations of inorganic phosphate, creatine phosphate and ATP remain constant within the accuracy limits of the method (Katz et al. 1989). These data suggest that the adenylate system does not regulate electron transport and oxygen consumption under in vivo conditions, although respiratory control by the adenylate system can be demonstrated in suspensions of isolated mitochondria.

1.9.3. Ca^{2+}

As described above in section 1.8, several possibilities exist for the regulation of mitochondrial NADH production in the myocardium. The main emphasis is on the adjustment of tricarboxylic acid cycle flux by calcium. The NAD-linked isocitrate and 2-oxoglutarate dehydrogenases are sensitive to activation by calcium at concentrations observed in the mitochondrial matrix. A correlation between intracellular calcium and oxygen consumption has been observed (Wu et al. 1992), but experiments with the calcium sensitizer levosimendan failed to distinguish between energy consumption or calcium elevation as the cause of enhanced oxygen consumption (Todaka et al. 1996). When the work output of isolated hearts perfused with buffer containing medium chain fatty acids was altered by changes in extracellular calcium, the mitochondrial redox state and cellular energy state data obtained by whole organ fluorometry and ^{31}P NMR, respectively, supported the notion that oxygen consumption is mainly determined by energy expenditure (Vuorinen et al. 1995b).

1.9.4. Regulation by ATP, ADP and Inorganic Phosphate

1.9.4.1. Mitochondrial ADP Availability

Oxidative phosphorylation is dependent on mitochondrial availability of ADP and inorganic phosphate, which are transported by the ADP/ATP translocator and phosphate translocator, respectively. The former is electrogenic because of a difference in the net charges of ATP and ADP, and becomes linked to the mitochondrial membrane potential. The phosphate translocator is a phosphate$^-$/OH$^-$ exchanger and thus is coupled to transmembrane $\Delta_p H$ (Palmieri et al. 1993).

The ADP/ATP translocase is an abundant protein in mitochondria, but atractyloside administration demonstrates its potential ability to regulate the rate of oxidative phosphorylation at least in isolated mitochondria (Groen et al. 1982). However, a detailed analysis and modeling of skeletal muscle mitochondria (Korzeniewski and Mazat 1996) according to the principles of Kacser (1983) yields a flux control coefficient of 0.14 for the ADP/ATP translocator; the flux control coefficient is defined as the ratio of the fractional change of flux to the fractional change in the concentration of the enzyme, with the sum of the coefficients of the pathway being unity (Kacser and Burns 1973).

1.9.4.2. ADP/ATP Feedback

The signal from the cytosolic energy state can be conveyed to the mitochondria through several modes, such as the concentration of ADP or ATP, the [ATP]/[ADP] ratio, and the [ATP]/[ADP]·[P_i] ratio. In some cases, an index of available high energy phosphate bonds, namely, the "energy charge" (EC) = ½(2·[ATP] + [ADP])/([ATP] + [ADP] + [AMP]), (Atkinson 1968) has been used to express the energy status even though it does not reflect the thermodynamic quantity of free energy available from ATP hydrolysis. The concentration of free ADP in heart muscle cells is very low (30-90 μM), compared with the free concentration of ATP (about 5 mM), or inorganic phosphate (around 1 mM), so that during fluctuations within the in vivo range of the ATP/ADP ratio the concentration of ATP remains almost constant.

The variation of oxygen consumption upon changes in work output of the isolated working heart preparation (Williamson et al., 1976) has been used to model the metabolic state of the mitochondria/cytosol/myofibril system (Vendelin et al. 2000) relative to oxidative phosphorylation, ADP/ATP translocase, phosphate translocase, mitochondrial and cytosolic creatine kinases and adenylate kinase. It was found that a model assuming equilibrium satisfactorily described the state of ATP, CrP and Cr, but the concentration of ADP changed in a concentration-dependent, non-equilibrium manner indicating the existence of intracellular concentration gradients. The data also revealed a reasonable fit to a Michaelis-Menten type hyperbolic relationship between cytosolic ADP and the rate of oxygen consumption (Vendelin et al. 2000).

The extreme sensitivity of the rate of oxidative phosphorylation to ADP, resulting in a nearly constant value for the ATP/ADP ratio is difficult to explain. It has been found that the regulation of oxidative phosphorylation does not follow simple Michaelis-Menten kinetics but is at least second order with respect to ADP (Jeneson et al. 1996). Thus, there must be an amplification mechanism to increase the sensitivity of mitochondria to cytosolic ADP. To test this hypothesis, an analysis was performed using curve fitting which was unbiased toward the specific mechanism of the reactions involved. A satisfactory fit for the dependence of the oxidative phosphorylation rate on ADP concentration was obtained when an arbitrary Hill equation was used, yielding an n value of 3.5. This finding supported the concept that an amplification mechanism sensitizes the mitochondria to ADP (Jeneson et al. 1996) and argues in favor of an allosteric regulatory mechanism.

The parallel activation of ATP-consuming and ATP-yielding pathways by calcium has, indeed, been a popular explanation for the relative constancy of the ATP/ADP ratio during workload transitions in heart muscle. Korzeniewsky et al. (2005) have modeled oxidative phosphorylation in different muscles and found that the closest fit to the experimental data is obtained when it is presumed that an "each-step activation" occurs. This means that a common signaling molecule influences all contributors of the pathway, i.e. upon activation of muscle ATP consumption, all individual enzyme complexes of the oxidative phosphorylation system become simultaneously activated (Korzeniewski et al. 2005). However, it is difficult to conceive the identity of the messenger. The experimental data indicate that a source of hysteresis exists in the system, such that the activated state of ATP synthesis is not immediately shut off after cessation of ATP consuming muscle stimulation, giving rise to an overshoot of creatine phosphate and undershoot of inorganic phosphate. According to Korzeniewski (1998), the best fit of the experimental data assumes that the enhanced rate of ATP synthesis decays with a time constant of 5 min.

The flux control coefficients of the complexes of the oxidative phosphorylation system and the related metabolite translocases have been estimated in skeletal muscle mitochondria, by fitting a theoretical model to the experimental data (Korzeniewski and Mazat 1996). In the resting state (state 4) control was largely imposed by the leakage of protons from the inner membrane, while in the intermediate state (state 3.5), control was largely exerted by the ATP consuming processes themselves. During the maximal rate, rate of oxidative phosphorylation (state 3) the control coefficient of complex III was 0.26 when the ADP/ATP carrier had a coefficient of 0.14 and the ATP synthase 0.16 (Korzeniewski and Mazat 1996). The titration of rat heart mitochondria with oligomycin has yielded a flux control coefficient of 0.04-0.09 for the F_1F_o-ATPase (Ylitalo et al. 2001).

1.9.4.3. Near-Equilibria in the Respiratory Chain

The classical measurements of the redox potential of the matrix NADH/NAD$^+$ couple and cytochrome c, as well as the free energy change of cytosolic ATP hydrolysis in isolated liver cells have demonstrated that near-equilibrium prevails (Wilson et al., 1974). The same relationship has been found in the isolated perfused rat heart (Hassinen and Hiltunen 1975). An extended network of redox- and free-energy near-equilibria (Hassinen and Hiltunen 1975) prevails in the isolated perfused rat heart (Hassinen 1986), including the redox span across mitochondrial complexes I and III, the ADP/ATP translocase, F_1F_o-ATPase and $\Delta\mu_H^+$ (Kauppinen et al. 1980; Kauppinen et al. 1983; Nuutinen et al. 1981; Nuutinen 1984). Recent data (Liimatta et al. 2004) obtained by organ spectrophotometry of isolated perfused hearts from myoglobin knock-out mice also argue that the dominant regulation exists at the level of the respiratory chain.

Mathematical modeling of these equilibria based on the kinetic properties of cytochrome oxidase, which obeys first order kinetics, predicts that oxygen consumption is proportional to the cytosolic [ATP]/[ADP]·[P_i] ratio (Wilson et al. 1977). The equilibria minimize the entropy increase and enhance the energy conservation efficiency of oxidative phosphorylation. This paradigm contradicts simple Michaelis-Menten kinetics for ADP regulation of cell respiration but is in accord with the classical notion that in a metabolic pathway the "rate-limiting" step is out of equilibrium and controls flux through the subsequent steps of the respiratory chain.

However, it should be borne in mind that a single rate-limiting step does not exist. All steps contribute, although the distribution of control is uneven (Korzeniewski and Mazat 1996). The control attributed to individual steps can be described by their flux control coefficient.

1.9.5. Allosteric and Covalent Regulation of the Respiratory Chain

1.9.5.1. Complex I

The 42 kD (NDUFA10) subunit of complex I is phosphorylated at some tyrosine residues that remain phosphorylated even during isolation of the complex (Schilling et al., 2005). Moreover, the 18 kD (AQDQ) subunit in the iron-sulfur protein fraction of complex I is phosphorylated in vitro by mitochondrial protein kinase-A (Papa et al. 1996). The 18 kD subunit is coded by the nuclear NDUFS4 gene, and its mutation results in complex I deficiency when homozygous. A 5 bp duplication in the NDUF4 gene destroys the phosphorylation site and abolishes the cAMP-dependent activation of complex I in fibroblast cultures. A deficiency of the 18 kD subunit due to a mutation leading to translation termination a few residues after the mitochondrial targeting sequence results in a defect in the assembly of complex I (Papa et al. 2002). The 18 kD complex I subunit can be dephosphorylated by a mitochondrial serine phosphatase which is inhibited by calcium in contrast to the mitochondrial PDH phosphatase which is activated by calcium (Signorile et al. 2002).

Cyclic AMP and PKA have been found in mitochondria, and the A-kinase anchoring proteins (AKAPs) which are variant splicing products of the AKAP1 gene have tag for targeting AKAPs and their PKA complexes to the mitochondrial outer membrane (for a review, see Feliciello et al. 2005). Cytosolic cAMP does not penetrate the mitochondrial inner membrane (Muller and Bandlow 1987), so that its signal probably reaches the mitochondrion by means of a "transductosome" composed of a protein conglomerate assembled by AKAP (Feliciello et al. 2005).

1.9.5.2. Complex IV

The multiplicity and overlap of the regulatory possibilities of cytochrome oxidase is promoted by the allosteric regulation of mammalian cytochrome oxidase. There are seven high affinity ATP binding sites and ten ADP binding sites in bovine heart cytochrome oxidase. ATP binds to the matrix domain of the subunit IV and acts as an allosteric inhibitor of the enzyme (Arnold and Kadenbach 1997). ATP is also bound to the heart isomorph of the VIa subunit, decreasing the proton pumping efficiency (H^+/e^- ratio) of the enzyme (Frank and Kadenbach 1996). On the other hand, ADP bound to the heart type VIa subunit acts as an allosteric activator of the enzyme (Anthony et al. 1993). It is noteworthy that this is specific for the heart enzyme, so that the liver enzyme is not activated by ADP.

Cytochrome c oxidase is also subject to covalent modification by means of protein kinases. It has been shown that a cyclic AMP dependent protein kinase exists in bovine heart mitochondria that is capable of phosphorylating some mitochondrial proteins (Technikova-Dobrova et al. 2001; Papa et al. 1999). Allosteric inhibition of cytochrome c oxidase by ATP is reversibly switched on by cyclic AMP-dependent protein phosphorylation of subunits II and Vb (Bender and Kadenbach 2000). Also subunit IV is phosphorylated, but the identity of the protein kinase responsible for that is not known (Steenaart and Shore 1997).

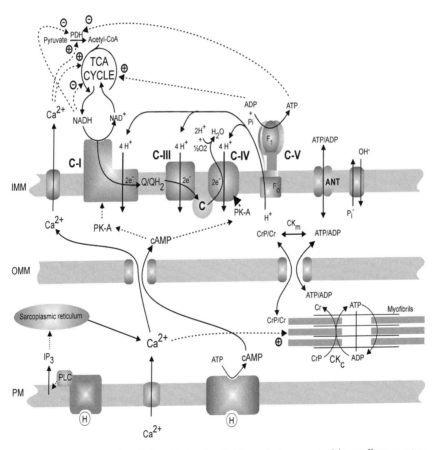

FIGURE 1.1. Metabolic interlinks of mitochondrial respiratory control in cardiomyocytes. Abbreviations: C-I, complex I; C-III, complex III; C-IV, complex IV; C-V, complex V; ANT, ADP/ATP exchange carrier; TCA, tricarboxylic acid; CK_m, mitochondrial creatine kinase; CK_c, cytosolic creatine kinase; PK-A, cAMP-dependent protein kinase; PLC phospholipase C. The solid arrows show metabolite transport and the dashed arrows depict regulation.

1.10. Concluding Remarks

In heart muscle, a tissue of continually high energy consumption, the regulation of electron transfer in the mitochondrial respiratory chain is multifaceted (Figure 1.1). It is evident that several organizational levels contribute to the rate adjustments. The level of dominance is dependent on the stimulus resulting from enhanced ATP consumption. It is apparent that when the primary stimulus is external (humoral) and activates a second messenger system, control is distributed because of the simultaneous activation of substrate-level metabolism, which enhances the production of reducing equivalents to feed into the respiratory chain, and the ATP consuming reactions, which through ADP stimulates the respiratory chain. Under conditions in which the stimulus is non-humoral, control by the adenylate system becomes dominant. The regulation is subtle and the apparent sensitivity of mitochondrial respiration to ADP is extremely high, and some modeling studies indicate that some kind of amplification system exists. This may be based on the simultaneous activation of all of the enzyme complexes of the respiratory chain, although the identity of the messenger remains unknown.

References

Abrahams JP, Leslie AG, Lutter R, and Walker JE (1994) Structure at 2.8 Å resolution of F_1-ATPase from bovine heart mitochondria. Nature 370: 621–628

Ala-Rämi A, Ylihautala M, Ingman P, Hassinen IE (2005) Influence of calcium-induced workload transitions and fatty acid supply on myocardial substrate selection. Metabolism 54: 410–420

Anderson S, Bankier AT, Barrell BG, de Bruijn MH, Coulson AR, Drouin J, Eperon IC, Nierlich DP, Roe BA, Sanger F, Schreier PH, Smith AJ, Staden R, Young IG (1981) Sequence and organization of the human mitochondrial genome. Nature 290: 457–465

Anthony G, Reimann A, Kadenbach B (1993) Tissue-specific regulation of bovine heart cytochrome-c oxidase activity by ADP via interaction with subunit VIa. Proc Natl Acad Sci USA 90: 1652–1656

Arnold S, Kadenbach B (1997) Cell respiration is controlled by ATP, an allosteric inhibitor of cytochrome-c oxidase. Eur. J Biochem 249: 350–354

Atkinson DE (1968) The energy charge of the adenylate pool as a regulatory parameter. Interaction with feedback modifiers. Biochemistry 7: 4030–4034

Beckmann JD, Frerman FE (1985) Electron-transfer flavoprotein-ubiquinone oxidoreductase from pig liver: purification and molecular, redox, and catalytic properties. Biochemistry 24: 3913–3921

Belke DD, Wang LC, Lopaschuk GD (1998) Acetyl-CoA carboxylase control of fatty acid oxidation in hearts from hibernating Richardson's ground squirrels. Biochim Biophys Acta 1391: 25–36

Bender E, Kadenbach B (2000) The allosteric ATP-inhibition of cytochrome c oxidase activity is reversibly switched on by cAMP-dependent phosphorylation. FEBS Lett 466: 130–134

Bessman SP, Geiger PJ (1981) Transport of energy in muscle: the phosphorylcreatine shuttle. Science 211: 448–452

Brandt U (1997) Proton-translocation by membrane-bound NADH:ubiquinone-oxidoreductase (complex I) through redox-gated ligand conduction. Biochim Biophys Acta 1318: 79–91

Cabezon E, Arechaga I, Jonathan P, Butler G, Walker JE (2000) Dimerization of bovine F_1-ATPase by binding the inhibitor protein, IF_1. J Biol Chem 275: 28353–28355

Chance B, Leigh JS Jr, Clark BJ, Maris J, Kent J, Nioka S, Smith D (1985) Control of oxidative metabolism and oxygen delivery in human skeletal muscle: a steady-state analysis of the work/energy cost transfer function. Proc Natl Acad Sci U S A 82: 8384–8388

Channer KS, Channer JL, Campbell MJ, Rees JR (1988) Cardiomyopathy in the Kearns-Sayre syndrome. Br Heart J 59: 486–490

Denton RM, Randle PJ, Bridges BJ, Cooper RH, Kerbey AL, Pask HT, Severson DL, Stansbie D, Whitehouse S (1975) Regulation of mammalian pyruvate dehydrogenase. Mol Cell Biochem 9: 27–53

Dutton PL, Moser CC, Sled VD, Daldal F, Ohnishi T (1998) A reductant-induced oxidation mechanism for complex I. Biochim Biophys Acta 1364: 245–257

Dyck JR, Barr AJ, Barr RL, Kolattukudy PE, Lopaschuk GD (1998) Characterization of cardiac malonyl-CoA decarboxylase and its putative role in regulating fatty acid oxidation. Am J Physiol 275: H2122–H2129

Ehrlich RS, Colman RF (1982) Interrelationships among nucleotide binding sites of pig heart NAD-dependent isocitrate dehydrogenase. J Biol Chem 257: 4769–4774

Feliciello A, Gottesman ME, Avvedimento EV (2005) cAMP-PKA signaling to the mitochondria: protein scaffolds, mRNA and phosphatases. Cell Signal 17: 279–287

Flemming D, Hellwig P, Friedrich T (2003) Involvement of tyrosines 114 and 139 of subunit NuoB in the proton pathway around cluster N2 in Escherichia coli NADH:ubiquinone oxidoreductase. J Biol Chem 278: 3055–3062

Forsander OA (1970) Effects of ethanol on metabolic pathways. In: Tremoliers J, editor. International encyclopedia of paharmacology and therapeutics. New York: Pergamon Press pp. 117–135

Frank V, Kadenbach B (1996) Regulation of the H^+/e^- stoichiometry of cytochrome c oxidase from bovine heart by intramitochondrial ATP/ADP ratios. FEBS Lett 382: 121–124

Friedrich T, Weiss H (1997) Modular evolution of the respiratory NADH:ubiquinone oxidoreductase and the origin of its modules. J Theor Biol 187: 529–540

Garofano A, Zwicker K, Kerscher S, Okun P, Brandt U (2003) Two aspartic acid residues in the PSST-homologous NUKM subunit of complex I from Yarrowia lipolytica are essential for catalytic activity. J Biol Chem 278: 42435–42440

Green DW, Murray HN, Sleph PG, Wang FL, Baird AJ, Rogers WL, Grover GJ (1998) Preconditioning in rat hearts is independent of mitochondrial F_1F_0 ATPase inhibition. Am J Physiol 274: H90–H97

Griffiths EJ, Halestrap AP (1993) Pyrophosphate metabolism in the perfused heart and isolated heart mitochondria and its role in regulation of mitochondrial function by calcium. Biochem J 290: 489–495

Groen AK, Wanders RJ, Westerhoff HV, Van der MR, Tager JM (1982) Quantification of the contribution of various steps to the control of mitochondrial respiration. J Biol Chem 257: 2754–2757

Guenebaut V, Schlitt A, Weiss H, Leonard K, Friedrich T (1998) Consistent structure between bacterial and mitochondrial NADH:ubiquinone oxidoreductase (complex I). J Mol Biol 276: 105–112

Gupte SS, Hackenbrock CR (1988) Multidimensional diffusion modes and collision frequencies of cytochrome c with its redox partners. J Biol Chem 263: 5241–5247

Hassinen I, Ito K, Nioka S, Chance B (1990) Mechanism of fatty acid effect on myocardial oxygen consumption. A phosphorus NMR study. Biochim Biophys Acta 1019: 73–80

Hassinen IE (1986) Mitochondrial respiratory control in the myocardium. Biochim Biophys Acta 853: 135–151

Hassinen IE, Hiltunen K (1975) Respiratory control in isolated perfused rat heart. Role of the equilibrium relations between the mitochondrial electron carriers and the adenylate system. Biochim Biophys Acta 408: 319–330

Hassinen IE, Vuokila PT (1993) Reaction of dicyclohexylcarbodiimide with mitochondrial proteins. Biochim Biophys Acta 1144: 107–124

Heineman FW, Balaban RS (1990) Control of mitochondrial respiration in the heart in vivo. Ann Rev Physiol 52: 523–542

Holt PJ, Morgan DJ, Sazanov LA (2003) The location of NuoL and NuoM subunits in the membrane domain of the Escherichia coli complex I: implications for the mechanism of proton pumping. J Biol Chem 278: 43114–43120

Honkakoski PJ, Hassinen IE (1986) Sensitivity to NN'-dicyclohexylcarbodi-imide of proton translocation by mitochondrial NADH:ubiquinone oxidoreductase. Biochem J 237: 927–930

Houstek J, Cannon B, Lindberg O (1975) Glycerol-3-phosphate shuttle and its function in intermediary metabolism of hamster brown-adipose tissue. Eur J Biochem 54: 11–18

Iwata S, Lee JW, Okada K, Lee JK, Iwata M, Rasmussen B, Link TA, Ramaswamy S, Jap BK (1998) Complete structure of the 11-subunit bovine mitochondrial cytochrome bc1 complex. Science 281: 64–71

Iwata S, Ostermeier C, Ludwig B, Michel H (1995) Structure at 2.8 Å resolution of cytochrome c oxidase from Paracoccus denitrificans. Nature 376: 660–669

Jeneson JA, Wiseman RW, Westerhoff HV, Kushmerick MJ (1996) The signal transduction function for oxidative phosphorylation is at least second order in ADP. J Biol Chem 271: 27995–27998

Kacser H (1983) The control of enzyme systems in vivo: elasticity analysis of the steady state. Biochem Soc Trans 11: 35–40

Kacser H, Burns JA (1973) The control of flux. Symp Soc Exp Biol 27: 65–104

Kato-Yamada Y, Noji H, Yasuda R, Kinosita K Jr, Yoshida M (1998) Direct observation of the rotation of epsilon subunit in F1-ATPase. J Biol Chem 273: 19375–19377

Katz LA, Swain JA, Portman MA, Balaban RS (1989) Relation between phosphate metabolites and oxygen consumption of heart in vivo. Am J Physiol 256: H265–H274

Kauppinen RA, Hiltunen JK, Hassinen IE (1980) Subcellular distribution of phosphagens in isolated perfused rat heart. FEBS Lett 112: 273–276

Kauppinen RA, Hiltunen JK, Hassinen IE (1983) Mitochondrial membrane potential, transmembrane difference in the NAD^+ redox potential and the equilibrium of the glutamate-aspartate translocase in the isolated perfused rat heart. Biochim Biophys Acta 725: 425–433

Kervinen M, Patsi J, Finel M, Hassinen IE (2004) A pair of membrane-embedded acidic residues in the NuoK subunit of Escherichia coli NDH-1, a counterpart of the ND4L subunit of the mitochondrial complex I, are required for high ubiquinone reductase activity. Biochemistry 43: 773–781

Klingenberg M (1970) Localization of the glycerol-phosphate dehydrogenase in the outer phase of the mitochondrial inner membrane. Eur J Biochem 13: 247–252

Kornberg H (1966) Anaplerotic sequences and their role in metabolism. Essays Biochem 1–31

Korzeniewski B (1998) Regulation of ATP supply during muscle contraction: theoretical studies. Biochem J 330: 1189–1195

Korzeniewski B, Mazat JP (1996) Theoretical studies on the control of oxidative phosphorylation in muscle mitochondria: application to mitochondrial deficiencies. Biochem J 319: 143–148

Korzeniewski B, Noma A, Matsuoka S (2005) Regulation of oxidative phosphorylation in intact mammalian heart in vivo. Biophys Chem 116: 145–157

Kupriyanov VV, Ya SA, Ruuge EK, Kapel'ko VI, Yu ZM, Lakomkin VL, Smirnov VN, Saks VA (1984) Regulation of energy flux through the creatine kinase reaction in vitro and in perfused rat heart. 31P-NMR studies. Biochim Biophys Acta 805: 319–331

Kurki S, Zickermann V, Kervinen M, Hassinen I, Finel M (2000). Mutagenesis of three conserved Glu residues in a bacterial homologue of the ND1 subunit of complex I affects ubiquinone reduction kinetics but not inhibition by dicyclohexylcarbodiimide. Biochemistry 39: 13496–13502

LaNoue KF, Williamson JR (1971) Interrelationships between malate-aspartate shuttle and citric acid cycle in rat heart mitochondria. Metabolism 20: 119–140

Lardy HA, Wellman H (1952) Oxidative phosphorylations; role of inorganic phosphate and acceptor systems in control of metabolic rates. J Biol Chem 195: 215–224

Leisey JR, Grotyohann LW, Scott DA, Scaduto RC Jr (1993) Regulation of cardiac mitochondrial calcium by average extramitochondrial calcium. Am J Physiol 265: H1203–H1208

Liimatta EV, Gödecke A, Schrader J, Hassinen IE (2004) Regulation of cellular respiration in myoglobin-deficient mouse heart. Mol Cell Biochem 256-257: 201–208

MacDonald MJ (1981) High content of mitochondrial glycerol-3-phosphate dehydrogenase in pancreatic islets and its inhibition by diazoxide. J Biol Chem 256: 8287–8290

MacGowan GA, Du C, Glonty V, Suhan JP, Koretsky AP, Farkas DL (2001) Rhod-2 based measurements of intracellular calcium in the perfused mouse heart: cellular and subcellular localization and response to positive inotropy. J Biomed Opt 6: 23–30

Manneschi L, Federico A (1995) Polarographic analyses of subsarcolemmal and intermyofibrillar mitochondria from rat skeletal and cardiac muscle. J Neurol Sci 128: 151–156

McCormack JG, Denton RM (1990) The role of mitochondrial Ca^{2+} transport and matrix Ca^{2+} in signal transduction in mammalian tissues. Biochim Biophys Acta 1018: 287–291

McGarry JD, Foster DW (1979) In support of the roles of malonyl-CoA and carnitine acyltransferase I in the regulation of hepatic fatty acid oxidation and ketogenesis. J Biol Chem 254: 8163–8168

Midgley PJ, Rutter GA, Thomas AP, Denton RM (1987) Effects of Ca^{2+} and Mg^{2+} on the activity of pyruvate dehydrogenase phosphate phosphatase within toluene-permeabilized mitochondria. Biochem J 241: 371–377

Mitchell P (1975) The protonmotive Q cycle: a general formulation. FEBS Lett 59: 137–139

Muller G, Bandlow W (1987) cAMP-dependent protein kinase activity in yeast mitochondria. Z. Naturforsch [C] 42: 1291–1302

Nichols BJ, Rigoulet M, Denton RM (1994) Comparison of the effects of Ca^{2+}, adenine nucleotides and pH on the kinetic properties of mitochondrial NAD^+-isocitrate dehydrogenase and oxoglutarate dehydrogenase from the yeast Saccharomyces cerevisiae and rat heart. Biochem J 303: 461–465

Nuutinen EM, Hiltunen K, Hassinen IE (1981) The glutamate dehydrogenase system and the redox state of mitochondrial free nicotinamide adenine dinucleotide in myocardium. FEBS Lett 128: 356–360

Nuutinen EM (1984) Subcellular origin of the surface fluorescence of reduced nicotinamide nucleotides in the isolated perfused rat heart. Basic Res Cardiol 79: 49–58

Ohnishi T, Salerno JC (2005) Conformation-driven and semiquinone-gated proton-pump mechanism in the NADH-ubiquinone oxidoreductase (complex I). FEBS Lett 579:4555–4561

Oster G, Wang H (2003) Rotary protein motors. Trends Cell Biol 13: 114–121

Ostro MJ, Fondy TP (1977) Isolation and characterization of multiple molecular forms of cytosolic NAD-linked glycerol-3-phosphate dehydrogenase from normal and neoplastic rabbit tissues. J Biol Chem 252: 5575–5583

Palmieri F, Bisaccia F, Capobianco L, Dolce V, Fiermonte G, Iacobazzi V, Zara V (1993) Transmembrane topology, genes, and biogenesis of the mitochondrial phosphate and oxoglutarate carriers. J Bioenerg Biomembr 25: 493–501

Papa S, Sardanelli AM, Cocco T, Speranza F, Scacco SC, Technikova-Dobrova Z (1996) The nuclear-encoded 18 kDa (IP) AQDQ subunit of bovine heart complex I is phosphorylated by the mitochondrial cAMP-dependent protein kinase. FEBS Lett 379: 299–301

Papa S, Sardanelli AM, Scacco S, Petruzzella V, Technikova-Dobrova Z, Vergari R, Signorile A (2002) The NADH: ubiquinone oxidoreductase (complex I) of the mammalian respiratory chain and the cAMP cascade. J Bioenerg Biomembr 34: 1–10

Papa S, Sardanelli AM, Scacco S, Technikova-Dobrova Z (1999) cAMP-dependent protein kinase and phosphoproteins in mammalian mitochondria. An extension of the cAMP-mediated intracellular signal transduction. FEBS Lett 444: 245–249

Penna C, Pagliaro P, Rastaldo R, Di Pancrazio F, Lippe G, Gattullo D, Mancardi D, Samaja M, Losano G, Mavelli I (2004) F0F1 ATP synthase activity is differently modulated by coronary reactive hyperemia before and after ischemic preconditioning in the goat. Am J Physiol Heart Circ Physiol 287: H2192–H2200

Peters SJ (2003) Regulation of PDH activity and isoform expression: diet and exercise. Biochem Soc Trans 31: 1274–1280

Peuhkurinen KJ, Hassinen IE (1982) Pyruvate carboxylation as an anaplerotic mechanism in the isolated perfused rat heart. Biochem J 202: 67–76

Ragan CI (1987) Structure of NADH-ubiquinone reductase (complex I). Curr Top Bioenerg 15: 1–35

Ravindran S, Radke GA, Guest JR, Roche TE (1996) Lipoyl domain-based mechanism for the integrated feedback control of the pyruvate dehydrogenase complex by enhancement of pyruvate dehydrogenase kinase activity. J Biol Chem 271: 653–662

Reichmann H, Vogler L, Seibel P (1996) Ragged red or ragged blue fibers. Eur Neurol 36: 98–102

Riva A, Tandler B, Loffredo F, Vazquez E, Hoppel C (2005) Structural differences in two biochemically defined populations of cardiac mitochondria. Am J Physiol Heart Circ Physiol 289: H868–H872

Ruzicka FJ, Beinert H (1977) A new iron-sulfur flavoprotein of the respiratory chain. A component of the fatty acid beta oxidation pathway. J Biol Chem 252: 8440–8445

Sazanov LA, Peak-Chew SY, Fearnley IM, Walker JE (2000) Resolution of the membrane domain of bovine complex I into subcomplexes: implications for the structural organization of the enzyme. Biochemistry 39: 7229–7235

Schagger H (2002) Respiratory chain supercomplexes of mitochondria and bacteria. Biochim Biophys Acta 1555: 154–159

Schilling B, Aggeler R, Schulenberg B, Murray J, Row RH, Capaldi RA, Gibson BW (2005) Mass spectrometric identification of a novel phosphorylation site in subunit NDUFA10 of bovine mitochondrial complex I. FEBS Lett 579: 2485–2490

Schulz H (1991) Beta oxidation of fatty acids. Biochim Biophys Acta 1081: 109–120

Siess EA, Kientsch-Engel RI, Wieland OH (1984) Concentration of free oxaloacetate in the mitochondrial compartment of isolated liver cells. Biochem J 218: 171–176

Signorile A, Sardanelli AM, Nuzzi R, Papa S (2002) Serine (threonine) phosphatase(s) acting on cAMP-dependent phosphoproteins in mammalian mitochondria. FEBS Lett 512: 91–94

Steenaart NA, Shore GC (1997) Mitochondrial cytochrome c oxidase subunit IV is phosphorylated by an endogenous kinase. FEBS Lett 415: 294–298

Sundqvist KE, Heikkilä J, Hassinen IE, Hiltunen JK (1987) Role of $NADP^+$-linked malic enzymes as regulators of the pool size of tricarboxylic acid-cycle intermediates in the perfused rat heart. Biochem J 243: 853–857

Sundqvist KE, Hiltunen JK, Hassinen IE (1989) Pyruvate carboxylation in the rat heart. Role of biotin-dependent enzymes. Biochem J 257: 913–916

Technikova-Dobrova Z, Sardanelli AM, Speranza F, Scacco S, Signorile A, Lorusso V, Papa S (2001) Cyclic adenosine monophosphate-dependent phosphorylation of mammalian mitochondrial proteins: enzyme and substrate characterization and functional role. Biochemistry 40: 13941–13947

Todaka K, Wang J, Yi GH, Stennett R, Knecht M, Packer M, Burkhoff D (1996) Effects of levosimendan on myocardial contractility and oxygen consumption. J Pharmacol Exp Ther 279: 120–127

Tsukihara T, Aoyama H, Yamashita E, Tomizaki T, Yamaguchi H, Shinzawa-Itoh K, Nakashima R, Yaono R, Yoshikawa S (1996) The whole structure of the 13-subunit oxidized cytochrome c oxidase at 2.8 A. Science 272: 1136–1144

Ueno H, Suzuki T, Kinosita K Jr, Yoshida M (2005) ATP-driven stepwise rotation of F_oF_1-ATP synthase. Proc Natl Acad Sci USA 102: 1333–1338

Vendelin M, Kongas O, Saks V (2000) Regulation of mitochondrial respiration in heart cells analyzed by reaction-diffusion model of energy transfer. Am J Physiol Cell Physiol 278: C747–C764

Vuorinen K, Ylitalo K, Peuhkurinen K, Raatikainen P, Ala-Rämi A, Hassinen IE (1995a) Mechanisms of ischemic preconditioning in rat myocardium. Roles of adenosine, cellular energy state, and mitochondrial F_1F_0-ATPase. Circulation 91: 2810–2818

Vuorinen KH, Ala-Rämi A, Yan Y, Ingman P, Hassinen IE (1995b) Respiratory control in heart muscle during fatty acid oxidation. Energy state or substrate-level regulation by Ca^{2+}? J Mol Cell Cardiol 27: 1581–1591

Weinstein ES, Benson DW, Fry DE (1986) Subpopulations of human heart mitochondria. J Surg Res 40: 495–498

Wieland OH, Portenhauser R (1974) Regulation of pyruvate-dehydrogenase interconversion in rat-liver mitochondria as related to the phosphorylation state of intramitochondrial adenine nucleotides. Eur J Biochem 45: 577–588

Wikström M (1984) Two protons are pumped from the mitochondrial matrix per electron transferred between NADH and ubiquinone. FEBS Lett 169: 300–304

Wikström M (1984) Pumping of protons from the mitochondrial matrix by cytochrome oxidase. Nature 308: 558–560

Williamson JR, Ford C, Illingworth J, Safer B (1976) Coordination of citric acid cycle activity with electron transport flux. Circ Res 38: I39–I51

Wilson DF, Owen CS, Holian A (1977) Control of mitochondrial respiration: a quantitative evaluation of the roles of cytochrome c and oxygen. Arch Biochem Biophys 182: 749–762

Wilson DF, Stubbs M, Veech RL, Erecinska M, Krebs HA (1974) Equilibrium relations between the oxidation-reduction reactions and the adenosine triphosphate synthesis in suspensions of isolated liver cells. Biochem J 140: 57–64

Wu ST, Kojima S, Parmley WW, Wikman-Coffelt J (1992) Relationship between cytosolic calcium and oxygen consumption in isolated rat hearts. Cell Calcium 13: 235–247

Ylitalo K, Ala-Rämi A, Vuorinen K, Peuhkurinen K, Lepojärvi M, Kaukoranta P, Kiviluoma K, Hassinen IE (2001) Reversible ischemic inhibition of F_1F_o-ATPase in rat and human myocardium. Biochim Biophys Acta 1504: 329–339

Zeviani M, Tiranti V, Piantadosi C (1998) Mitochondrial disorders. Medicine (Baltimore) 77: 59–72

2
Regulation of Fatty Acid Oxidation of the Heart

Clifford D. L. Folmes and Gary D. Lopaschuk

2.1. Introduction

The heart has very high energy demands and requires a large amount of ATP in order to maintain contraction and ionic homeostasis (Opie 1998). To meet this high demand, the heart acts as a metabolic omnivore, metabolizing a variety of carbon substrates, including carbohydrates (glucose, lactate, and pyruvate), fatty acids, and ketone bodies (Neely and Morgan 1974; King and Opie 1998; Stanley et al. 2005). Under normal aerobic conditions, the heart preferentially metabolizes fatty acids, which contribute between 60% and 80% of the required ATP (Lopaschuk et al. 1994a; Stanley and Chandler 2002), with carbohydrates contributing the residual 20% to 40%. While there are many similarities in fatty acid oxidation in various tissues, the regulation of fatty acid oxidation in the heart can differ dramatically from regulation in tissues like liver, skeletal muscle and kidney. This chapter will not attempt to compare these differences, but instead will concentrate on the regulation of fatty acid oxidation specifically in the heart. Although peroxisomal β-oxidation also occurs in the heart and may play an important role in providing acetyl-CoA for the production of malonyl-CoA, this chapter will focus on mitochondrial fatty acid oxidation. In addition, fatty acids can also undergo α- or θ-oxidation as well; however the existence of these pathways in the heart has yet to be established. Various disease states are associated with a dysregulation of myocardial metabolism; this chapter will review this dysregulation and will introduce the use of metabolic modulators as a novel pharmacological approach to treating heart disease.

2.1.1. Overview of Myocardial Fatty Acid Metabolism

An overview of the pathway for fatty acid oxidation is found in Figure 2.1. The major sources of fatty acids for the myocardium are fatty acids bound to albumin and triglycerides in lipoproteins. Free fatty acids are removed from these sources in the coronary vasculature, and transported across the endothelium into the interstitial space by an as of yet unknown mechanism.

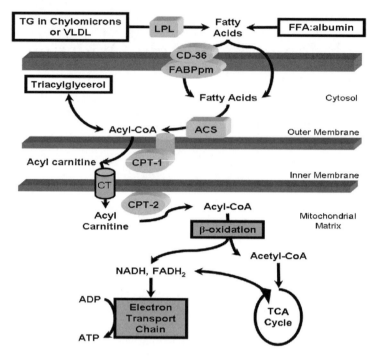

FIGURE 2.1. Schematic overview of myocardial fatty acid metabolism. ACS, acyl-CoA synethase; CPT-1, carnitine palmitoyltransferase-1: CPT-2, carnitine palmitoyltransferase-2; CT, carnitine acyltranslocase; FABPpm, plasma membrane fatty acid binding protein; FFA, free fatty acid; LPL, lipoprotein lipase.

These fatty acids then enter the cardiomyocytes by either passive diffusion or a carrier-mediated process. These fatty acids are transported in the cytoplasm by fatty acid binding proteins and are activated to long-chain fatty acyl-CoAs, a step required for their further oxidation. The acyl moieties are then transported across the impermeable inner mitochondrial membrane by carnitine dependent transport complex. Once in the mitochondrial matrix acyl-CoAs enter the β-oxidation spiral and for each successive cycle they are reduced by two carbon units and produce one acetyl-CoA, one NADH and one $FADH_2$. Acetyl-CoA is further oxidized in the Krebs cycle, which results in the liberation of two CO_2, three NADH, and one $FADH_2$. These reduced electron carriers enter the electron transport chain, the hydrogens are transferred to oxygen to produce water, and their energy is used for ATP synthesis. The remainder of the chapter will outline the regulation of this process and its dysregulation in disease.

2.2. Fatty Acid Supply to the Cardiomyocyte

In healthy individuals circulating plasma fatty acid levels can range from 0.2 to 0.8 mM during the course of a day. However, in pathological states which are associated with a metabolic stress, such as diabetes and ischemia, circulating fatty acid levels can far exceed these normal levels (>1.0 mM) (Lopaschuk et al. 1994b). Plasma fatty acid concentration is dependent on their release from triacylglycerols (TGs) in the adipocytes, which is dependent on the balance of enzymatic activity of hormone-sensitive lipase for TG degradation and glycerolphosphate acyltransferase for TG synthesis (Lopaschuk et al. 1994a). As the name implies, hormone-sensitive lipase is regulated by hormones, including catecholamines which activate the enzyme, and insulin which inhibits the enzyme. Thus pathological states such as ischemic heart disease which increases β-adrenergic stimulation, results in elevated plasma fatty acid concentration. This is in contrary to the fed state, where insulin levels are elevated and fatty acid levels are suppressed. These fluctuations in plasma fatty acid levels are important as the rate of fatty acid uptake into the cardiomyocyte is very dependent on the concentration of nonesterified fatty acids supplied in the coronary vasculature (Bing et al. 1954; Wisneski et al. 1987; Lopaschuk et al. 1994a). However due to the hydrophobic nature of free fatty acids, they are never found in the free form and must be associated with hydrophilic moieties such as proteins, carnitine, or coenzyme A, to increase their solubility in the aqueous environment of the plasma or cytoplasm. In the plasma they are found in their nonesterified form bound to high affinity binding sites on albumin, or contained as TGs within TG-rich lipoproteins, chylomicrons and very-low-density lipoproteins (VLDL) (Richieri and Kleinfeld 1995). Even though the molar concentration of fatty acids in lipoproteins is an order of magnitude greater than that of fatty acids bound to albumin, it is believed that albumin fatty acids are the primary source of energy for the heart (Hamilton and Kamp 1999). However TGs from these lipoproteins cannot cross the plasma membrane, and must first be hydrolyzed by lipoprotein lipase (LPL) to free fatty acids. Studies by Hauton et al. have determined the contribution of fatty acids from albumin, VLDL and chylomicrons to fatty acid oxidation (Hauton et al. 2001). These authors showed that TG from VLDL or chylomicrons could support cardiac function; however they were of secondary importance when in the presence of fatty acid bound to albumin. In addition the presence of free fatty acids suppressed utilization of TG from chylomicrons but not from VLDL. Further studies by this group have shown that TG from chylomicrons are oxidized and contribute to the maintenance of mechanical function, while TG from VLDL is mostly incorporated into tissue lipids (Niu et al. 2004).

Cardiac muscle is the tissue with the highest LPL expression, likely due to tissue-specific regulatory elements in the LPL gene promoter region that allows for differential regulation (Enerback and Gimble 1993; Fielding and Frayn 1998). LPL is bound to the luminal side of the capillary endothelial cells and the outer surface of the cardiomyocyte by charge interactions with heparin sulfate, a constituent of the glycocalyx (Merkel et al. 2002a; Augustus et al. 2003; Yagyu et al. 2003; Yokoyama et al. 2004). In vitro observations have shown that both active and inactive LPL can anchor lipoproteins to the cell surface, thereby increasing the efficiency of lipolysis (Merkel et al. 1998a; Merkel et al. 2002b). Mice with a cardiac specific deletion of LPL survive, and exhibit increased circulating TGs, decreased uptake of lipoprotein derived fatty acids and a suppressed expression of fatty acid oxidation genes, likely due to a downregulation of a PPAR mediated pathway (Merkel et al. 1998b; Augustus et al. 2004). In addition, transgenic mice with cardiac restricted expression of LPL are able to maintain normal circulating lipid levels, suggesting that the heart is an important site for the homeostasis of serum lipid levels (Levak-Frank et al. 1999; Augustus et al. 2003). Following the hydrolysis of lipoprotein TGs, the free fatty acids bind to albumin, until they are transported to the interstitial space by an unknown concentration gradient dependent mechanism. Uptake of fatty acids into the cardiomyocyte occurs by two mechanisms: passive diffusion and a protein-mediated process (Van der Vusse et al. 2000).

2.3. Fatty Acid Uptake

2.3.1. Passive Diffusion

As fatty acids have low solubility in aqueous environments and have high solubility in lipids, they partition into phospholipid bilayers (the kinetics of which is reviewed elsewhere (Hamilton and Kamp 1999)). Therefore when fatty acids dissociate from albumin, they readily adsorb into the plasma membrane despite being in their ionized form at physiological pH (where their pKa is 4.5) (Hamilton et al. 1994). Movement of ionized long-chain fatty acid (LCFA) anions from one leaflet of the membrane to the other is slow (Gutknecht 1988; Schmider et al. 2000), whereas the movement of non-ionized LCFAs is rapid. This process is thus accelerated due to the fact that the pKa of ionization of LCFAs is 7.6 in the phospholipid bilayer. It has been proposed that this so called "flip-flop" model includes several steps: 1) dissociation of fatty acid from albumin; 2) adsorption of fatty acid to the membrane; 3) protonation; 4) flip-flop within the membrane; and 5) desorption into the cytosolic space. (as reviewed (Hamilton and Kamp 1999; Schaffer 2002)).

2.3.2. Protein Mediated Fatty Acid Uptake

Several lines of evidence support the idea of facilitated fatty acid transport. Studies in isolated adipocytes where fatty acids were not esterified, showed that permeation could be distinguished from metabolism, and supply of fatty acids by albumin did not limit uptake (Richieri and Kleinfeld 1995). Fatty acid uptake produced nonlinear saturation kinetics as a function of fatty acid concentration and produced a Km of 7 nM, a value expected for a carrier-mediated process in the range of the physiological concentration of fatty acid (Richieri and Kleinfeld 1995). These results were reproduced in cardiac myocytes by Stremmel and Berk (Stremmel and Berk 1986). Competition for the transporter was observed in competition assays between long chain fatty acids, such as oleate and palmitate (Abumrad et al. 1981; Sorrentino et al. 1996). Another study making use of fluorescent fatty acid derivatives showed that greater than 87% of the fluorescent fatty acids remained unesterified inside the cell, therefore the competition could not be explained by metabolism (Storch et al. 1991). Another line of evidence is that fatty acid uptake is sensitive to protein modifying agents, such as phloretin which completely blocks membrane flux of fatty acids, DIDS (4,4'-di-isothiocyanostilbene-2,2'-sulphonate) which produces reversible inhibition of transport and antibodies raised against CD36 inhibited uptake (Abumrad et al. 1981; Abumrad et al. 1984; Harmon et al. 1992). For a more comprehensive review of this evidence see the review by Abumrad et al. (1998).

2.3.3. CD36/Fatty Acid Translocase

FAT/CD36 was originally identified as a fatty acid transporter based on its binding of two fatty acid uptake inhibitors, DIDS and sulpho-N-succinimidyl FA esters (Harmon and Abumrad 1993; Abumrad et al. 1998). This 88 kDa transmembrane protein is abundantly expressed in fatty acid metabolizing tissues such as skeletal and cardiac muscle, and appears to be the major fatty acid transporter found in the heart (Vannieuwenhoven et al. 1995; Greenwalt et al. 1995; Pelsers et al. 1999). Evidence for a role of CD36 in fatty acid uptake comes both from experimental models and human studies. Transgenic mice that overexpress FAT/CD36 in muscle have reduced circulating fatty acid levels and fatty acid oxidation is increased only during muscle contraction, suggesting that fatty acid uptake is only accelerated when there is a metabolic demand (Ibrahimi et al. 1999). Conversely, CD36 null mice have increased circulating fatty acids and tissue uptake of an iodinated fatty acid analog, [^{125}I]BMIPP (15-(p-iodophenyl)-3(R,S)-methylpentadecanoic acid) is reduced in heart, skeletal muscle and adipose tissue (Febbraio et al. 1999; Coburn et al. 2000). If hearts from CD36 knockout mice are perfused as isolated working hearts, a 40-60% reduction in fatty acid oxidation occurs, which is compensated for by an increase in glucose oxidation (Kuang et al. 2004). In addition, humans with mutations in CD36 have lower rates of uptake of [^{123}I]BMIPP compared to normal people, suggesting that CD36 also partially regulates cardiac fatty acid uptake in humans (Kintaka et al. 2002). Taken together, these results suggest that CD36 is an important regulator of fatty acid uptake in the heart.

Electrically stimulated contractions of cardiomyocytes in suspension results in increased fatty acid uptake, and this is completely blocked by sulfo-N-succinimidyloleate, a specific inhibitor of CD36 (Luiken et al., 2001a). This increase in flux could be a result of either an increase in sarcolemmal abundance of CD36 or an increase in intrinsic transport activity (Luiken et al. 2004a). In similar studies performed in skeletal muscle, the apparent maximum velocity of transport is increased independent of changes in Km. This finding suggests that the increase in fatty acid uptake is dependent on an increase in total transporter in the sarcolemma (Bonen et al. 2000). Subsequent fractionation of cardiac myocytes showed that under basal conditions CD36 is localized at both the sarcolemma membrane and in intracellular stores (Luiken et al. 2002).

Contraction or an increase in energy demand was shown to induce a translocation of CD36 from an intracellular pool to the plasma membrane, in a similar mode of regulation as the GLUT-4 transporter (Luiken et al. 2001a; Luiken et al. 2003b; Luiken et al. 2004b). To date several signaling proteins that may play a role in the contraction induced translocation of CD36 have been identified, including AMP-activated protein kinase (AMPK) (Luiken et al. 2003b), protein kinase C α, δ or ε or extracellular signal-related kinases (Luiken et al. 2004a). Surprisingly a similar phenomenon of increased CD36 translocation to the sarcolemmal membrane is observed in response to insulin stimulation (Luiken et al., 2001a). Despite these interesting results in isolated cell systems, it has yet to be demonstrated if translocation of CD36 occurs in the intact heart and thus it remains to be determined that translocation of CD36 regulates cardiac fatty acid oxidation under physiological conditions.

2.3.4. Plasma Membrane Fatty Acid Binding Protein

The plasma membrane fatty acid binding protein (FABPpm) was originally identified based on its binding of radiolabelled oleate (Stremmel et al. 1985). FABPpm is a 40 kDa protein that is loosely associated with the plasma membrane. The evidence supporting a role of FABPpm in fatty acid transport was initially based on an observation that treatment with an antibody raised against this protein results in a reduction in fatty acid uptake in a variety of cells (Berk et al. 1990; Zhou et al. 1992; Luiken et al. 1997). Additional evidence stems from expression of FABPpm in Xenopus laevis oocytes and 3T3 fibroblasts, which resulted in an increase in fatty acid uptake rates (Zhou et al. 1992; Zhou et al. 1994). A variety of conditions show concomitant changes in FABPpm protein expression and fatty acid utilization, including differentiation of 3T3-L1 cells, fasting, endurance training, diabetes and obesity (Zhou et al. 1992; Turcotte et al. 1997; Kiens et al. 1997; Berk et al. 1997; Luiken et al. 2001b).

2.3.5. Fatty Acid Transport Proteins

Fatty acid transport protein (FATP) was discovered as a fatty acid transporter by use of an expression cloning strategy in 2T2-L1 adipocytes (Schaffer and Lodish 1994). Expression of this 63 kDa integral membrane protein in a stable fibroblast cell line results in a 3- to 4-fold increase in long-chain fatty acid transport (Schaffer and Lodish 1994). The cloning of this first FATP led to the discovery of a family of evolutionarily conserved proteins, encoded by at least five different genes in the mouse and displaying different tissue-specific expression patterns (Anderson and Welsh 1990). FATP1 is the major isoform found in heart and skeletal muscle (Hirsch et al. 1998; Binnert et al. 2000; Pohl et al. 2000). However the role of this protein in fatty acid uptake is controversial. Bonen et al. (2002) found an inverse relationship between FATP1 protein expression in the plasma membrane and fatty acid uptake, however this may be due to the use of a poor FATP1 antibody in those studies. At the protein level, FATP1 has 40%

amino acid homology with very-long chain acyl-CoA synthetase (VLACS), and FATP1 has been reported to be a VLACS (Uchiyama et al. 1996; Kiens et al. 1997; Coe et al. 1999). For a more in depth review of this literature please see a review by Brinkmann et al. (2002). It has yet to be determined whether FATP1 has separate fatty acid uptake and VLACS activities.

2.3.6. Synergistic Action of Transporters

It has been proposed that CD36 and FABPpm may cooperate to transport fatty acids across the plasma membrane (Glatz et al. 1997; Glatz and Storch 2001; Bonen et al. 2002). Indirect evidence from studies in giant sarcolemmal vesicles suggest that fatty acid uptake may be inhibited by either sulpho-N-succinimidyl oleate or an anti-FABPpm antiserum (Luiken et al. 1999; Turcotte et al. 2000). However when these agents are applied together they do not result in a further decrease in fatty acid uptake (Luiken et al. 1999).

2.3.7. Very-Low-Density Lipoprotein (VLDL) Receptor

The VLDL receptor was originally cloned from a rabbit heart cDNA library and showed striking similarity with the LDL receptor cDNA (Takahashi et al. 1992). The heart is the site with the most abundant VLDL receptor expression and indirect evidence suggests that it may play a role in the uptake of fatty acids. A study in Balb/c mice showed that VLDL receptor expression increases with fasting and they (Kwok et al. 1997) proposed a role of the VLDL receptor in delivery of fatty acids as a source of fuel. Additionally these fasting mice also produced an increase in mRNA of other fatty acid metabolism proteins such as acyl-CoA synthetase and LCAD, and an increase of remnant lipoprotein particles (Kamataki et al. 2002). Studies are still required to elucidate the exact role of the VLDL receptor on cardiac fatty acid metabolism.

2.4. Cytosolic Transport and Activation of Fatty Acids

2.4.1. Fatty Acid Binding Proteins (FABP)

FABPs are the cytoplasmic equivalent of albumin. While albumin transports fatty acids between organs, FABPs are required for the bulk transportation of long-chain fatty acids to specific intracellular targets, such as the mitochondrial outer membrane. FABPs are a family of low molecular weight proteins (14–15 kDa) with at least 13 members, with slightly differing amino acid sequence. These protein were first identified as having a high affinity for fatty acids (Ockner et al. 1972), however their role in fatty acid metabolism was not determined until recently. Hearts from mice with a disruption of the gene encoding H-FABPc (the heart isoform of the cytosolic FABP) show a 50% reduction in rates of palmitate uptake. In addition, fatty acid utilization by cardiomyocytes isolated from H-FABPc deficient mouse hearts are suppressed by 45% despite the fact that the capacity for fatty acid oxidation in heart homogenates does not differ from wild-type animals (Schaap et al. 1998; Binas et al. 1999). A study of skeletal muscle giant vesicles from H-FABP$^{-/-}$ mice demonstrated a similar reduction in fatty acid uptake, but uptake in vesicles from H-FABP$^{+/-}$ is not impaired, suggesting that H-FABP plays a permissive rather than regulatory role in fatty acid uptake (Luiken et al. 2003a). Evidence suggests that fatty acids are desorbed from the inner leaflet of the sarcolemma by collisional interaction of muscle FABPc with the membrane, which differs from liver FABPc whereby the fatty acids are released into the water layer adjacent to the membrane before they bind to FABPc (Storch and Thumser 2000). It has yet to be determined whether FABPc can physically interact with either fatty acyl-CoA sythetase or with the phospholipids of the outer mitochondrial membrane where this enzyme resides. However this interaction would help to aid in the transport of cytoplasmic fatty acids to the enzyme, which activates the fatty acids before they can undergo fatty acid oxidation in the mitochondria.

2.4.2. Activation of Fatty Acids

Fatty acids are activated by esterification to fatty acyl-CoAs by the action of a family of fatty acyl-CoA synthetases (ACS), which differ in their chain length specificity and their subcellular location (Knudsen et al. 1993). The heart isoform of long-chain acyl-CoA synthase is localized to the aqueous cytoplasmic face of the outer mitochondrial membrane (Waku 1992). This conversion to a more reactive CoA thioester is a prerequisite for the subsequent catabolic process. Several lines of evidence suggest that ACS plays a role in the uptake of fatty acids. In adipocytes ACS1 is an integral membrane protein, and it has been shown to localize to the plasma membrane and codistribute with FATP1 (Schaffer 2002). Disruption of yeast orthologs of ACS1, FAA1 and FAA4, results in a restriction of fatty acid uptake and utilization (Faergeman and Knudsen 1997; Choi and Martin 1999). A transgenic mouse model that has increased ACS1 expression in the heart produces marked cardiac TG accumulation, even when serum lipid levels are normal (Chiu et al. 2001). Despite the essential role ACS plays in the catabolism of fatty acids, their regulatory role in fatty acid metabolism has yet to be elucidated.

Once activated, fatty acyl-CoAs can bind to another cytoplasmic binding protein, acyl-CoA-binding protein (ACBP). This protein copurifies with FABP and has been identified in a number of tissues including heart (Mikkelsen and Knudsen 1987). It has been suggested that this protein may regulate fatty acid metabolism by the vectorial delivery of fatty acyl moieties to distinct metabolic pathways (Faergeman and Knudsen 1997). In addition, these proteins may buffer the concentration of long-chain acyl-CoA esters, which would relieve the end product inhibition of acetyl-CoA carboxylase (ACC) and adenine nucleotide translocase (Rasmussen et al. 1993). These long-chain fatty acyl-CoAs can then either be esterified to TGs by glycerolphosphate acyltransferase

(Coleman et al. 2000) or transported into the mitochondria for subsequent β-oxidation (Lopaschuk et al. 1994a).

2.5. Triacylglycerol as a Source of Fatty Acids

TG stores in the myocardium can be very important endogenous sources of fatty acids for fatty acid oxidation (Crass 1977; Paulson and Crass 1982; Saddik and Lopaschuk 1992). In the unphysiological condition of an isolated heart being perfused in the absence of exogenous fatty acids, the heart can meet 50% of its energy requirements by mobilizing its TG stores. However as concentrations of fatty acids delivered to the heart increase, the contribution of these TG pools to myocardial oxidative metabolism decrease; i.e. during perfusion with a high concentration of fatty acids, TG fatty acids still contribute approximately 11% of the total cardiac ATP production (Crass 1972; Saddik and Lopaschuk 1991). This suggests that myocardial triglycerides are a mobilizable endogenous substrate source due to a constant and rapid turnover of the pool. Approximately 10% of fatty acids taken into the heart are first cycled through the TG pool before they undergo oxidation. This TG turnover can be accelerated by adrenergic stimulation (Crass 1977; Paulson and Crass 1982; Stanley et al. 1997; Goodwin et al. 1998; Goodwin and Taegtmeyer 2000), uncontrolled diabetes (Saddik and Lopaschuk 1994) and during reperfusion of the heart following ischemia (Paulson and Crass 1982; Saddik and Lopaschuk 1992).

2.6. Regulation of Fatty Acid Transport into the Mitochondria

2.6.1. Carnitine Transport System

β-oxidation of fatty acids occurs primarily in the mitochondria, however a small portion also occurs in the peroxisomes (Schulz 1994; Kunau et al. 1995). Before long-chain acyl-CoAs can undergo oxidation, they first must be transported into the mitochondrial matrix. The inner mitochondrial membrane is not freely permeable to acyl-CoA, most likely to maintain separate cytosolic and mitochondrial CoA pools, therefore a carnitine-dependent transport system is required (Lopaschuk et al. 1994a; Kerner and Hoppel 2000). This system consists of three components: carnitine palmitoyl transferase 1 (CPT1), carnitine acyltranslocase (CT) and carnitine palmitoyl transferase 2 (CPT2), please see Figure 2.2. First CPT1, found on the inner surface of the outer mitochondrial membrane (Murthy and Pande 1987), catalyzes the conversion of long-chain acyl-CoA to long-chain acylcarnitine. Secondly CT exchanges long-chain acylcarnitine for free carnitine across the inner mitochondrial membrane. Lastly, CPT2 regenerates long-chain acyl-CoA in the mitochondrial matrix. Of these three enzymes, CPT1 is the major site of regulation in controlling the uptake of fatty acids into the mitochondria (Kerner and Hoppel 2000).

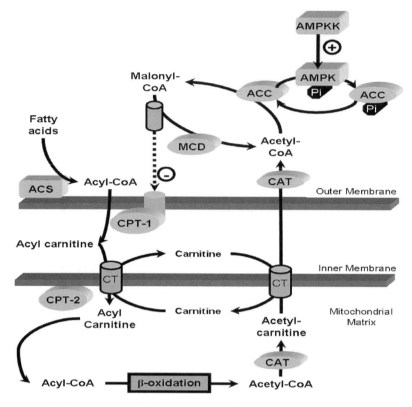

FIGURE 2.2. Regulation of fatty acid transport into the mitochondria. ACC, acetyl-CoA carboxylase; ACS, acyl-CoA synethase; AMPK, AMP-activated protein kinase; AMPKK, AMP-activated protein kinase kinase; CAT, carnitine acetyltransferase; CPT-1, carnitine palmitoyltransferase-1: CPT-2, carnitine palmitoyltransferase-2; CT, carnitine acyltranslocase; FFA, free fatty acid; MCD, malonyl-CoA decarboxylase.

2.6.2. Malonyl-CoA Regulation of Long-Chain Acyl-CoA Transport into the Mitochondria

Fatty acid oxidation is tightly controlled at a number of steps, but the uptake of fatty acyl-CoAs into the mitochondria is arguably one of the most important. CPT1 is the major site of regulation as it is strongly inhibited by malonyl-CoA, a compound that binds on the cytosolic side of the enzyme (McGarry et al. 1983; Lopaschuk et al. 1994a; Zammit et al. 1997; Kerner and Hoppel 2000). The heart expresses two isoforms of CPT1 (CPT1β (82 kDa) and CPT1α (88kDa)) with the 82 kDa isoform predominating. The CPT1β isoform is 30-fold more sensitive to malonyl-CoA inhibition that the CPT1α liver isoform (McGarry et al. 1983; Weis et al. 1994; McGarry and Brown 1997). Therefore, CPT1 activity in the heart is between 10 and 50 times more sensitive to malonyl-CoA than in the liver. This makes malonyl-CoA a key endogenous inhibitor of fatty acid oxidation in the heart, as malonyl-CoA levels are inversely correlated with fatty acid oxidation rates. (Saddik et al. 1993; Kudo et al. 1995; Stanley et al. 1996; Hall et al. 1996a; Abu-Elheiga et al. 2001).

The half-life of malonyl-CoA in the heart is approximately 1.25 minutes (Reszko et al. 2001), making both the synthesis and degradation of malonyl-CoA important in the control of its steady state level. Malonyl-CoA is synthesized by the carboxylation of acetyl-CoA by acetyl-CoA carboxylase (ACC) (Thampy 1989; Kudo et al. 1995; Hall et al. 1996a; Hall et al. 1996b) and degraded to acetyl-CoA and carbon dioxide in the cytosol and mitochondria by malonyl-CoA decarboxylase (MCD) (Kim and Kolattukudy 1978; Dyck et al. 1998; Saha et al. 2000; Kerner and Hoppel 2002; Dyck et al. 2004)

2.6.3. Acetyl-CoA Carboxylase Regulation of Malonyl-CoA

Two different isoform of ACC have been identified, a 265 kDa isoform (ACC265 or ACCα) and a 280 kDa isoform (ACC280 or ACCβ); both isoforms are expressed in the heart, although the prominent form is ACC280 and it appears to contribute most of the heart ACC activity (Bai et al. 1986; Ha et al. 1996; Widmer et al. 1996; Dyck et al. 1999). It has been speculated that ACC280 is directly involved in the regulation of fatty acid oxidation because it is found in tissue with high oxidative capacity and is in close association with the mitochondrial membrane that would allow malonyl-CoA produced to directly inhibit CPT-1 activity (Saddik et al. 1993; Ha et al. 1996; Lee and Kim 1999; Abu-Elheiga et al. 2001). Isolated working heart studies from our laboratory (Saddik et al. 1993; Lopaschuk et al. 1994a; Kudo et al. 1995; Kudo et al. 1996; Makinde et al. 1997) provided direct evidence that ACC is an important regulator of fatty acid oxidation. This has been confirmed in isolated myocytes as well (Awan and Saggerson 1993).

ACC activity is dependent on the supply of its substrate, acetyl-CoA, however most of the acetyl-CoA in cardiomyocytes resides in the mitochondria (Idell-wenger et al. 1978). This is indirect evidence that acetyl-CoA can be derived from the export of mitochondrial acetyl-CoA as acetylcarnitine by the CT (Lysiak et al. 1986; Saddik et al. 1993) or from citrate via the ATP-citrate lyase reaction (Saha et al. 1997; Poirier et al. 2002). Recently, studies making use of ^{13}C-labelled substrates and isotopomer analysis of malonyl-CoA provided direct evidence that malonyl-CoA is derived from acetyl-CoA produced by peroxisomal β-oxidation (Reszko et al. 2004). This study also suggested that long-chain fatty acids underwent partial oxidation in the peroxisomes, and the released products, C12 and C14 acyl-CoAs, were subsequently oxidized in the mitochondria (Reszko et al. 2004). A more in depth review of the regulation of mammalian ACC has been published (Munday 2002).

A major regulator of ACC activity in the hearts and skeletal muscle is AMPK, which phosphorylates three serine residues in ACC265 (serines 79, 1200, and 1215), however only phosphorylation status at ser79 corresponds with inactivation of the enzyme (Davies et al. 1990; Ha et al. 1994). Due to an N-terminal extension in ACC280 the equivalent phosphorylation site is ser219 (Scott et al. 2002). AMPK is a heterotrimeric serine/threonine kinase which consists of one catalytic subunit (α) and two regulatory subunits (β and γ); the most prominent heart isoforms are the α2, β2, and γ1/2 (Stapleton et al. 1996; Dyck et al. 1996; Gao et al. 1996; Thornton et al. 1998). AMPK was originally discovered as a regulator of fatty acid and cholesterol biosynthesis (Carling et al. 1987; Carling et al. 1989). It is now known that AMPK acts as a key molecular regulator of energy metabolism as it functions as an "energy conservation agent" in the liver by inhibiting anabolic processes when ATP levels are depleted (Carling et al. 1994; Carling et al. 1997). However, in skeletal and heart muscle, its action is to up-regulate fatty acid oxidation during occasions of increased energy demand via the phosphorylation and inactivation of ACC. Thus AMPK is a metabolic sensor in these tissues, which monitors the ratios of AMP/ATP and Cr/PCr, via a mechanism involving the direct allosteric modification of the enzyme by AMP, the activation of an upstream kinase (AMPKK), and the inhibition of protein phosphatase 2C activity (Hawley et al., 1995; Ponticos et al. 1998). Despite the discovery of the identity of two of the AMPKKs, LKB1 (Hawley et al. 2003) and Ca calmodulin dependent kinase kinase (Hurley et al. 2005), it appears that there are additional AMPKKs which mediate the ischemic induced activation of AMPK (Altarejos et al. 2005). Our laboratory has shown that in isolated working rat hearts, not only does ACC copurify with the α2 catalytic subunit of AMPK, it can phosphorylate and inactivate both isoforms of ACC, resulting in an almost complete loss of activity (Kudo et al. 1996; Dyck et al. 1999). Many studies have shown a correlation between increased AMPK activity, decreased ACC activity and increased fatty acid oxidation in working rat hearts (Kudo et al. 1995; Kudo et al. 1996; Makinde et al. 1997; Makinde et al. 1998).

2.6.4. *Malonyl-CoA Decarboxylase Regulation of Malonyl-CoA*

MCD catalyzes the decarboxylation of malonyl-CoA, and thus has been proposed to play an important role in the regulation of myocardial fatty acid oxidation. Generally in situations where MCD activity is elevated, malonyl-CoA content is low, which results in elevated rates of fatty acid oxidation (Dyck et al. 1998; Campbell et al. 2002; Dyck et al. 2004). Despite this correlative data, there is a controversy as to where MCD is localized in the cell. As only cytosolic malonyl-CoA is able to inhibit CPT-1, it would be logical that MCD would reside on or near the mitochondrial membrane, however others have reported it to be localized to the mitochondria, peroxisomes or in the cytosol (Alam and Saggerson 1998; FitzPatrick et al. 1999; Sacksteder et al. 1999; Dyck et al. 2000). This differential localization may be due to two putative translational start sequences (Gao et al. 1999), one of which encodes a protein with a mitochondrial targeting sequence and a peroxisomal targeting sequence (SKL), the second of which codes for a shorter protein that lacks the mitochondrial targeting sequence. Despite this controversy, it still appears that MCD plays an important role in the regulation of fatty acid oxidation, and patients with MCD deficiencies have phenotypes consistent with alterations in fatty acid oxidation (Brown et al. 1984; Yano et al. 1997).

2.7. Regulation of β-Oxidation

Once acyl-CoAs have entered the mitochondria they undergo β-oxidation, which produces NADH and $FADH_2$ for the electron transport chain and acetyl-CoA for further catabolism in the Krebs cycle. β-oxidation requires four different steps as outlined in Figure 2.3, each of which is catalyzed by a specific enzyme with specificity for short-, medium- and long-chain acyl-CoA intermediates (Bing et al. 1954). The acyl-CoA undergoes repeated oxidation via these reactions until only acetyl-CoA units remain. The first step of β-oxidation consists of the conversion of acyl-CoA to Δ2-3-trans-enoyl-CoA in the presence of the FAD, and is catalyzed by acyl-CoA dehydrogenase. Enoyl-CoA hydratase then adds water across the double bond to produce L-3-hydroxyacyl-CoA. In the second dehydrogenation step of the pathway, L-3-hydroxyacyl-CoA dehydrogenase catalyzes the oxidation of L-3-hydroxyacyl-CoA to L-3-ketoacyl-CoA,

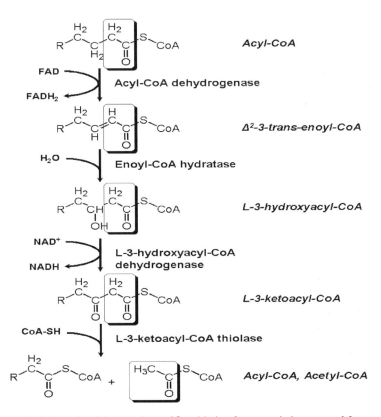

FIGURE 2.3. Schematic of the reactions of β-oxidation for even-chain saturated fatty acids.

with the involvement of NAD^+ as a cofactor. The final step of the pathway consists of the splitting of L-3-ketoacyl-CoA at the 2,3-position by 3-ketoacyl-CoA thiolase, in the presence of CoA, to produce one acetyl-CoA unit and an acyl-CoA that is 2 carbons shorter than the original acyl-CoA. This acyl-CoA feeds back to acyl-CoA dehydrogenase to start another cycle of β-oxidation, while acetyl-CoA goes to the Krebs cycle for further oxidation.

The oxidation of unsaturated fatty acids require additional enzymes. For odd numbered double bonds, oxidation occurs until a Δ3-cis-enoyl-CoA intermediate is produced, which is isomerized by Δ3, Δ2-enoyl-CoA isomerase to Δ2-trans-enoyl-CoA, which can be further oxidized. The oxidation of even numbered double bonds requires an additional step. First these fatty acids undergo oxidation until a Δ2-trans, Δ4-cis-dienoyl-CoA intermediate is produced. This intermediate is reduced by 2,4-dienoyl-CoA reductase to Δ3-trans-enoyl-CoA which is isomerized by Δ3, Δ2-enoyl-CoA isomerase to Δ2-trans-enoyl-CoA, which can be further oxidized. For a more comprehensive review of β-oxidation in the heart please see the reviews by Schulz (Schulz 1991; Schulz 1994).

β-oxidation is primarily dependent on the workload of the heart, as ratios of $NADH/NAD^+$ and acetyl-CoA/CoA decrease in response to an increase in workload (Oram et al. 1973; Neely et al. 1976). The decrease in acetyl-CoA/CoA ratio releases the suppression of L-3-ketoacyl-CoA thiolase activity and a decrease in matrix acetyl-CoA also activates acyl-CoA dehydrogenase (Oram et al. 1973). A decrease in workload would result in an increase in L-3-ketoacyl-CoA, which feeds back and inhibits acyl-CoA dehydrogenase (Olowe and Schulz 1980). Reports in liver suggest that the $NADH/NAD^+$ ratio regulates L-3-hydroxyacyl-CoA dehydrogenase in the presence of low acetyl-CoA levels, however it is unknown whether this is an important site of regulation in the heart (Latipaa et al. 1986).

Alternative methods of feedback regulation may exist in the heart. Elevated levels β-oxidation or flux through pyruvate dehydrogenase result in an increase in matrix acetyl-CoA. It has been proposed that this excess acetyl-CoA is transported out of the mitochondria by one of two mechanisms: 1) conversion to acetylcarnitine by carnitine acetyltransferase (CAT), transport across the membrane by CT, and conversion back to acetyl-CoA by a cytosolic CAT (Lysiak et al. 1986; Saddik et al. 1993) or 2) removal of citrate from the mitochondria by the tricarboxylic acid transporter and conversion to acetyl-CoA by the ATP-citrate lyase (Saha et al. 1997; Poirier et al. 2002). Each of these processes increases cytosolic acetyl-CoA levels, which has the potential to accelerate the production of malonyl-CoA, and therefore suppress fatty acid oxidation as shown in Figure 2.2.

Recently a mechanism to augment fatty acid oxidation has been proposed that involves mitochondrial thioesterase 1 (MTE1) and uncoupling protein 3 (UCP3) (Himms-Hagen and Harper 2001). MTE1 catalyzes the hydrolysis of fatty acyl-CoA molecules to nonesterified fatty acid anions and free CoA, while UCP3 are classically known for uncoupling ATP synthesis from oxidative metabolism, but may act as a fatty acid anion transporter (Palou et al. 1998; Samec et al. 1998). This hypothesis suggests that in the presence of high rates of fatty acid oxidation, free CoA may become limiting for CPT-2 and 3-ketoacyl-CoA thiolase, and hence could suppress the flux thru β-oxidation. Therefore hydrolysis of acyl-CoAs by MTE1 will liberate free CoA to allow for continued β-oxidation, and the fatty acid anion generated would be transported out of the mitochondria by UCP3 and be regenerated to acyl-CoA in the cytosol (Stavinoha et al. 2004). However this hypothesis has yet to be examined in the intact heart.

2.8. Interregulation of Carbohydrate and Fatty Acid Metabolism

Rates of glucose and fatty acid oxidation are coordinately regulated in the heart, with the rate of fatty acid oxidation being the primary physiological regulator of glucose oxidation. Elevated rates of fatty acid oxidation can inhibit pyruvate dehydrogenase (PDH) activity, the rate limiting enzyme of glucose oxidation. This occurs due to an increase in mitochondrial NADH/NAD$^+$ and acetyl-CoA/CoA ratios which directly interferes with PDH activity, as well as activating the PDH kinase which phosphorylates and inhibits PDH. In addition, glucose oxidation may regulate fatty acid oxidation via the efflux of acetyl-CoA from the mitochondria and its conversion to malonyl-CoA (as described previously). Therefore inhibition of fatty acid oxidation results in an increase in glucose and lactate uptake, as well as oxidation by lowering mitochondrial acetyl-CoA and NADH levels and relieving the inhibition of PDH, and by decreasing citrate levels and inhibition of phosphofructokinase-1, the rate limiting step of glycolysis. The functional consequences in the switch in substrate usage will be discussed in section 2.11.

2.9. Transcriptional Control of Fatty Acid Oxidation Enzymes

The regulation of the expression of proteins that are involved in the myocardial catabolism of fatty acid is not well understood. Except for genes that encode thirteen subunits of the respiratory complexes which are found in circular mitochondria DNA (Scarpulla 2002), all the proteins required for fatty acid uptake and oxidation are encoded by nuclear genes. Recent studies have provided insight that many cardiac genes are regulated by the ligand-activated transcription factors, peroxisome proliferators-activated receptors (PPARs) (Berger and Moller 2002; Huss and Kelly 2004). Three PPAR isoforms have been discovered: PPARα, PPARβ/δ and PPARβ/δ, and their functional specificity is achieved by isoform-specific tissue distribution, or by differences in ligand affinity or cofactor interactions (Braissant et al. 1996; Escher and Wahli 2000). All of these PPAR isoforms are activated when fatty acids bind, and form heterodimers with retinoid X receptors, and then bind to specific PPAR response elements

(PPREs) which are located within the promoter regions of many metabolic genes. Binding of fatty acids and eicosanoids increase the activity of the heterodimers, therefore they act as lipid sensors, which increase the facility for fatty acid metabolism in response to greater exposure to fatty acids (Berger and Moller 2002; Huss and Kelly 2004). Additionally PPARs also recruit transcriptional coactivators which are required to initiate target gene transcription, such as PPARγ coactivator 1 (PGC-1), which when overexpressed increases the mRNA production of metabolic genes and starts mitochondrial biogenesis (Vega et al. 2000; Lehman et al., 2000; Robyr et al. 2000). Once this complex binds to the PPRE, it increases the rate of transcription of the specific gene that it is bound to by recruitment of histone acetyltransferases, which increases access to the transcriptional start site (Barger and Kelly 2000), as shown in Figure 2.4.

PPARα is highly expressed in tissues that have high rates of fatty acid oxidation, such as heart, liver and skeletal muscle. At present the endogenous ligand for PPARα is unknown, however the fibrate class of hyperlipidemic drugs are synthetic PPARα ligands, and they increase the expression of fatty acid oxidation genes and the rates of fatty acid oxidation in cardiomyocytes (Gilde et al. 2003). PPARα regulated genes are involved in most steps of cardiac FA utilization including fatty acid uptake, activation to fatty acyl-CoA, transport into the mitochondria, and mitochondrial ß-oxidation. Evidence from PPARα knockout mouse studies show that fatty acid oxidation enzymes are downregulated and have suppressed fatty acid oxidation (Campbell et al. 2002). Mice with a cardiac specific overexpression of PPARα result in an upregulation of fatty acid utilization enzymes and elevated rates of fatty acid oxidation (Finck et al. 2002; Hopkins et al. 2003).

PPARβ/δ is also expressed in cardiomyocytes and PPARβ/δ selective ligands can increase mRNA for fatty acid oxidation genes and increase fatty acid oxidation in both neonatal and adult cardiomyocytes (Gilde et al. 2003; Cheng et al. 2004b). In addition cardiomyocyte deletion of PPARβ/δ downregulates mRNA and protein of fatty acid oxidation genes and reduces catabolism of fatty acids (Cheng et al. 2004a). PPARß activation can also rescue the expression of fatty acid oxidations enzymes that are reduced at baseline in PPAR knockout cardiomyocytes (DeLuca et al. 2000; Cheng et al. 2004b). It appears that PPARα and PPARβ/δ regulate a similar series of genes, however they are not functionally redundant and PPARβ/δ cannot completely compensate for PPARα in the knockout mice (Huss and Kelly 2004).

PPARγ is primarily expressed in adipocytes but is expressed at low levels in other tissues including the vascular wall, skeletal muscle, pancreatic β-cell, and heart. PPARγ is primarily a regulator of lipid storage and adipose formation, and does not appear to play a role in the regulation of fatty acid oxidation in the heart (Gilde et al. 2003; Lee et al. 2003). However a recent study showed that a cardiac specific knockout of PPARγ results in hypertrophic growth; however mRNA or protein expression of fatty acid utilization genes or rates of fatty acid oxidation were not measured in this study (Duan et al. 2005). This would suggest that PPARγ plays a role in the heart; however PPARγ can also have

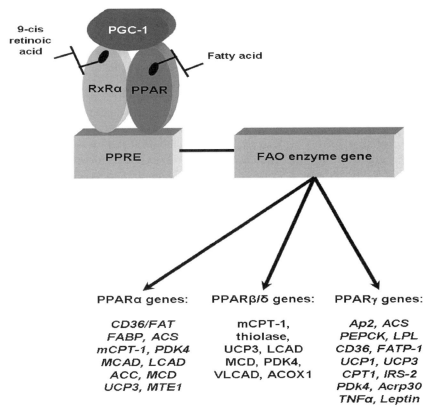

FIGURE 2.4. Regulation of expression of fatty acid metabolism genes by stimulation of peroxisome proliferators activated receptors. PPARα and PPARβ/δ directly modify genes in the heart, while most of PPARγ effects are extracardiac and occur in either the adipose tissue or skeletal muscle tissue. ACC, acetyl-CoA carboxylase; ACOX1, acyl-CoA oxidase 1; Acrp 30, adipocyte complement-related factor 30; ACS, acyl-CoA synthetase; FAT, fatty acid transporter; FATP-1, fatty acid transport protein 1; IRS-2, insulin receptor substrate 2; LCAD, long-chain acyl-CoA dehydrogenase; LPL, lipoprotein lipase; MCAD, medium-chain acyl-CoA dehydrogenase; MCD, malonyl-CoA decarboxylase, mCPT-1, muscle isoform of carnitine palmitoyltransferase-1; MTE1, mitochondrial thioesterase 1; PDK4, pyruvate dehydrogenase kinase 4; PEPCK, phosphoenolpyruvate carboxylase; thiolase, 3-ketoacyl-CoA thiolase; TNFα, tumor necrosis factor α; UCP1, uncoupling protein 1; UCP3, uncoupling protein 3; VLCAD, very-long-chain acyl-CoA dehydrogenase.

indirect effects on myocardial fatty acid oxidation via decreasing the fatty acid concentration to which the heart is exposed. For example, the insulin sensitizing agent the thiazolidinediones are PPARγ ligands, and their mode of action is to sequester fatty acids in the adipocytes, thus lowering circulating fatty acids and triglycerides (Berger and Moller 2002).

2.10. Alterations in Fatty Acid Oxidation in Disease

2.10.1. *Hypertrophy Induced Changes in Myocardial Fatty Acid Metabolism*

Cardiac hypertrophy has been associated with reduced pre- and post-ischemic function and is a risk factor for the development of congestive heart failure, myocardial infarction and sudden death (Taegtmeyer and Overturf 1988; Frohlich et al. 1992). Hypertrophied hearts have a reduction in the oxidation of long-chain fatty acids, however oxidation of octanoate, a medium chain fatty acid, is not different (el Alaoui-Talibi et al. 1992; Allard et al. 1994). This suggests that mitochondrial β-oxidation is not impaired in non-decompensated hypertrophied hearts, but some other component of long-chain fatty acid utilization is impaired. A number of alterations may explain the reduction of long-chain fatty acid oxidation in the hypertrophied heart. Spontaneously hypertensive rats have suppressed expression of cytosolic FABP and CD36, which would reduce fatty acid supply to the mitochondria (Aitman et al. 1999; Van der Vusse et al. 2000). In addition, long-chain fatty acid transport into the mitochondria may also be suppressed due to reduced carnitine concentrations in the hypertrophied heart, which could limit carnitine availability for CPT1 (Reibel et al. 1983; Allard et al. 1994; el Alaoui-Talibi et al. 1997). It has also been hypothesized that there is impaired interaction between acyl-CoA synthetase and CPT1 which would also suppress long-chain fatty acid transport into the mitochondria (Christian et al. 1998). Hypertrophied hearts which are showing signs of decompensation (heart failure) also have a reduction of mRNA expression of enzymes required for fatty acid oxidation, including long- and medium-chain acyl-CoA dehydrogenase and long-chain 3-hydroxyacyl-CoA dehydrogenase (Sack et al. 1996; Barger and Kelly 1999).

Alterations in fatty acid oxidation in heart failure may share similar characteristics of fatty acid oxidation alterations in hypertrophy. These changes will not be discussed here since the metabolic effect of heart failure has recently been extensively reviewed (Stanley et al. 2005). However, similar to hypertrophy, fatty acid oxidation appears to decrease as the severity of heart failure increases. These alterations in fatty acid utilization in the hypertrophied heart have profound effects on ischemic tolerance, as reviewed in (Sambandam et al. 2002).

2.10.2. *Diabetes Induced Changes in Myocardial Fatty Acid Oxidation*

Dramatic changes in energy metabolism occur in diabetes and are associated with impaired contractile function (Lopaschuk 1996). In uncontrolled diabetes fatty acid oxidation can provide between 90 and 100% of the heart's energy requirement (Randle 1986; Wall and Lopaschuk 1989). Even if diabetic rat hearts are perfused in the absence of fatty acids, glucose oxidation still only provides about 20% of the heart's total ATP requirement and endogenous triacylglyerol provides a major source of energy. These elevated rates of fatty acid oxidation can be partially explained by a deficiency in insulin signaling resulting in a suppression of glucose uptake. The primary reasons for the elevated rates appear to be an increased level of circulating fatty acids and triglycerides and a modification of the intracellular control of fatty acid oxidation. One site of regulation that appears to be important in diabetes is CPT1. It appears that sensitivity of heart CPT1 to malonyl-CoA does not change (Stanley et al. 1999). However, we have shown in streptozotocin diabetic rats (a model of type 1 diabetes), that AMPK activity is elevated which is associated with a significant decrease in ACC activity (Lopaschuk 1996). In addition MCD activity is significantly increased in the heart (Dyck et al. 1998; Sakamoto et al. 2000). Taken together these results suggest that a reduction in malonyl-CoA inhibition of CPT1 would account for elevated rates of fatty acid oxidation in the diabetic heart. Elevated levels of circulating fatty acids may also result in the activation of PPARα and transcription of fatty acid utilization genes (as review in (Young et al. 2002)). Indeed, transgenic mice with a cardiac specific overexpression of PPARα produce a cardiac phenotype that is similar to the phenotype caused by diabetes (Finck et al., 2002). When these mice were treated with a specific PPARα activator, many fatty acid utilization genes were upregulated including M-CPT1, FATP1, FACS1, CD36, UCP2 and UCP3 (Finck et al. 2002). Additionally PPARα also stimulates the induction of PDK4, which results in the phosphorylation and inhibition of PDH, thus further suppressing glucose oxidation. Overall similar results were also seen in hearts from type 2 diabetic animals as reviewed in (Carley and Severson 2005).

2.10.3. *Genetic Deficiencies in Fatty Acid Oxidation*

The last two decades have seen a steady increase in the identification of a number of mitochondrial defects and associated disease states in patients (Gregersen et al. 2001; Gregersen et al. 2004). Deficiencies in many different fatty acid utilization transporters and enzymes have now been discovered. One of the largest groups of enzymes with heritable mutations are the acyl-CoA dehydrogenases (Gregersen et al. 2004). Generally, defects in the catabolism of medium- or short-chain fatty acids are less likely to produce heart failure or cardiomyopathies than defects in long-chain fatty acid oxidation. One of the most common of these deficiencies is in medium chain acyl-CoA dehydrogenase (MCAD) (Tanaka et al. 1992). Over 90% of these cases involved a homozygous mutation at base pair 985 (A985G) of MCAD, which results in the replacement of a lysine with a glutamate at residue 304 that interferes with tetramer assembly and decreases protein stability (Kelly et al. 1992). Specific defects have also been found in longchain and very long-chain acyl-CoA dehydrogenases (Hale et al. 1985; Strauss et al. 1995).

Carnitine deficiencies are usually associated with severe cardiomyopathies and are typically dependent on defects in carnitine transport or carnitine translocase

(Engel and Angelini 1973). For a more in depth review of fatty acid oxidation defects please see the following reviews (Marin-Garcia and Goldenthal 2002; Gregersen et al. 2004). Although the direct myocardial metabolic effects of these defects have not been determined, it is believed that due to suppressed oxidation of fatty acids, there is an accumulation of intermediary metabolites of fatty acid oxidation, such as long-chain acyl CoAs. These intermediates have detergent-like properties which can modify membrane proteins and lipids, and lead to cardiac arrhythmias and conduction defects (Marin-Garcia and Goldenthal 2002). Additionally fatty acids have also been implicated in the induction of apoptosis (Marin-Garcia and Goldenthal 2002).

2.11. Fatty Acid Oxidation in the Ischemic and Reperfused Heart and Optimization of Fatty Acid Oxidation as a Therapeutic Approach to Treat Ischemic Heart Disease

Inadequate supply of oxygen during ischemia results in a reduction in overall oxidative metabolism associated with an impairment of ATP production, the magnitude of which is dependent on the severity of ischemia (Neely and Morgan 1974; Opie 1998). During severe ischemia, the primary source of ATP is glycolytic metabolism of glycogen-derived glucose because coronary blood flow is significantly reduced or even halted; thus substrate and oxygen supply to the heart is suppressed and all oxidative metabolism effectively ceases (Neely and Morgan 1974; Opie 1998). As pyruvate dehydrogenase, the rate limiting enzyme of glucose oxidation, is inhibited, the pyruvate derived from glycolysis is converted to lactate to preserve sufficient NAD^+ to sustain flux through glycolysis rather than being metabolized in the mitochondria. This uncoupling of glycolysis from glucose oxidation produces protons from glycolytic ATP hydrolysis, and the inability to remove these protons due to the reduction in coronary blood flow is associated with an intracellular acidosis (Liu et al. 1996a; Liu et al. 1996b; King and Opie 1998; Liu et al. 2002). Homeostasis of normal proton concentrations is dependent on a number of pathways, with two of them involving Na^+-dependent mechanisms: the Na^+-H^+ exchanger 1 (NHE1) (Karmazyn et al. 1999) and the Na^+-HCO_3^- cotransporter (NBC1, 3, or 4) (Sterling and Casey 2002). During normal aerobic perfusion excess intracellular sodium is removed by the Na^+-K^+ ATPase, however during ischemia and reperfusion this transporter is inactivated due to a drop in ATP and intracellular pH. Therefore during ischemia and reperfusion proton extrusion is linked to an intracellular Na^+ overload and activation of the reverse mode of the Na^+-Ca^{2+} exchanger (NCX) (Karmazyn and Moffat 1993), as shown in Figure 2.5. Activation of NCX results in the ischemia-induced Ca^{2+} overload that is associated with irreversible injury such as apoptosis and necrosis, and reversible injury such as arrhythmias and stunning. This ionic derangement is also detrimental because there is an increased need for ATP to correct these ionic imbalances, which shunts ATP away from contractile function and results in a decrease in contractile efficiency. Despite the fact that ATP is quickly replenished due to the rapid recovery of oxygen consumption and TCA cycle activity during reperfusion, mechanical function can only recover once Ca^{2+} levels have normalized (Lopaschuk et al. 1990; Benzi and Lerch 1992; Liu et al. 1996b; Liu et al. 2002).

During reperfusion of previously ischemic hearts, fatty acids can provide over 90% of the myocardium's ATP requirement (Lopaschuk et al. 1990). This excessive use of fatty acids is due to processes discussed earlier in this review, including an increase in substrate supply due to hormonal stimulation of lipolysis in adipose tissue (Oliver and Opie 1994; Lopaschuk et al. 1994b) and due to a decrease in myocardial malonyl-CoA levels and reduced inhibition of CPT-1 (Kudo et al. 1995; Dyck et al. 1999). The reduction in malonyl-CoA is due to the ischemia-induced activation of AMPK and phosphorylation and inactivation of ACC. MCD activity is preserved during reperfusion; therefore there is a reduction in malonyl-CoA levels, and a stimulation of fatty acid oxidation. These excessive rates of fatty acid oxidation have important consequences on glucose oxidation as these pathways are coordinately regulated (Randle et al. 1963). This phenomenon is of particular importance since elevated rates of fatty acid

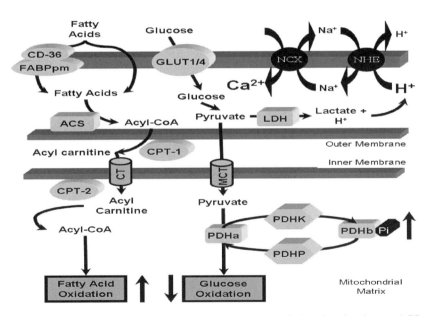

FIGURE 2.5. Metabolic effects of ischemia and reperfusion in the heart. ACS, acyl-CoA synethase; CPT-1, carnitine palmitoyltransferase-1: CPT-2, carnitine palmitoyltransferase-2; CT, carnitine acyltranslocase; FABPpm, plasma membrane fatty acid binding protein, GLUT1/4, glucose transporter 1 or 4; LDH, lactate dehydroganse; NCX, sodium/calcium exchanger; NHE, sodium/hydrogen exchanger; PDH, pyruvate dehydrogenase; PDHK, pyruvate dehydrogenase kinase; PDHP, pyruvate dehydrogenase phosphatase.

oxidation rise produce an allosteric activation of pyruvate dehydrogenase kinase, which phosphorylates and inactivates the pyruvate dehydrogase complex, thus inhibiting glucose oxidation (Neely and Morgan 1974). Therefore elevated fatty acid oxidation rates during reperfusion of ischemic hearts can exacerbate the uncoupling of glucose metabolism during reperfusion. The use of metabolic modulation to inhibit fatty acid oxidation and stimulate glucose oxidation is a novel therapeutic approach that can be utilized to improve cardiac efficiency (McCormack et al. 1996; Liu et al. 1996a; Kantor et al. 2000; Lopaschuk et al. 2003). Modification of myocardial metabolism is one of the key consequences of ischemia and reperfusion that can contribute to reversible or irreversible ischemia-reperfusion injury. An important therapeutic approach would therefore be to reduce fatty acid oxidation and improve the coupling of glucose metabolism. This would prevent the dysregulation of ion homeostasis and the associated impairment of contractile function. Several different agents have already been tested in experimental models and some are currently being used clinically. These agents can be divided into two classes: those that lower circulating free fatty acids and those that directly modify fatty acid oxidation or glucose oxidation.

Agents that lower circulating fatty acids include compounds such as β-adrenoceptor antagonists (β-blockers), glucose-insulin-potassium (GIK) infusions, and nicotinic acid. The hypothesized modes of action of these compounds is the reduction in circulating free fatty acids, which reduces fatty acid supply to the mitochondria and therefore decrease fatty acid oxidation rates and increase the aerobic disposal of pyruvate.

Another approach to lower fatty acid oxidation is to directly inhibit fatty acid oxidation. This approach has been shown with a class of piperazine derivatives, including trimetazidine and ranolazine. Trimetazidine is an inhibitor of long-chain 3-ketoacylcoenzyme A thiolase that reduces fatty acid oxidation and increases glucose oxidation via the Randle cycle (Kantor et al. 2000; Lopaschuk et al. 2003). It is the first widely used antianginal drug with a mechanism of action that can be attributed to the optimization of fatty acid oxidation. Ranolazine is an anti-anginal agent that also inhibits fatty acid oxidation (McCormack et al. 1996). Agents which accelerate glucose oxidation have also been shown to reduce ischemic injury. The prototype of this class, dichloroacetate (DCA), acts by way of the inhibition of PDH kinase, thus ultimately activating PDH and glucose oxidation (Stacpoole 1989). For a more in depth review of metabolic modulation please see these reviews ((Lopaschuk et al. 2002; Folmes et al. 2005).

2.12. Conclusions

The heart is dependent on the oxidation of fatty acids as a source of ATP for contraction and ion homeostasis. Therefore the use of fatty acids by the heart is a highly regulated process and is dependent on many different processes: substrate supply and uptake by the cardiomyocyte, transport and activation in the cytosol, transport into the mitochondria and finally β-oxidation of fatty acids. Despite the complexity of this regulation, it ensures a continual supply of acyl-CoA for oxidation. However, the detrimental effects of dysregulation of fatty acid utilization are readily apparent in various forms of disease. As a result the use of pharmacological modulators of myocardial metabolism is an exciting new approach for the treatment of heart disease.

Acknowledgements. CDLF is a trainee of the Alberta Heritage Foundation for Medical Research, Canadian Institute of Health Research and Tomorrow's Research Cardiovascular Health Professionals (TORCH). GDL is a Medical Scientist of the Alberta Heritage Foundation for Medical Research.

References

Abu-Elheiga L, Matzuk MM, Abo-Hashema KA, Wakil SJ (2001) Continuous fatty acid oxidation and reduced fat storage in mice lacking acetyl-CoA carboxylase 2. Science 291: 2613–6

Abumrad N, Harmon C, Ibrahimi A (1998) Membrane transport of long-chain fatty acids: evidence for a facilitated process. J Lipid Res 39: 2309–2318

Abumrad NA, Park JH, Park CR (1984) Permeation of long-chain fatty acid into adipocytes. Kinetics, specificity, and evidence for involvement of a membrane protein. J Biol Chem 259: 8945–8953

Abumrad NA, Perkins RC, Park JH, Park CR (1981) Mechanism of long chain fatty acid permeation in the isolated adipocyte. J Biol Chem 256: 9183–9191

Aitman TJ, Glazier AM, Wallace CA, Cooper LD, Norsworthy PJ, Wahid FN, Al-Majali KM, Trembling PM, Mann CJ, Shoulders CC, Graf D, St LE, Kurtz TW, Kren V, Pravenec M, Ibrahimi A, Abumrad NA, Stanton LW, Scott J (1999) Identification of Cd36 (Fat) as an insulin-resistance gene causing defective fatty acid and glucose metabolism in hypertensive rats. Nat Genet 21: 76–83

Alam N, Saggerson ED (1998) Malonyl-CoA and the regulation of fatty acid oxidation in soleus muscle. Biochem J 334 (Pt 1), 233–41

Allard MF, Schonekess BO, Henning SL, English DR, Lopaschuk GD (1994) Contribution of oxidative metabolism and glycolysis to ATP production in hypertrophied hearts. Am J Physiol 267: 742–50

Altarejos JY, Taniguchi M, Clanachan AS, Lopaschuk GD (2005) Myocardial ischemia differentially regulates LKB1 and an alternate 5'-AMP-activated protein kinase kinase. J Biol Chem 280: 183–90

Anderson MP, Welsh MJ (1990) Fatty acids inhibit apical membrane chloride channels in airway Epithelia. PNAS 87: 7334–7338

Augustus AS, Kako Y, Yagyu H, Goldberg IJ (2003) Routes of FA delivery to cardiac muscle: modulation of lipoprotein lipolysis alters uptake of TG-derived FA. Am J Physiol Endocrinol Metab 284: E331–E339

Augustus A, Yagyu H, Haemmerle G, Bensadoun A, Vikramadithyan RK, Park SY, Kim JK, Zechner R, Goldberg IJ (2004) Cardiac-specific knock-out of lipoprotein lipase alters plasma lipoprotein triglyceride metabolism and cardiac gene expression. J Biol Chem 279: 25050–25057

Awan MM, Saggerson ED (1993) Malonyl-CoA metabolism in cardiac myocytes and its relevance to the control of fatty acid oxidation. Biochem J 295 (Pt 1): 61–6

Bai DH, Pape ME, Lopez-Casillas F, Luo XC, Dixon JE, Kim KH (1986) Molecular cloning of cDNA for acetyl-coenzyme A carboxylase. J Biol Chem 261: 12395–12399

Barger PM, Kelly DP (1999) Fatty acid utilization in the hypertrophied and failing heart: molecular regulatory mechanisms. Am J Med Sci 318: 36–42

Barger PM, Kelly DP (2000) PPAR Signaling in the control of cardiac energy metabolism. Trends Cardiovasc Med 10: 238–245

Benzi RH, Lerch R (1992) Dissociation between contractile function and oxidative metabolism in postischemic myocardium. Attenuation by ruthenium red administered during reperfusion. Circ Res 71: 567–76

Berger J, Moller DE (2002) The mechanisms of action of PPARs. Annu Rev Med 53: 409–435

Berk PD, Wada H, Horio Y, Potter BJ, Sorrentino D, Zhou S, Isola LM, Stump D, Kiang C, Thung S (1990) Plasma membrane fatty acid-binding protein and mitochondrial Glutamic- Oxaloacetic Transaminase of rat liver are related. PNAS 87: 3484–3488

Berk PD, Zhou SL, Kiang CL, Stump D, Bradbury M, Isola LM (1997) Uptake of long chain free fatty acids is selectively up-regulated in adipocytes of Zucker Rats with genetic obesity and non-insulin-dependent diabetes mellitus. J Biol Chem 272: 8830–8835

Binas BERT, Danneberg HEIK, McWhir JI, Mullins LIN, Clark AJ (1999) Requirement for the heart-type fatty acid binding protein in cardiac fatty acid utilization. FASEB J 13: 805–812

Bing RJ, Siegel A, Ungar I, Gilbert M (1954).Metabolism of the Human Heart .2. Studies on fat, ketone and amino acid metabolism. Am J Med 16: 504–515

Binnert C, Koistinen HA, Martin G, Andreelli F, Ebeling P, Koivisto VA, Laville M, Auwerx J, Vidal H (2000) Fatty acid transport protein-1 mRNA expression in skeletal muscle and in adipose tissue in humans. Am J Physiol Endocrinol Metab 279: E1072–E1079

Bonen A, Luiken JJ, Glatz J F (2002) Regulation of fatty acid transport and membrane transporters in health and disease. Mol Cell Biochem 239: 181–192

Bonen A, Luiken JJFP, Arumugam Y, Glatz JFC, Tandon NN (2000) Acute Regulation of fatty acid uptake involves the cellular redistribution of fatty acid translocase. J Biol Chem 275: 14501–14508

Braissant O, Foufelle F, Scotto C, Dauca M, Wahli W (1996) Differential expression of peroxisome proliferator-activated receptors (PPARs): tissue distribution of PPAR-alpha, -beta, and -gamma in the adult rat. Endocrinology 137: 354–366

Brinkmann JF, Abumrad NA, Ibrahimi A, van d V, Glatz JF (2002) New insights into long-chain fatty acid uptake by heart muscle: a crucial role for fatty acid translocase/CD36. Biochem J 367: 561–570

Brown GK, Scholem RD, Bankier A, Danks DM (1984) Malonyl coenzyme A decarboxylase deficiency. J Inherit Metab Dis. 7: 21–26

Campbell FM, Kozak R, Wagner A, Altarejos JY, Dyck JRB, Belke DD, Severson DL, Kelly DP, Lopaschuk GD (2002) A role for peroxisome proliferator-activated receptor alpha (PPAR alpha) in the control of cardiac malonyl-CoA levels - Reduced fatty acid oxidation rates and increased glucose oxidation rates in the hearts of mice lacking PPAR alpha are associated with higher concentrations of maloncyl-CoA and reduced expression of malonyl-CoA decarboxlase. J Biol Chem 277: 4098–4103

Carley AN, Severson DL (2005) Fatty acid metabolism is enhanced in type 2 diabetic hearts. Biochim Biophys Acta 1734: 112–26

Carling D, Aguan K, Woods A, Verhoeven AJ, Beri RK, Brennan CH, Sidebottom C, Davison MD, Scott J (1994) Mammalian AMP-activated protein kinase is homologous to yeast and plant protein kinases involved in the regulation of carbon metabolism. J Biol Chem 269: 11442–8

Carling D, Clarke PR, Zammit VA, Hardie DG (1989) Purification and characterization of the AMP-activated protein kinase. Copurification of acetyl-CoA carboxylase kinase and 3-hydroxy-3-methylglutaryl-CoA reductase kinase activities. Eur J Biochem 186: 129–36

Carling D, Woods A, Thornton C, Cheung PC, Smith FC, Ponticos M, Stein SC (1997) Molecular characterization of the AMP-activated protein kinase and its role in cellular metabolism. Biochem Soc Trans 25: 1224–8

Carling D, Zammit VA, Hardie DG (1987) A common bicyclic protein kinase cascade inactivates the regulatory enzymes of fatty acid and cholesterol biosynthesis. FEBS Lett 223: 217–222

Cheng LH, Ding GL, Qin QH, Huang Y, Lewis W, He N, Evans RM, Schneider MD, Brako FA, Xiao Y, Chen YQE, Yang QL (2004a) Cardiomyocyte-restricted peroxisome proliferator-activated receptor-delta deletion perturbs myocardial fatty acid oxidation and leads to cardiomyopathy. Nat Med 10: 1245–1250

Cheng L, Ding G, Qin Q, Xiao Y, Woods D, Chen YE, Yang Q (2004b) Peroxisome proliferator-activated receptor [delta] activates fatty acid oxidation in cultured neonatal and adult cardiomyocytes. Biochem Biophys Res Commun 313: 277–286

Chiu HC, Kovacs A, Ford DA, Hsu FF, Garcia R, Herrero P, Saffitz JE, Schaffer JE (2001) A novel mouse model of lipotoxic cardiomyopathy. J Clin Invest 107: 813–822

Choi JY, Martin CE (1999) The Saccharomyces cerevisiae FAT1 Gene Encodes an Acyl-CoA Synthetase That Is Required for Maintenance of Very Long Chain Fatty Acid Levels. J Biol Chem 274: 4671–4683

Christian B, Zainab EAT, Mireille M, Josef M (1998) Palmitate oxidation by the mitochondria from volume-overloaded rat hearts. Mol Cell Biochem 180: 117–128

Coburn CT, Knapp FF Jr, Febbraio M, Beets AL, Silverstein RL, Abumrad NA (2000) Defective uptake and utilization of long chain fatty acids in muscle and adipose tissues of CD36 knockout mice. J Biol Chem 275: 32523–32529

Coe NR, Smith AJ, Frohnert BI, Watkins PA, Bernlohr DA (1999) The fatty acid transport protein (FATP1) Is a very long chain acyl-CoA synthetase. J Biol Chem 274: 36300–36304

Coleman RA, Lewin TM, Muoio DM (2000) Physiological and nutritional regulation of enzymes of triacylglycerol synthesis. Annu Rev Nutr 20: 77–103

Crass MF (1972) Exogenous substrate effects on endogenous lipid-metabolism in working rat-heart. Biochim Biophys Acta 280: 71–81

Crass MF (1977) Regulation of triglyceride-metabolism in isotopically pre-labeled perfused heart. Fed Proc 36: 1995–1999

Davies SP, Sim AT, Hardie DG (1990) Location and function of three sites phosphorylated on rat acetyl-CoA carboxylase by the AMP-activated protein kinase. Eur J Biochem 187: 183–190

DeLuca JG, Doebber TW, Kelly LJ, Kemp RK, Molon-Noblot S, Sahoo SP, Ventre J, Wu MS, Peters JM, Gonzalez FJ, Moller DE (2000) Evidence for Peroxisome Proliferator-Activated Receptor (PPAR)alpha -independent peroxisome proliferation: Effects of PPARgamma /delta-specific agonists in PPARα -null mice. Mol Pharmacol 58: 470–476

Duan SZ, Ivashchenko CY, Russell MW, Milstone DS, Mortensen RM (2005) Cardiomyocyte-specific knockout and agonist of Peroxisome Proliferator-Activated Receptor-{gamma} Both induce cardiac hypertrophy in mice. Circ Res 97: 372–379

Dyck JR, Berthiaume LG, Thomas PD, Kantor PF, Barr AJ, Barr R, Singh D, Hopkins TA, Voilley N, Prentki M, Lopaschuk GD (2000) Characterization of rat liver malonyl-CoA decarboxylase and the study of its role in regulating fatty acid metabolism. Biochem J 350 Pt 2:599–608

Dyck JR, Cheng JF, Stanley WC, Barr R, Chandler MP, Brown S, Wallace D, Arrhenius T, Harmon C, Yang G, Nadzan AM, Lopaschuk GD (2004) Malonyl coenzyme a decarboxylase inhibition protects the ischemic heart by inhibiting fatty acid oxidation and stimulating glucose oxidation. Circ Res 94: 78–84

Dyck JR., Kudo N, Barr AJ, Davies SP, Hardie DG, Lopaschuk GD (1999) Phosphorylation control of cardiac acetyl-CoA carboxylase by cAMP-dependent protein kinase and 5'-AMP activated protein kinase. Eur J Biochem 262: 184–90

Dyck JRB, Barr AJ, Barr RL, Kolattukudy PE, Lopaschuk GD (1998) Characterization of cardiac malonyl-CoA decarboxylase and its putative role in regulating fatty acid oxidation. Am J Physiol Heart Circ Physiol 44: H2122–H2129

Dyck JRB, Gao G, Widmer J, Stapleton D, Fernandez CS, Kemp BE, Witters LA (1996) Regulation of 5'-AMP-activated Protein Kinase Activity by the Noncatalytic beta and gamma Subunits. J Biol Chem 271: 17798–17803

el Alaoui-Talibi Z, Guendouz A, Moravec M, Moravec J (1997) Control of oxidative metabolism in volume-overloaded rat hearts: effect of propionyl-L-carnitine. Am J Physiol Heart Circ Physiol 272: H1615–H1624

el Alaoui-Talibi Z, Landormy S, Loireau A, Moravec J (1992) Fatty acid oxidation and mechanical performance of volume-overloaded rat hearts. Am J Physiol Heart Circ Physiol 262: H1068–H1074

Enerback S, Gimble JM (1993) Lipoprotein-Lipase gene-expression - physiological regulators at the transcriptional and posttranscriptional level. Biochim Biophys Acta 1169: 107–125

Engel AG, Angelini C (1973) Carnitine deficiency of human skeletal muscle with associated lipid storage myopathy: a new syndrome. Science 179: 899–902

Escher P, Wahli W (2000) Peroxisome proliferator-activated receptors: insight into multiple cellular functions. Mutat Res 448: 121–38

Faergeman NJ, Knudsen J (1997) Role of long-chain fatty acyl-CoA esters in the regulation of metabolism and in cell signalling. Biochem J 323: 1–12

Febbraio M, Abumrad NA, Hajjar DP, Sharma K, Cheng W, Pearce SF, Silverstein RL (1999) A null mutation in Murine CD36 reveals an important role in fatty acid and lipoprotein metabolism. J Biol Chem 274: 19055–19062

Fielding BA, Frayn KN (1998) Lipoprotein lipase and the disposition of dietary fatty acids. Br J Nutr 80: 495–502

Finck BN, Lehman JJ, Leone TC, Welch MJ, Bennett MJ, Kovacs A, Han X, Gross RW, Kozak R, Lopaschuk GD, Kelly DP (2002) The cardiac phenotype induced by PPARα overexpression mimics that caused by diabetes mellitus. J Clin Invest 109: 121–30

FitzPatrick DR, Hill A, Tolmie JL, Thorburn DR, Christodoulou J (1999) The molecular basis of malonyl-CoA decarboxylase deficiency. Am J Hum Genet 65: 318–326

Folmes CD, Clanachan AS, Lopaschuk, GD (2005) Fatty acid oxidation inhibitors in the management of chronic complications of atherosclerosis. Curr Atheroscler Rep 7: 63–70

Frohlich ED, Apstein C, Chobanian AV, Devereux RB, Dustan HP, Dzau V, Fauad-Tarazi F, Horan MJ, Marcus M, Massie B (1992) The heart in hypertension. N Engl J Med 327: 998–1008

Gao G, Fernandez CS, Stapleton D, Auster AS, Widmer J, Dyck JRB, Kemp BE, Witters LA (1996) Non-catalytic beta- and [IMAGE]-Subunit Isoforms of the 5'-AMP-activated Protein Kinase. J Biol Chem 271: 8675–8681

Gao J, Waber L, Bennett MJ, Gibson KM, Cohen JC (1999) Cloning and mutational analysis of human malonyl-coenzyme A decarboxylase. J. Lipid Res 40: 178–182

Gilde AJ, van der Lee KAJM, Willemsen PHM, Chinetti G, van der Leij FR, Van der Vusse GJ, Staels B, van Bilsen M (2003) Peroxisome proliferator-activated receptor (PPAR) alpha and PPAR beta/delta, but not PPAR gamma, modulate the expression of genes involved in cardiac lipid metabolism. Circ Res 92: 518–524

Glatz JF, Storch J (2001) Unravelling the significance of cellular fatty acid-binding proteins. Curr Opin Lipidol 12: 267–274

Glatz JF, van Nieuwenhoven FA, Luiken JJ, Schaap FG, van d V (1997) Role of membrane-associated and cytoplasmic fatty acid-binding proteins in cellular fatty acid metabolism. Prostaglandins Leukot. Essent. Fatty Acids 57: 373–378

Goodwin GW, Taegtmeyer H (2000) Improved energy homeostasis of the heart in the metabolic state of exercise. Am J Physiol Heart Circ Physiol 279: H1490–H1501

Goodwin GW, Taylor CS, Taegtmeyer H (1998) Regulation of energy metabolism of the heart during acute increase in heart work. J Biol Chem 273: 29530–29539

Greenwalt DE, Scheck SH, Rhinehart-Jones T (1995) Heart CD36 expression is increased in murine models of diabetes and in mice fed a high fat diet. J Clin. Invest 96: 1382–1388

Gregersen N, Andresen BS, Corydon MJ, Corydon TJ, Olsen RK, Bolund L, Bross P (2001) Mutation analysis in mitochondrial fatty acid oxidation defects: Exemplified by acyl-CoA dehydrogenase deficiencies, with special focus on genotype-phenotype relationship. Hum Mutat 18: 169–189

Gregersen N, Bross P, Andresen BS (2004) Genetic defects in fatty acid beta-oxidation and acyl-CoA dehydrogenases. Molecular pathogenesis and genotype-phenotype relationships. Eur J Biochem 271: 470–482

Gutknecht J (1988) Proton conductance caused by long-chain fatty acids in phospholipid bilayer membranes. J Membr Biol 106: 83–93

Ha J, Daniel S, Broyles SS, Kim KH (1994) Critical phosphorylation sites for acetyl-CoA carboxylase activity. J Biol Chem 269: 22162–22168

Ha J, Lee JK, Kim KS, Witters LA, Kim KH (1996) Cloning of human acetyl-CoA carboxylase-[beta] and its unique features. Proc Natl Acad Sci USA 93: 11466–11470

Hale DE, Batshaw ML, Coates PM, Frerman FE, Goodman SI, Singh I, Stanley CA (1985) Long-chain acyl coenzyme A dehydrogenase deficiency: an inherited cause of nonketotic hypoglycemia. Pediatr Res 19: 666–671

Hall JL, Lopaschuk GD, Barr A, Bringas J, Pizzurro RD, Stanley WC (1996a) Increased cardiac fatty acid uptake with dobutamine infusion in swine is accompanied by a decrease in malonyl CoA levels. Cardiovasc Res 32: 879–885

Hall JL, Stanley WC, Lopaschuk GD, Wisneski JA, Pizzurro RD, Hamilton CD, McCormack JG (1996b) Impaired pyruvate oxidation but normal glucose uptake in diabetic pig heart during dobutamine-induced work. Am J Physiol Heart Circ Physiol 40: H2320–H2329

Hamilton JA, Civelek VN, Kamp F, Tornheim K, Corkey BE (1994) Changes in internal pH caused by movement of fatty acids into and out of clonal pancreatic beta-cells (HIT). J Biol Chem 269: 20852–20856

Hamilton JA, Kamp F (1999) How are free fatty acids transported in membranes? Is it by proteins or by free diffusion through the lipids? Diabetes 48: 2255–2269

Harmon CM, Abumrad NA (1993) Binding of sulfosuccinimidyl fatty acids to adipocyte membrane proteins: isolation and amino-terminal sequence of an 88-kD protein implicated in transport of long-chain fatty acids. J Membr Biol 133: 43–49

Harmon CM, Luce P, Abumrad NA (1992) Labelling of an 88 kDa adipocyte membrane protein by sulpho-N-succinimidyl long-chain fatty acids: inhibition of fatty acid transport. Biochem Soc Trans 20: 811–813

Hauton D, Bennett MJ, Evans RD (2001) Utilisation of triacylglycerol and non-esterified fatty acid by the working rat heart: myocardial lipid substrate preference. Biochim Biophys Acta 1533: 99–109

Hawley SA, Boudeau J, Reid JL, Mustard KJ, Udd L, Makela TP, Alessi DD Hardie DG (2003) Complexes between the LKB1 tumor suppressor, STRADalpha/beta and MO25alpha/beta are upstream kinases in the AMP-activated protein kinase cascade. J Biol 2: 28

Hawley SA, Selbert MA, Goldstein EG, Edelman AM, Carling D, Hardie DG (1995) 5'-AMP activates the AMP-activated protein kinase cascade, and Ca2+/calmodulin activates the calmodulin-dependent protein kinase I cascade, via three independent mechanisms. J Biol Chem 270: 27186–91

Himms-Hagen J, Harper ME (2001) Physiological role of UCP3 may be export of fatty acids from mitochondria when fatty acid oxidation predominates: An hypothesis. Exp Biol Med 226: 78–84

Hirsch D, Stahl A, Lodish HF (1998) A family of fatty acid transporters conserved from mycobacterium to man. PNAS 95: 8625–8629

Hopkins TA, Sugden MC, Holness MJ, Kozak R, Dyck JR, Lopaschuk GD (2003) Control of cardiac pyruvate dehydrogenase activity in peroxisome proliferator-activated receptor-alpha transgenic mice. Am J Physiol Heart Circ Physiol 285: 270–6

Hurley RL, Anderson KA, Franzone JM, Kemp BE, Means AR, Witters LA (2005) The Ca^{2+}/Calmodulin-dependent protein kinase kinases are AMP-activated protein kinase kinases. J Biol Chem 280: 29060–6

Huss JM, Kelly DP (2004) Nuclear receptor signaling and cardiac energetics. Circ Res 95: 568–578

Ibrahimi A, Bonen A, Blinn WD, Hajri T, Li X, Zhong K, Cameron R, Abumrad N A (1999) Muscle-specific overexpression of FAT/CD36 enhances fatty acid oxidation by contracting muscle, reduces plasma triglycerides and fatty acids, and increases plasma glucose and insulin. J Biol Chem 274: 26761–26766

Idellwenger JA, Grotyohann LW, Neely JR (1978) Coenzyme-A and carnitine distribution in normal and ischemic hearts. J Biol Chem 253: 4310–4318

Kamataki A, Takahashi S, Masamura K, Iwasaki T, Hattori H, Naiki H, Yamada K, Suzuki J, Miyamori I, Sakai J (2002) Remnant lipoprotein particles are taken up into myocardium through VLDL receptor–a possible mechanism for cardiac fatty acid metabolism. Biochem Biophys Res Commun 293: 1007–1013

Kantor PF, Lucien A, Kozak R, Lopaschuk GD (2000) The antianginal drug trimetazidine shifts cardiac energy metabolism from fatty acid oxidation to glucose oxidation by inhibiting mitochondrial long-chain 3-ketoacyl coenzyme A thiolase.Circ Res 86: 580–588

Karmazyn M, Moffat MP (1993) Na^+/H^+ exchange and regulation of intracellular Ca^{2+}. Cardiovasc Res 27: 2079–2080

Karmazyn M, Gan XT, Humphreys RA, Yoshida H, Kusumoto K (1999) The Myocardial Na^+-H^+ exchange: structure, regulation, and its role in heart disease. Circ Res 85: 777–786

Kelly DP, Hale DE, Rutledge SL, Ogden ML, Whelan AJ, Zhang Z, Strauss AW (1992) Molecular basis of inherited medium-chain acyl-CoA dehydrogenase deficiency causing sudden child death. J Inherit Metab Dis 15: 171–80

Kerner J, Hoppel C (2000) Fatty acid import into mitochondria. Biochim Biophys Acta 1486: 1–17

Kerner J, Hoppel CL (2002) Radiochemical malonyl-CoA decarboxylase assay: Activity and subcellular distribution in heart and skeletal muscle. Anal Biochem 306: 283–289

Kiens B, Kristiansen S, Jensen P, Richter EA, Turcotte LP (1997) Membrane associated fatty acid binding protein (FABPpm) in human skeletal muscle is increased by endurance training. Biochem Biophys Res Commun 231: 463–465

Kim YS, Kolattukudy PE (1978) Purification and properties of malonyl-CoA decarboxylase from rat-liver mitochondria and its immunological comparison with enzymes from rat-brain, heart, and mammary-gland. Arch Biochem Biophys 190: 234–246

King LM, Opie LH (1998) Glucose and glycogen utilisation in myocardial ischemia–changes in metabolism and consequences for the myocyte. Mol Cell Biochem 180: 3–26

Kintaka T, Tanaka T, Imai M, Adachi I, Narabayashi I, Kitaura Y (2002) CD36 genotype and long-chain fatty acid uptake in the heart. Circ J 66: 819–825

Knudsen J, Mandrup S, Rasmussen JT, Andreasen PH, Poulsen F, Kristiansen K (1993) The function of acyl-CoA-binding protein (ACBP)/diazepam binding inhibitor (DBI). Mol Cell Biochem 123: 129–138

Kuang M, Febbraio M, Wagg C, Lopaschuk GD, Dyck JR (2004) Fatty acid translocase/CD36 deficiency does not energetically or functionally compromise hearts before or after ischemia. Circulation 109: 1550–7

Kudo N, Barr AJ, Barr RL, Desai S, Lopaschuk GD (1995) High-rates of fatty-acid oxidation during reperfusion of ischemic hearts are associated with a decrease in malonyl-CoA levels due to an increase in 5'-AMP-activated protein-kinase inhibition of acetyl-CoA carboxylase. J Biol Chem 270: 17513–17520

Kudo N, Gillespie JG, Kung L, Witters LA, Schulz R, Clanachan AS, Lopaschuk GD (1996) Characterization of 5'AMP-activated protein kinase activity in the heart and its role in inhibiting acetyl-CoA carboxylase during reperfusion following ischemia. Biochim Biophys Acta 1301: 67–75

Kunau WH, Dommes V, Schulz H (1995) Beta-oxidation of fatty acids in mitochondria, peroxisomes, and bacteria: A century of continued progress. Prog Lipid Res.34: 267–342

Kwok S, Singh-Bist A, Natu V, Kraemer FB (1997) Dietary regulation of the very low density lipoprotein receptor in mouse heart and fat. Horm. Metab Res 29: 524–529

Latipaa PM, Karki TT, Hiltunen JK, Hassinen IE (1986) Regulation of palmitoylcarnitine oxidation in isolated rat liver mitochondria. Role of the redox state of NAD(H). Biochim Biophys Acta 875: 293–300

Lee CH, Olson P, Evans RM (2003) Minireview: Lipid metabolism, metabolic diseases, and peroxisome proliferator-activated receptors. Endocrinology 144: 2201–2207

Lee JK, Kim KH (1999) Roles of acetyl-CoA carboxylase [beta] in muscle cell differentiation and in mitochondrial fatty acid oxidation. Biochem. Biophys Res Commun. 254: 657–660

Lehman JJ, Barger PM, Kovacs A, Saffitz JE, Medeiros DM, Kelly DP (2000) Peroxisome proliferator-activated receptor gamma coactivator-1 promotes cardiac mitochondrial biogenesis. J Clin Invest 106: 847–856

Levak-Frank S, Hofmann W, Weinstock PH, Radner H, Sattler W, Breslow JL, Zechner R (1999) Induced mutant mouse lines that express lipoprotein lipase in cardiac muscle, but not in skeletal muscle and adipose tissue, have normal plasma triglyceride and high-density lipoprotein-cholesterol levels. PNAS 96: 3165–3170

Liu B, Clanachan AS, Schulz R, Lopaschuk GD (1996a) Cardiac efficiency is improved after ischemia by altering both the source and fate of protons. Circ Res 79: 940–8

Liu B, el Alaoui-Talibi Z, Clanachan AS, Schulz R, Lopaschuk GD (1996b) Uncoupling of contractile function from mitochondrial TCA cycle activity and MVO2 during reperfusion of ischemic hearts. Am J Physiol 270: 72–80

Liu Q, Docherty JC, Rendell JC, Clanachan AS, Lopaschuk GD (2002) High levels of fatty acids delay the recovery of intracellular pH and cardiac efficiency in post-ischemic hearts by inhibiting glucose oxidation. J Am Coll Cardiol 39: 718–25

Lopaschuk GD (1996) Abnormal mechanical function in diabetes: relationship to altered myocardial carbohydrate/lipid metabolism. Coron Artery Dis 7: 116–23

Lopaschuk GD, Barr R, Thomas PD, Dyck JRB (2003) Beneficial effects of trimetazidine in ex vivo working ischemic hearts are due to a stimulation of glucose oxidation secondary to inhibition of long-chain 3-ketoacyl coenzyme A thiolase. Circ Res 93: E33–E37

Lopaschuk GD, Belke DD, Gamble J, Itoi T, Schonekess BO (1994a) Regulation of fatty-acid oxidation in the mammalian heart in health and disease. Biochim Biophys Acta 1213: 263–276

Lopaschuk GD, Collins-Nakai R, Olley PM, Montague TJ, Mcneil G, Gayle M, Penkoske P, Finegan BA (1994b) Plasma fatty acid levels in infants and adults after myocardial ischemia. Am Heart J 128: 61–7

Lopaschuk GD, Rebeyka IM, Allard MF (2002) Metabolic modulation: a means to mend a broken heart. Circulation 105: 140–2

Lopaschuk GD, Spafford MA, Davies NJ, Wall SR (1990) Glucose and palmitate oxidation in isolated working rat hearts reperfused after a period of transient global ischemia. Circ Res 66:546–53

Luiken JJ, Coort SL, Koonen DP, Bonen A, Glatz JF (2004a) Signalling components involved in contraction-inducible substrate uptake into cardiac myocytes. Proc Nutr Soc 63: 251–258

Luiken JJ, Koonen DP, Coumans WA, Pelsers MM, Binas B, Bonen A, Glatz JF (2003a) Long-chain fatty acid uptake by skeletal muscle is impaired in homozygous, but not heterozygous, heart-type-FABP null mice. Lipids 38: 491–496

Luiken JJ, van Nieuwenhoven FA, America G, Van der Vusse GJ, Glatz JF (1997) Uptake and metabolism of palmitate by isolated cardiac myocytes from adult rats: involvement of sarcolemmal proteins. J Lipid Res 38: 745–758

Luiken JJFP, Coort SLM, Koonen DPY, van der Horst DJ, Bonen A, Zorzano A, Glatz JFC (2004b) Regulation of cardiac long-chain fatty acid and glucose uptake by translocation of substrate transporters. Pflugers Arch 448: 1–15

Luiken JJFP, Coort SLM, Willems J, Coumans WA, Bonen A, Van der Vusse GJ, Glatz JFC (2003b) Contraction-induced fatty acid translocase/CD36 translocation in rat cardiac myocytes is mediated through AMP-activated protein kinase signaling. Diabetes 52: 1627–1634

Luiken JJFP, Willems J, Van der Vusse GJ, Glatz JFC (2001a) Electrostimulation enhances FAT/CD36-mediated long-chain fatty acid uptake by isolated rat cardiac myocytes. Am J Physiol Endocrinol Metab 281: E704–E712

Luiken JJFP, Arumugam Y, Dyck DJ, Bell RC, Pelsers MML, Turcotte LP, Tandon NN, Glatz JFC, Bonen A (2001b) Increased rates of fatty acid uptake and plasmalemmal fatty acid transporters in Obese Zucker Rats. J Biol Chem 276: 40567–40573

Luiken JJFP, Koonen DPY, Willems J, Zorzano A, Becker C, Fischer Y, Tandon NN, van der Vusse GJ, Bonen A, Glatz JFC (2002) Insulin stimulates long-chain fatty acid utilization by rat cardiac myocytes through cellular redistribution of FAT/CD36. Diabetes 51: 3113–3119

Luiken JJFP, Turcotte LP, Bonen A (1999) Protein-mediated palmitate uptake and expression of fatty acid transport proteins in heart giant vesicles. J Lipid Res 40, 1007–1016

Lysiak W, Toth PP, Suelter CH, Bieber LL (1986) Quantitation of the efflux of acylcarnitines from rat-heart, brain, and liver-mitochondria. J Biol Chem 261: 3698–3703

Makinde AO, Gamble J, Lopaschuk GD (1997) Upregulation of 5'-AMP-activated protein kinase is responsible for the increase in myocardial fatty acid oxidation rates following birth in the newborn rabbit. Circ Res 80: 482–489

Makinde AO, Kantor PF, Lopaschuk GD (1998) Maturation of fatty acid and carbohydrate metabolism in the newborn heart. Mol Cell Biochem 188: 49–56

Marin-Garcia J, Goldenthal MJ (2002) Fatty acid metabolism in cardiac failure: biochemical, genetic and cellular analysis. Cardiovasc Res 54: 516–527

McCormack JG, Barr RL, Wolff AA, Lopaschuk GD (1996) Ranolazine stimulates glucose oxidation in normoxic, ischemic, and reperfused ischemic rat hearts. Circulation 93: 135–42

McGarry JD, Brown NF (1997) The mitochondrial carnitine palmitoyltransferase system. From concept to molecular analysis. FEBS Journal 244: 1–14

McGarry JD, Mills SE, Long CS, Foster DW (1983) Observations on the affinity for carnitine, and malonyl-CoA sensitivity, of carnitine palmitoyltransferase-I in animal and human-tissues - Demonstration of the presence of malonyl-CoA in non-hepatic tissues of the rat. Biochem J 214: 21–28

Merkel M, Eckel RH, Goldberg IJ (2002a) Lipoprotein lipase: genetics, lipid uptake, and regulation. J Lipid Res 43: 1997–2006

Merkel M, Heeren J, Dudeck W, Rinninger F, Radner H, Breslow JL, Goldberg IJ, Zechner R, Greten H (2002b) Inactive lipoprotein lipase (LPL) alone increases selective cholesterol ester uptake in vivo, whereas in the presence of active LPL it also increases triglyceride hydrolysis and whole particle lipoprotein uptake. J Biol Chem 277: 7405–7411

Merkel M, Kako Y, Radner H, Cho IS, Ramasamy R, Brunzell JD, Goldberg IJ, Breslow JL (1998a) Catalytically inactive lipoprotein lipase expression in muscle of transgenic mice increases very low density lipoprotein uptake: Direct evidence that lipoprotein lipase bridging occurs in vivo. Proc Natl Acad Sci USA 95: 13841–13846

Merkel M, Weinstock PH, Chajek-Shaul T, Radner H, Yin BY, Breslow JL, Goldberg IJ (1998b) Lipoprotein lipase expression exclusively in liver - A mouse model for metabolism in the neonatal period and during cachexia. J Clin Invest 102: 893–901

Mikkelsen J, Knudsen J (1987) Acyl-CoA-binding protein from cow. Binding characteristics and cellular and tissue distribution. Biochem J 248: 709–714

Munday MR (2002) Regulation of mammalian acetyl-CoA carboxylase. Biochem Soc Trans 30: 1059–1064

Murthy MSR, Pande SV (1987) Malonyl-CoA binding site and the overt carnitine palmitoyltransferase activity reside on the opposite sides of the outer mitochondrial membrane. PNAS 84: 378–382

Neely JR, Morgan JE (1974) Relationship between carbohydrate metabolism and energy balance of heart muscle. Ann Rev Physiol 36: 413–459

Neely JR, Whitmer M, Mochizuki S (1976) Effects of mechanical activity and hormones on myocardial glucose and fatty acid utilization. Circ Res 38: I22–I30

Niu YG, Hauton D, Evans RD (2004) Utilization of triacylglycerol-rich lipoproteins by the working rat heart: routes of uptake and metabolic fates. J Physiol (Lond) 558: 225–237

Ockner RK, Manning JA, Poppenhausen RB, Ho WK (1972) A binding protein for fatty acids in cytosol of intestinal mucosa, liver, myocardium, and other tissues. Science 177: 56–58

Oliver MF, Opie LH (1994) Effects of glucose and fatty acids on myocardial ischaemia and arrhythmias. Lancet 343: 155–158

Olowe Y, Schulz H (1980) Regulation of thiolases from pig heart. Control of fatty acid oxidation in heart. Eur J Biochem 109: 425–429

Opie LH (1998) Heart Physiology, From Cell to Circulation, Lippincott-Raven

Oram JF, Bennetch SL, Neely JR (1973) Regulation of Fatty Acid Utilization in Isolated Perfused Rat Hearts. J Biol Chem 248: 5299–5309

Palou A, Pico C, Bonet ML, Oliver P (1998) The uncoupling protein, thermogenin. Int. J Biochem Cell Biol. 30: 7–11

Paulson DJ, Crass MF (1982) Endogenous Triacylglycerol Metabolism in Diabetic Heart. Am J Physiol 242: 1084–1094

Pelsers MM, Lutgerink JT, Nieuwenhoven FA, Tandon NN, van der Vusse GJ, Arends JW, Hoogenboom HR, Glatz JF (1999) A sensitive immunoassay for rat fatty acid translocase (CD36) using phage antibodies selected on cell transfectants: abundant presence of fatty acid translocase/CD36 in cardiac and red skeletal muscle and up-regulation in diabetes. Biochem J 337: 407–414

Pohl J, Fitscher BA, Ring A, Ihl-Vahl R, Strasser RH, Stremmel W (2000). Fatty acid transporters in plasma membranes of cardiomyocytes in patients with dilated cardiomyopathy. Eur J Med Res 5: 438–442

Poirier M, Vincent G, Reszko AE, Bouchard B, Kelleher JK, Brunengraber H, Des Rosiers C (2002) Probing the link between citrate and malonyl-CoA in perfused rat hearts. Am J Physiol Heart Circ Physiol 283: H1379–H1386

Ponticos M, Lu QL, Morgan JE, Hardie DG, Partridge TA, Carling D (1998) Dual regulation of the AMP-activated protein kinase provides a novel mechanism for the control of creatine kinase in skeletal muscle. Embo J 17: 1688–99

Randle PJ (1986). Fuel Selection in Animals. Biochem Soc Trans 14: 799–806

Randle PJ, Garland PB, Newsholme EA, Hales CN (1963) Glucose Fatty-Acid Cycle - Its Role in Insulin Sensitivity and Metabolic Disturbances of Diabetes Mellitus. Lancet 1: 785–789

Rasmussen JT, Rosendal J, Knudsen J (1993) Interaction of acyl-CoA binding protein (ACBP) on processes for which acyl-CoA is a substrate, product or inhibitor. Biochem J 292: 907–913

Reibel DK, Uboh CE, Kent RL (1983) Altered coenzyme A and carnitine metabolism in pressure-overload hypertrophied hearts. Am J Physiol Heart Circ Physiol 244: H839–H843

Reszko AE, Kasumov T, David F, Jobbins KA, Thomas KR, Hoppel CL, Brunengraber H, Rosiers CD (2004) Peroxisomal fatty acid oxidation is a substantial source of the acetyl moiety of malonyl-CoA in rat heart. J Biol. Chem 279: 19574–19579

Reszko AE, Kasumov T, Comte B, Pierce BA, David F, Bederman IR, Deutsch J, Des Rosiers C, Brunengraber H (2001) Assay of the concentration and 13C-isotopic enrichment of malonyl-coenzyme A by gas chromatography-mass spectrometry. Anal Biochem 298: 69–75

Richieri GV, Kleinfeld AM (1995) Unbound free fatty acid levels in human serum. J Lipid Res 36: 229–240

Robyr D, Wolffe AP, Wahli W (2000) Nuclear hormone receptor coregulators in action: Diversity for shared tasks. Mol Endocrinol 14: 329–347

Sack MN, Rader TA, Park SH, Bastin J, Mccune SA, Kelly DP (1996) Fatty acid oxidation enzyme gene expression is downregulated in the failing heart. Circulation 94: 2837–2842

Sacksteder KA, Morrell JC, Wanders RJA, Matalon R, Gould SJ (1999) MCD encodes peroxisomal and cytoplasmic forms of malonyl-CoA decarboxylase and is mutated in malonyl-CoA decarboxylase deficiency. J Biol Chem 274: 24461–24468

Saddik M, Gamble J, Witters LA, Lopaschuk GD (1993) Acetyl-CoA carboxylase regulation of fatty-acid oxidation in the heart. J Biol Chem 268: 25836–25845

Saddik M, Lopaschuk GD (1991) Myocardial triglyceride turnover and contribution to energy substrate utilization in isolated working rat hearts. J Biol Chem 266: 8162–70

Saddik M, Lopaschuk GD (1992) Myocardial triglyceride turnover during reperfusion of isolated rat hearts subjected to a transient period of global ischemia. J Biol Chem 267: 3825–31

Saddik M, Lopaschuk GD (1994) Triacylglycerol turnover in isolated working hearts of acutely diabetic rats. Can J Physiol Pharmacol 72: 1110–9

Saha AK, Schwarsin AJ, Roduit R, Masse F, Kaushik V, Tornheim K, Prentki M, Ruderman NB (2000) Activation of malonyl-CoA decarboxylase in rat skeletal muscle by contraction and the AMP-activated protein kinase activator 5-aminoimidazole-4-carboxamide-1-beta-D-ribofuranoside. J Biol Chem 275: 24279–24283

Saha AK, Vavvas D, Kurowski TG, Apazidis A, Witters LA, Shafrir E, Ruderman NB (1997) Malonyl-CoA regulation in skeletal muscle: Its link to cell citrate and the glucose-fatty acid cycle. Am J Physiol Endocrinol Metab 35: E641–E648

Sakamoto J, Barr RL, Kavanagh KM, Lopaschuk GD (2000) Contribution of malonyl-CoA decarboxylase to the high fatty acid oxidation rates seen in the diabetic heart. Am J Physiol Heart Circ Physiol 278: 1196–204

Sambandam N, Lopaschuk GD, Brownsey RW, Allard MF (2002) Energy metabolism in the hypertrophied heart. Heart Fail Rev 7: 161–73

Samec S, Seydoux J, Dulloo AG (1998) Role of UCP homologues in skeletal muscles and brown adipose tissue: mediators of thermogenesis or regulators of lipids as fuel substrate? FASEB J 12: 715–724

Scarpulla RC (2002) Nuclear activators and coactivators in mammalian mitochondrial biogenesis. Biochim Biophys Acta 1576: 1–14

Schaap FG, van der Vusse GJ, Glatz JFC (1998) Fatty acid-binding proteins in the heart. Mol Cell Biochem 180: 43–51

Schaffer JE, Lodish HF (1994) Expression cloning and characterization of a novel adipocyte long chain fatty acid transport protein. Cell 79: 427–436

Schaffer JE (2002) Fatty acid transport: the roads taken. Am J Physiol Endocrinol Metab 282: E239–E246

Schmider W, Fahr A, Blum HE, Kurz G (2000) Transport of heptafluorostearate across model membranes. Membrane transport of long-chain fatty acid anions I. J. Lipid Res 41: 775–787

Schulz H (1991) Beta oxidation of fatty acids. Biochim. Biophys. Acta 1081: 109–120

Schulz H (1994) Regulation of Fatty-Acid Oxidation in Heart. J Nutr 124: 165–171

Scott JW, Norman DG, Hawley SA, Kontogiannis L, Hardie DG (2002) Protein kinase substrate recognition studied using the recombinant catalytic domain of AMP-activated protein kinase and a model substrate. J Mol Biol 317: 309–23

Sorrentino D, Stump DD, Van Ness K, Simard A, Schwab AJ, Zhou SL, Goresky CA, Berk PD (1996) Oleate uptake by isolated hepatocytes and the perfused rat liver is competitively inhibited by palmitate. Am J Physiol Gastrointest Liver Physiol 270: G385–G392

Stacpoole PW (1989) The pharmacology of dichloroacetate. Metabolism 38: 1124–1144

Stanley WC, Lopaschuk GD, Kivilo KM (1999) Alterations in myocardial energy metabolism in streptozotocin diabetes. In: Experimental Models of Diabetes, ed. M.N.JH CRC Press, 19–38

Stanley WC, Chandler MP (2002) Energy metabolism in the normal and failing heart: potential for therapeutic interventions. Heart Fail Rev 7: 115–30

Stanley WC, Lopaschuk GD, Hall JL, McCormack JG (1997) Regulation of myocardial carbohydrate metabolism under normal and ischaemic conditions. Potential for pharmacological interventions. Cardiovasc Res 33: 243–57

Stanley WC, Recchia FA, Lopaschuk GD (2005) Myocardial substrate metabolism in the normal and failing heart. Physiol Rev 85: 1093–129

Stanley W C, Hernandez LA, Spires D, Bringas J, Wallace S, McCormack JG (1996) Pyruvate dehydrogenase activity and malonyl CoA levels in normal and ischemic swine myocardium: Effects of dichloroacetate. J Mol Cell Cardiol 28: 905–914

Stapleton D, Mitchelhill KI, Gao G, Widmer J, Michell BJ, Teh T, House CM, Fernandez CS, Cox T, Witters LA, Kemp BE (1996) Mammalian AMP-activated protein kinase subfamily. J Biol Chem 271: 611–4

Stavinoha MA, RaySpellicy JW, Essop MF, Graveleau C, Abel ED, Hart-Sailors M L, Mersmann HJ, Bray MS, Young ME (2004) Evidence for mitochondrial thioesterase 1 as a peroxisome proliferator-activated receptor-alpha-regulated gene in cardiac and skeletal muscle. Am J Physiol Endocrinol Metab 287: E888–E895

Sterling D, Casey JR (2002) Bicarbonate transport proteins. Biochem Cell Biol 80: 483–497

Storch J, Lechene C, Kleinfeld AM (1991) Direct determination of free fatty acid transport across the adipocyte plasma membrane using quantitative fluorescence microscopy. J Biol Chem 266: 13473–13476

Storch J, Thumser AEA (2000) The fatty acid transport function of fatty acid-binding proteins. Biochim Biophys Acta 1486: 28–44

Strauss AW, Powell CK, Hale DE, Anderson MM, Ahuja A, Brackett JC, Sims HF (1995) Molecular basis of human mitochondrial very-long-chain acyl-CoA dehydrogenase deficiency causing cardiomyopathy and sudden death in childhood. PNAS 92: 10496–10500

Stremmel W, Berk PD (1986) Hepatocellular influx of [^{14}C]oleate reflects membrane transport rather than intracellular metabolism or binding. PNAS 83: 3086–3090

Stremmel W, Strohmeyer G, Borchard F, Kochwa S, Berk PD (1985) Isolation and partial characterization of a fatty acid binding protein in rat liver plasma membranes. PNAS 82: 4–8

Taegtmeyer H, Overturf ML (1988) Effects of moderate hypertension on cardiac function and metabolism in the rabbit. Hypertension 11: 416–426

Takahashi S, Kawarabayasi Y, Nakai T, Sakai J, Yamamoto T (1992) Rabbit very low density lipoprotein receptor: A low density lipoprotein density receptor-like protein with distinct ligand specificity. PNAS 89: 9252–9256

Tanaka K, Yokota I, Coates PM, Strauss AW, Kelly DP, Zhang Z, Gregersen N, Andresen BS, Matsubara Y, Curtis D, (1992) Mutations in the medium chain acyl-CoA dehydrogenase (MCAD) gene. Hum Mutat 1: 271–279

Thampy KG (1989) Formation of Malonyl Coenzyme-A in Rat-Heart - Identification and purification of an isozyme of acetyl-coenzyme-A carboxylase from rat-heart. J Biol Chem. 264: 17631–17634

Thornton C, Snowden MA, Carling D (1998) Identification of a novel AMP-activated protein kinase beta subunit isoform that is highly expressed in skeletal muscle. J Biol Chem 273: 12443–50

Turcotte LP, Swenberger JR, Tucker MZ, Yee AJ, Trump G, Luiken JJ, Bonen A (2000) Muscle palmitate uptake and binding are saturable and inhibited by antibodies to FABP(PM). Mol Cell Biochem 210: 53–63

Turcotte LP, Srivastava AK, Chiasson JL (1997) Fasting increases plasma membrane fatty acid-binding protein (FABPPM) in red skeletal muscle. Mol Cell Biochem 166: 153–158

Uchiyama A, Aoyama T, Kamijo K, Uchida Y, Kondo N, Orii T, Hashimoto T (1996) Molecular cloning of cDNA encoding rat very long-chain acyl-CoA synthetase. J Biol Chem 271: 30360–30365

Van der Vusse GJ, van Bilsen M, Glatz JFC (2000) Cardiac fatty acid uptake and transport in health and disease. Cardiovasc Res 45: 279–293

Vannieuwenhoven FA, Verstijnen CPHJ, Abumrad NA, Willemsen PHM, Vaneys GJJM, Vandervusse GJ, Glatz JFC (1995) Putative membrane fatty acid translocase and cytoplasmic fatty-acid-binding protein are co-expressed in rat heart and skeletal muscles. Biochem Biophys Res Commun 207: 747–752

Vega RB, Huss JM, Kelly DP (2000) The coactivator PGC-1 cooperates with peroxisome proliferator-activated receptor alpha in transcriptional control of nuclear genes encoding mitochondrial fatty acid oxidation enzymes. Mol Cell Biol 20: 1868–1876

Waku K (1992) Origins and fates of fatty acyl-CoA esters. Biochim Biophys Acta 1124: 101–111

Wall SR, Lopaschuk GD (1989) Glucose oxidation rates in fatty acid-perfused isolated working hearts from diabetic rats. Biochim Biophys Acta 1006: 97–103

Weis BC, Cowan AT, Brown N, Foster D W, McGarry JD (1994) Use of a selective inhibitor of liver carnitine palmitoyltransferase-I (Cpt-I) allows quantification if its contribution to total CPT-I activity in rat-heart - evidence that the dominant cardiac CPT-I isoform is identical to the skeletal-muscle enzyme. J Biol Chem 269: 26443–26448

Widmer J, Fassihi KS, Schlichter SC, Wheeler KS, Crute BE, King N, Nutile-McMenemy N, Noll WW, Daniel S, Ha J, Kim KH, Witters LA (1996) Identification of a second human acetyl-CoA carboxylase gene. Biochem J 316: 915–922

Wisneski JA, Gertz EW, Neese RA, Mayr M (1987) Myocardial-Metabolism of Free Fatty-Acids - Studies with ^{14}C-labeled substrates in humans. J Clin Invest 79: 359–366

Yagyu H, Chen GP, Yokoyama M, Hirata K, Augustus A, Kako Y, Seo T, Hu YY, Lutz EP, Merkel M, Bensadoun A, Homma S, Goldberg IJ (2003) Lipoprotein lipase (LpL) on the surface of cardiomyocytes increases lipid uptake and produces a cardiomyopathy. J Clin Invest 111: 419–426

Yano S, Sweetman L, Thorburn DR, Mofidi S, Williams JC (1997) A new case of malonyl coenzyme A decarboxylase deficiency presenting with cardiomyopathy. Eur J Pediatr 156: 382–383

Yokoyama M, Yagyu H, Hu YY, Seo T, Hirata K, Homma S, Goldberg IJ (2004) Apolipoprotein B production reduces lipotoxic cardiomyopathy - Studies in heart-specific lipoprotein lipase transgenic mouse. J Biol Chem 279: 4204–4211

Young, ME, McNulty P, Taegtmeyer H (2002) Adaptation and maladaptation of the heart in diabetes: Part II - Potential mechanisms. Circulation 105: 1861–1870

Zammit VA, Fraser F, Orstorphine CG (1997) Regulation of mitochondrial outer-membrane carnitine palmitoyltransferase (CPT I): Role of membrane-topology. Adv Enzyme Regul 37: 295–317

Zhou SL, Stump D, Isola L, Berk PD (1994) Constitutive expression of a saturable transport system for non-esterified fatty acids in Xenopus laevis oocytes. Biochem J 297: 315–319

Zhou SL, Stump D, Sorrentino D, Potter BJ, Berk PD (1992) Adipocyte differentiation of 3T3-L1 cells involves augmented expression of a 43-kDa plasma membrane fatty acid-binding protein. J Biol Chem 267: 14456–14461

Zhou, S. L., Stump, D, Sorrentino, D, Potter, B J, and Berk, P D (1992). Adipocyte differentiation of 3T3-L1 cells involves augmented expression of a 43-kDa plasma membrane fatty acid-binding protein. J. Biol. Chem. 267, 14456–14461.

3
Regulation of Mitochondrial Fuel Handling by the Peroxisome Proliferator-Activated Receptors

Mary C. Sugden and Mark J. Holness

3.1. Introduction

Within the body, lipids are essential for many aspects of cellular processes, and the PPARS are sensors of lipids that act to regulate gene expression. As lipid metabolism and glucose homeostasis are intrinsically related, these transcription factors exert a profound influence on the selection of metabolic fuels for ATP generation via mitochondrial oxidation. This chapter therefore focuses on the role of the PPARs as key coordinators of the selection of metabolic fuels used for mitochondrial ATP production.

In man, the nuclear hormone receptor family at present consists of 48 mammalian transcription factors that are related in terms of evolution, function and structure. These transcription factors regulate a variety of genes that control development, inflammation and nutrient metabolism in a range of tissues and organs (reviewed in Berkenstam and Gustafsson 2005), and can be divided into 3 groups. The first consists of classical high-affinity receptors for steroid and thyroid hormones. The second group consists of low-affinity receptors that bind non-classical ligands, including a wide range of metabolic intermediates including fatty acids (FA) and cholesterol derivatives, and are implicated in the regulation of lipid metabolism. The third group consists of "orphan" receptors, such as the estrogen-related receptor α (ERRα), for which neither endogenous nor synthetic ligands have yet been identified.

The low-affinity nuclear receptors include the peroxisome proliferator-activated receptors (PPARs). Other important low-affinity receptors include the RAR-related orphan receptors (which bind cholesterol or retinoic acid), the Liver X receptor (LXR) subfamily (LXRα and LXRβ), which are activated by oxidised cholesterol derivatives (oxysterols), and the Farnesoid X receptor (FXR), which binds bile acids. The retinoid X receptor (RXR) is a common binding partner for the PPARs. LXRs regulate cholesterol homeostasis in liver and adipose tissue and "reverse cholesterol transport" from peripheral tissues to liver, and are also increasingly recognized to be involved in inflammation and atherosclerosis. FXR regulates hepatic intracellular bile acid and cholesterol levels by regulating the expression of genes involved in cholesterol catabolism and biliary cholesterol secretion.

Within the body, lipids are essential for many aspects of cellular processes, including oxidative substrates for ATP production and, as such, are integral to mitochondrial function. The PPARs act as lipid sensors. Through the actions of lipids - or their derivatives - to activate the PPARs, lipids influence their metabolic fates in an interactive manner which is fundamental to metabolic fuel homeostasis under conditions of variable lipid availability - either dietary or released from stored triglyceride (TAG). As lipid metabolism and glucose homeostasis are intrinsically related, as initially described by the pioneering work of Randle and his colleagues (Randle et al. 1963); reviewed in Randle (1998), these transcription factors exert a profound influence on not only lipid, but also glucose handling. The PPARs and their roles as key coordinators of metabolic fuel oxidation will form the major focus of this review. The roles of the PPARs in atherosclerosis and inflammation are described in detail elsewhere (Israelian-Konaraki and Reaven 2005; Staels 2005).

3.2. PPARs: General Function in Relation to Tissue Distribution

PPARα (also known as NR1C1) was originally identified as a target for hepatocarcinogens that elicited peroxisome proliferation in the liver of rodents (Issemann and Green, 1990). These effects are not observed in man. Subsequent studies identified two further PPAR receptor isoforms: PPARδ (also known as PPARβ, NR1C2 or FA-activated receptor, FAAR) and PPARγ (also known as NR1CR3). Subsequently, PPARs were demonstrated to be closely implicated in mediating adaptive metabolic responses to changes in systemic metabolic fuel availability, in particular lipids (reviewed in Kliewer et al. 2001). More recently, the involvement of PPARs has been identified in a diverse range of physiological and pathological processes. The differing biological actions of the PPAR isoforms reflect differential tissue expression patterns.

PPARα, a lipo-oxidative transcription factor, is most abundantly expressed in a range of tissues with a high capacity for FA oxidation and which have a high energy demand related to their physiological functions: these include cardiac myocytes, hepatocytes, kidney cortex and pancreatic islets (Desvergne and Wahli 1999; Kersten et al. 1999; Sugden et al. 2001). PPARα is also present in monocytes/macrophages, the artery wall, vascular smooth muscle cells and endothelial cells (Duval et al. 2002). PPARα promotes a programme of lipid-induced activation of genes encoding proteins involved in FA uptake, β-oxidation, FA transport into peroxisomes and ω-oxidation of unsaturated FAs

(Braissant et al. 1996; Wolfrum et al. 2001) and is established to be critical to adaptations to fasting.

PPARγ, a lipogenic transcription factor, has a more limited tissue distribution. It is highly expressed in white and brown adipocytes, the placenta, cells of the immune system (macrophages, monocytes), and the endocrine pancreas (Braissant et al. 1996; Dubois et al. 2000; Lupi et al. 2004; Zhou et al. 1998). PPARγ promotes adipocyte differentiation, regulates intracellular lipid accumulation, induces genes involved in lipogenesis (Rosen et al. 1999), alters the secretion of adipokines, and regulates genes involved in or enhancing insulin signalling (Gray et al. 2005; Mueller et al. 2002). Thus, while PPARγ does not have a direct effect on mitochondrial fuel selection, it influences partitioning of available dietary lipid between adipose tissue and oxidative tissues. Altered lipid delivery to oxidative tissues such as heart, liver and oxidative skeletal muscle has two effects: it influences competition between FA and glucose as oxidative fuel by virtue of their relative availability and also may influence insulin action within non-adipose tissues should lipid delivery be excessive. In muscle, the strong relationship between PPARγ and altered lipid handling is accompanied by increased intracellular concentrations of lipid metabolites, such as acyl-CoA and diacylglycerol, which appear to impair insulin signalling by activating a serine kinase cascade. This culminates in serine phosphorylation of components of the insulin signalling pathway and a resultant impairment of their activation by tyrosine phosphorylation (see e.g. Roden et al. 1996).

PPARδ is expressed ubiquitously, but is found only at low levels in liver. Of the three PPAR isoforms, the function of PPARδ is the least well defined, except perhaps in skeletal muscle. Emerging research has demonstrated distinct tissue-specific patterns of regulation of FA utilisation by PPARδ, with a common theme being the promotion of oxidative metabolism (Fredenrich and Grimaldi 2005; Grimaldi 2005; Lee et al. 2003; Luquet et al. 2005). Thus, PPARδ acts as a central regulator of FA breakdown, as well as influencing the intrinsic oxidative capability of skeletal muscle.

3.3. PPARs: Domain Structure and Regulation of Transcriptional Activity

As with all nuclear receptors, the PPARs have a modular structure comprising a ligand-independent transcriptional activation domain (AF-1) in the N-terminal region, zinc finger motifs, a hinge region important for cofactor docking, a conserved C-terminal ligand binding domain and an AF2 domain that binds to co-repressors and co-activators (Figure 3.1).

Each of the PPAR subtypes bind to DNA as obligate heterodimers with a second nuclear receptor, RXR (Kliewer et al. 1992) (Figure 3.1). In their inactive state (in the absence of ligand), the PPARs can repress transcription by recruiting accessory proteins (co-repressors) and recruiting histone deacetylase. Activation

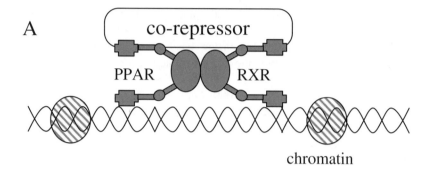

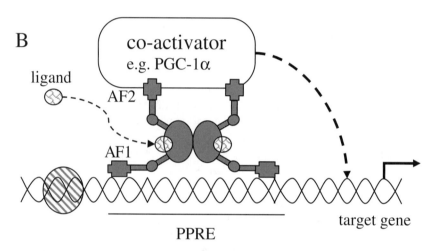

FIGURE 3.1. Interaction between PPARs, RXRs, co-repressors and co-activators in relation to the regulation of gene expression Binding of the co-repressor to the PPAR/RXR dimer in the absence of ligand blocks gene expression (panel A). Ligand binding leads to a conformational change promoting release of the co-repressor and binding of the co-activator resulting in gene transcription (panel B).

through ligand binding modifies local chromatin structure, inducing a conformational change leading to dissociation of co-repressors and recruitment of protein complexes containing histone acetyl transferase activity that enhance gene transcription (co-activators) (Figure 3.1). Several transcriptional coactivators, including PPARγ coactivator-1 (PGC-1), the biological importance of which is discussed later in this review, bind and enhance PPAR activity (Puigserver and Spiegelman 2003).

The RXRs can be activated by 9-cis retinoic acid, an endogenous vitamin A derivative. RXR heterodimers formed with PPARs can be activated by RXR and PPAR agonists either independently or together, in which case they elicit synergistic activation. The individual roles of each of the RXR isoforms (α, β, γ) (Bastien and Rochette-Egly 2004) which complex with the PPARs to form the heterodimeric PPAR:RXR complex has not yet been determined (Berger and Moller 2002). However, like the PPARs, the RXR isoforms are characterised by a tissue-specific distribution: in rat heart, RXRγ and RXRβ are most abundant, whereas RXRα is only marginally detected (Mangelsdorf et al. 1992). The preferred DNA-binding site, referred to as the DR-1 motif, is the same for each of the PPAR/RXR heterodimer pairs (Kliewer et al. 2001). Specificity of function may be conferred by mechanisms that impart subtype-specific regulation of PPAR activity, including differences in binding affinity for endogenous ligands (Forman et al. 1997; Lin et al. 1999) or through phosphorylation of their transcriptional activation domains (Barger et al. 2000; Barger et al. 2001; Forman et al. 1997; Lazennec et al. 2000). PPARγ contains a consensus mitogen-activated protein kinase phosphorylation site which, when phosphorylated, decreases the ligand binding affinity of PPARγ (Shao et al. 1998) and negatively regulates transcriptional activity (Adams et al., 1997; Camp et al. 1999; Hu et al. 1996; Zhang et al. 1996). Docking of coactivators is generally regulated by a conformational change. Some of the most interesting of the transcriptional coactivators include PGC-1 and PGC-2. PGC-1α, in particular, is a key regulator of cellular energy metabolism: it is enriched in metabolically active tissues including heart (reviewed in Finck et al. 2005).

The activated PPAR:RXR complexes bind to specific response elements (PPREs) in the promoters of the various target genes. These are comprised of a consensus sequence (AGGTCA), the NR hexameric core recognition motif, that can be configured as a single element or two tandem repeats separated by one nucleotide. Binding to the PPREs modifies transcription, leading to an altered complement of enzymes in the cell, with a resultant alteration in cellular functional characteristics. Through this mechanism, each PPAR can, once activated, confer distinct properties of lipid handling to the tissues in which it is found, with an overall systemic lipid-lowering effect.

3.4. Pharmacological PPAR Ligands

The importance of the nuclear hormone receptors in general was revealed when it was discovered that they represented the targets of a number of widely-prescribed drugs. It was discovered that PPARα and PPARγ are the molecular targets for major classes of drugs used to ameliorate atherogenic hypertriglyceridemia and insulin resistance, both of which are commonly used in the treatment of type 2 diabetes. The fibrate class of drugs (clofibrate, gemfibrozil, fenofibrate, bezafibrate and ciprofibrate) have been in clinical use for over 40 years. These drugs target PPARα. Synthetic ligands for PPARα, in addition to the fibrate drugs, include phthalate ester plasticers, herbicides, and leukotriene D4 receptor antagonists. PPARγ is the molecular target for the insulin-sensitising thiazolidinediones (TZDs) (e.g., rosiglitazone, pioglitazone). Some of the side effects of the TZDs include weight gain through effects of PPARγ on adipose tissue (described below) and current research is evaluating newer PPARγ ligands, whose mode of binding differs from the TZDs or which are selected on the basis of attenuated regulation of lipogenic genes (Shi et al. 2005). A new class of novel PPAR-targeted compounds termed dual-acting PPAR agonists, which include the compound ragaglitizar, specifically activate both PPARγ and PPARα. Synthetic PPARδ ligands include GW501516 and GW0742, which activate PPARδ with a 1000-fold selectivity over other PPAR isotypes (Sznaidman et al. 2003). There is also current pharmaceutical interest in PPAR-pan agonists (which target all three PPAR isoforms): these generate enhanced insulin sensitivity, mild suppression of circulating triglycerides, reductions in low-density lipoprotein (LDL) cholesterol, and increased high density lipoprotein (HDL) cholesterol. A recent review has described emerging concepts in the use of PPAR agonists for treating metabolic and cardiovascular disease (Staels and Fruchart 2005).

3.5. Physiological PPAR Ligands

The PPARs are vitally important sensors of the lipid status of the body and of individual tissues or organs of the body. Although the complete range of physiological ligands that regulate the activities of each of the PPAR isoforms still remains to be established, it is now known that the PPARs are activated by binding of naturally-occurring endogenous or dietary FA or lipid derivatives. Known natural ligands of PPARα include polyunsaturated FAs and FA derivatives, including eicosanoids and leukotriene B4 (a lipoxygenase product of arachidonic acid). Recently the ω-hydroxylated derivatives of 14,15-epoxyeicosatrienoic acid, another derivative of arachidonic acid, has been demonstrated to activate PPARα (Cowart et al. 2002): other EETs and their derivatives may also function as other natural ligands. Oleoylthanolamide, a naturally occurring lipid that is structurally similar to the endogenous cannabinoid anandamide, also activates PPARα (Fu et al., 2003). New data also indicate that, in liver, newly-synthesised fat (but not pre-existing fat) may be an endogenous PPARα activator (Chakravarthy et al. 2005). These studies used a mouse model with liver-specific inactivation of the gene encoding the lipogenic enzyme FA synthase (FAS). When normal chow was substituted by a high carbohydrate-zero fat diet, these mice showed a fasting metabolic profile (hypoglycemia, high FA, lowered insulin, low leptin), and a fatty liver: this resembles the phenotype of fasting PPARα null mice. This phenotype was reversed by the provision of high-fat diet. Activation of PPARα also "rescued" the phenotype. This suggests that in the absence of dietary fat, FA synthesised de novo through FAS ("new fat") can act as a PPARα ligand (and upregulate FA oxidation). It was also suggested that dietary FA transported to the liver via chylomicrons could represent another source of "new" fat. However, it was proposed that FA released from stored TAG in adipose tissue, "old fat", could not activate hepatic PPARα as effectively. It was proposed that new fat and old fat access separate intracellular compartments, each containing a different pool of PPARα, and these pools cannot be accessed by the FA derived from the bulk of intrahepatic TAG (most of which originate from adipose-tissue derived FA). One pool of PPARα is activated by both newly synthesised and dietary FA (and upregulates glucose synthesis and FA oxidation); the other pool is activated only by biosynthetic FA, and this pool additionally activates cholesterol balance. The possibility was suggested that a lipid-binding protein specific for new fat shuttles FA to nuclear PPAR or that PPARα shuttles between the nucleus and cytoplasm via chaperones (e.g. hsp90) that promote an interaction with new fat. VLDLs produced by the liver and chylomicrons containing dietary TAG are targeted to heart and skeletal muscle, generating a local supply of FA and monoacylglycerol via lipoprotein lipase (LPL). Physiological TAG hydrolysis mediated by LPL might represent a link between lipid metabolism and PPARα activation in tissues, such as heart, where there is less de novo fat synthesis than in liver a finding relevant to the effects of provision of high-fat diets, particularly those rich in very-long-chain FAs and polyunsaturated FAs, which can induce FA oxidation via PPARα. Lipolysis of TAG-rich lipoproteins (VLDL) generates PPAR ligands in endothelial cells (Ziouzenkova et al. 2003). Cardiac specific knockout of LPL elevates plasma and cardiac TAG and lowers expression of the PPARα target gene pyruvate dehydrogenase kinase (PDK) 4 (Augustus et al. 2004). Whether or not LPL activity in adipose tissue generates natural agonists for PPARγ has not been investigated, but it is known that PPARγ is activated by specific prostanoids, such as 15-deoxy-$\Delta^{12,14}$ prostaglandin J2. Activation of PPARγ by components of oxidised LDL (9-hydroxyoctadecadienoic acid, 13- hydroxyoctadecadienoic acid) enhances macrophage lipid accumulation through induction of a scavenger receptor/transporter (FA translocase/CD36) (Lee et al. 2004). PPARδ is activated by long-chain FA, both saturated and unsaturated, and by prostacyclin produced by the conversion of polyunsaturated FAs (Amri et al. 1995). Again, it is not yet known whether PPARδ activation is coupled with LPL activity in oxidative skeletal muscle.

3.6. The Role of PPARα in the Regulation of Fuel Handling

The actions of PPARα in heart and liver reflect the biological roles of these tissues: the heart as a continually-contracting pump, the liver as a major fuel producer and transformer. A recent study has demonstrated that, during neonatal development of the heart, all three PPAR isoforms are upregulated almost in parallel (Steinmetz et al. 2005). In parallel, cardiac metabolism switches from glucose to lipid as the preferred energy substrate to generate ATP (Makinde et al.1998). This transition is accompanied by changes in the activity and expression levels of several enzymes and regulators involved in FA metabolism, indicating a functional bias towards the lipo-oxidative PPARs, namely PPARs -α and -β.

3.6.1. Cardiac Fuel Utilisation

In the adult, the heart uses most available fuels to sustain a continuous and enormous demand for energy to fulfill its function (Stanley and Chandler 2002). Although pyruvate derived from either glycolysis or lactate that enters the cell via the monocarboxylate transporter (MCT) can be converted to acetyl-CoA via the pyruvate dehydrogenase (PDH) complex for entry into the tricarboxylic acid (TCA) cycle (Figure 3.2), lipids are thought to constitute the predominant energy substrate for the adult heart, even in the well-fed state. FAs account for 70% of total ATP production in the normal heart; this proportion increases to >90% in

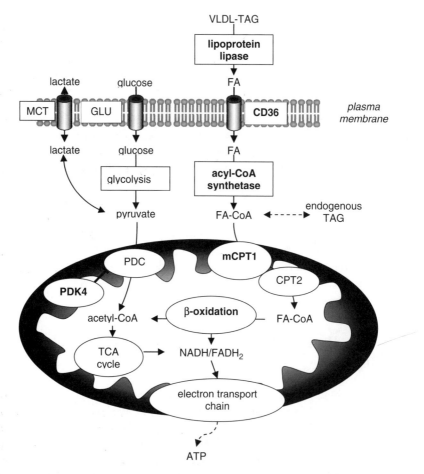

FIGURE 3.2. Regulation of cardiac glucose and lipid utilisation by PPARα. Sites regulated by PPARα are shown in bold. Abbreviations: AQP, aquaporin; CPT, carnitine palmitoyl-transferase; GLUT, glucose transporter; FA, fatty acid; LPL, lipoprotein lipase; MCT, monocarboxylate transporter; PDC, pyruvate dehydrogenase complex; PDK4, pyruvate dehydrogenase kinase 4; VLDL-TAG, very-low-density lipoprotein-triacylglycerol.

diabetes (Saddik and Lopaschuk 1991). In addition, cardiac FA oxidation is very high upon reperfusion after ischemia (Kantor et al., 1999). Acetyl-CoA can be generated from a range of lipid sources: circulating non-esterified FA (NEFA); endogenous (stored) TAG; TAG in circulating very-low-density lipoproteins (VLDL); or circulating ketone bodies. In perfused rat hearts, a considerable rate of intramyocellular TAG breakdown supplies FA for post-ischemic fuel oxidation, and TAG lipolysis contributes about 20% of total FA oxidation even when hearts are perfused with exogenous FA (Saddik and Lopaschuk 1992).

3.6.2. Regulation of Cardiac Lipid Utilisation by PPARα

The importance of PPARα in the adult heart appears relatively limited when the dietary or endogenous lipid supply is restricted, but PPARα assumes an important role in cardiac lipid management when lipid delivery to the heart is increased e.g., by fasting or by increasing dietary fat. In these situations, PPARα activation enhances the expression of genes encoding many proteins involved in FA transport, FA activation and FA oxidation (Figure 3.2): examples of PPARα-activated genes include muscle carnitine palmitoyltransferase 1 (m-CPT1) (which is crucial for the transport of long-chain FA into the mitochondria) and the β-oxidation enzymes medium-chain acyl-CoA dehydrogenase (MCAD) and long-chain acyl-CoA dehydrogenase (LCAD). Increased expression of these genes allows increased mitochondrial FA oxidation. Thus, PPARα deficient (null) mice provided with free access to a high-carbohydrate low-fat diet do not show any obvious cardiac abnormalities, but starvation or high-fat feeding results in significant perturbations in cardiac lipid metabolism as a consequence of a failure of the heart to induce FA β-oxidation pathways (Djouadi et al. 1998; Kersten et al. 1999; Leone et al. 1999), including abnormal accumulation of neutral lipid (Djouadi et al. 1998; Leone et al. 1999). Absence of PPARα decreases mitochondrial FA by approximately 70% (Aoyama et al. 1998; Campbell et al. 2002; Watanabe et al. 2000) due to reduced constitutive and/or inducible expression of several FA catabolic enzymes (Campbell et al. 2002).

3.6.3. Regulation of Cardiac Pyruvate Handling by PPARα

Although PPARα is closely involved in the control of lipid handling, it also exerts a direct and vital influence on glucose utilisation at the level of the pyruvate dehydrogenase complex (PDC). Regulation of PDC activity is important since PDC inactivation is crucial for glucose conservation when glucose is scarce, whereas adequate PDC activity is required to allow both ATP and FA production from glucose. The regulatory kinase pyruvate dehydrogenase kinase (PDK) 4 is an important negative regulator of glucose oxidation, functioning via post-translational modification of PDC (Sugden and Holness 2003) (Figure 3.2). Phosphorylation of E1 of PDC by PDK4 (and others of the PDK family) causes inactivation, sparing glucose and favouring FA oxidation. High saturated fat feeding increases cardiac PDK4 expression and PPARα activation acts synergistically with high-fat feeding to enhance cardiac PDK4 expression (Holness et al. 2002b). This essentially forces cardiac FA oxidation, despite available carbohydrate, and thereby optimises clearance of incoming lipid, particularly as the activity of this specific PDK isoform is increased by the increasing $NADH/NAD^+$ concentration ratios that accompany FA β-oxidation (Bowker-Kinley et al. 1998) (reviewed in Sugden and Holness 2003). In starvation, enhanced cardiac expression of PDK4 blocks glucose oxidation and again forces FA oxidation to maintain the ATP supply for cardiac contraction (Sugden and Holness 2003), this time in the face of a diminished supply of glucose. Increased expression of cardiac PDK4 in starvation is again linked to PPARα signalling, as well as to cardiac FA supply (Holness et al. 2002b; Wu et al., 2001). The involvement of lipid signalling via PPARα in the regulation of cardiac PDK4 expression, as well as that of FA oxidation enzymes, extends the operation of the classical glucose-FA cycle to the gene level and the newly-emerging field of nutrigenomics (Gillies 2003; Mutch et al. 2005).

3.6.4. Regulation of Hepatic Nutrient Handling by PPARα

The liver is capable of either oxidising, or storing, FA at high rates - although not simultaneously. This metabolic versatility underlies its central position in whole-body lipid handling during the absorptive phase and after fasting. Incoming FA, derived from the diet or adipocyte TAG, can be re-esterified within the liver to provide the TAG component of VLDL: this function can be increased when hepatic FA delivery is higher than that required to fuel hepatic ATP requirements. During starvation, FAs generated by adipocyte lipolysis are taken up by the liver and oxidised to form the water-soluble lipid-derived fuels, the ketone bodies (Figure 3.3). The ketone bodies are avidly used as metabolic fuels by skeletal muscle and brain in starvation, reducing their requirement for glucose as metabolic fuel. PPARα regulates genes encoding regulatory proteins enhancing each stage of hepatic FA utilisation: FA uptake into the hepatocyte across the plasma membrane, FA retention within the cell and its total or partial mitochondrial oxidation, leading to ATP production and ketone body formation respectively (Figure 3.3). It is essential that the capacity for hepatic FA oxidation be up-regulated when lipid delivery is increased by a diet high in fat and also in starvation, where there is increased generation of FA derived from the hydrolysis of adipose-tissue TAG. These roles are best illustrated by the abnormal hepatic responses to starvation seen in PPARα–deficient mice (Kersten et al. 1999; Leone et al. 1999; Sugden et al. 2002). These animals exhibit an impaired ability to up-regulate hepatic FA oxidation in response to a high-fat diet, despite increases in FA supply. Secondary to impaired FA oxidation, PPARα deficient mice become hypoglycemic when fasted in part due to inadequate generation of ATP to fuel hepatic gluconeogenesis (Figure 3.3). Hypoglycemia also arises out of the necessity of heart and oxidative skeletal muscle to continue to use

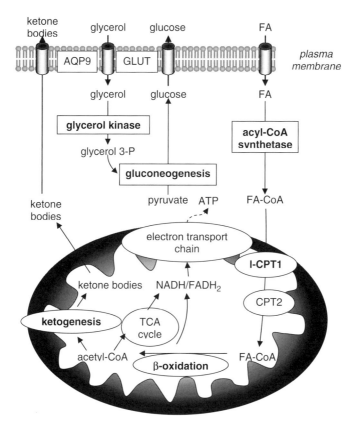

FIGURE 3.3. Regulation of hepatic glucose and lipid utilisation by PPARα. Sites regulated by PPARα are shown in bold. Abbreviations: AQP, aquaporin; GLUT, glucose transporter; CPT, carnitine palmitoyltransferase; FA, fatty acid.

glucose at high rates, rather than to be able to switch to FA or ketone bodies as alternative energy fuels. Fasting hypoketonemia in PPARα null mice reflects a failure to enhance expression of the key mitochondrial enzyme in ketogenesis, mitochondrial HMG-CoA synthase (Le May et al. 2000).

In response to lipolytic stimulation (exercise and fasting), adipocytes release not only hydrolysed FA, but also glycerol into the circulation. Glycerol is used for hepatic gluconeogenesis, a process involving its conversion to glycerol 3-phosphate by hepatic glycerol kinase (Figure 3.4). The contribution of glycerol to glucose production depends on the nutritional state: it may vary from 5% postprandially in humans (Baba et al. 1995), and approx 50% in the postabsorptive state in rodents (Peroni et al. 1995) to 20% in humans (Baba et al. 1995; Jensen et al. 2001) and >90% rodents after prolonged fasting (Peroni et al. 1995). It has been proposed that PPARα governs hepatic glycerol metabolism: the expression of several genes involved in the hepatic metabolism of glycerol

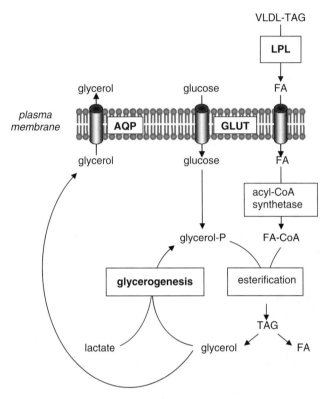

FIGURE 3.4. Actions of PPARγ on the adipocyte. Sites regulated by PPARγ are shown in bold. Abbreviations: AQP, aquaporin; GLUT, glucose transporter; FA, fatty acid; LPL, lipoprotein lipase.

(e.g. glycerol kinase, cytosolic and mitochondrial glycerol 3-phosphate dehydrogenase) is enhanced by fasting in wild type mice but not PPARα null mice (Patsouris et al. 2004). Furthermore, administration of pharmacological PPARα agonists lowers plasma glycerol concentrations (Patsouris et al. 2004).

Under conditions of PPARα stimulation by the non-β-oxidisable FA tetradecylthioacetic acid, the liver functions as a major drain for circulating FA (Madsen et al. 2002). Furthermore, the fibrates in current clinical use primarily act on the liver in humans; hence, the TAG-lowering action of fibrates and related PPARα-targeted drugs in patients with hyperlipidemia is ascribed to increased FA clearance via oxidation primarily, but not exclusively in liver. However, new compounds are currently being developed to target PPARα in tissues other than the liver. Treatment of obese hypertriglyceridemic monkeys with a selective PPARα agonist increases LPL activity in muscle, with a reciprocal decrease in adipose tissue. Thus PPARα agonists suppress the uptake of TAGs in adipose tissue, while increasing uptake in muscle for oxidation (Bodkin et al. 2003).

3.7. Regulation of Lipid Storage by PPARγ

The human PPARγ gene contains alternative transcription start sites resulting in three splice variants that produce three mRNA species in man (PPARγ1, 2 and 3) and two in mice (PPARγ1 and 2) (Fajas et al. 1997; Fajas et al. 1998). These mRNAs yield two distinct protein products: PPARγ1and 3 mRNA encode an identical protein, PPARγ1, while PPARγ2 encodes a protein containing an additional 30 amino acids, PPARγ2. PPARγ1 is expressed in many tissues; PPARγ2 is expressed predominantly in adipose tissue (Auboeuf et al. 1997). Conditioned PPARγ knockouts have confirmed an essential role of PPARγ in adipocyte differentiation and survival (Kubota et al. 1999; Rosen et al. 1999). Homozygous PPARγ knockout mice die at embryonic day 10 from impaired placental development (Barak et al. 1999), highlighting that PPARγ is also vital for nutrient and oxygen transfer to the fetus. PPARγ interacts with a number of ligand-dependent co-activator complexes, including the CBP/p300 complex (Gelman et al. 1999), steroid receptor c-activator (SRC1) (Zhu et al, 1996), and PPARγ coactivators PGC1 and PGC2, resulting in a range of biological activities.

The ability of adipocytes to store FA as TAG in the fed state provides animals with a fuel store for use in time of need. In the fed state, insulin increases adipocyte TAG storage by augmenting adipocyte glucose uptake (allowing the production of glycerol 3-phosphate for FA esterification to form TAG), by direct stimulation of enzymes involved in the esterification of incoming FA, mainly generated locally via adipocyte LPL, and by effects to block TAG lipolysis. Adipocyte production of glycerol 3-phosphate is enhanced by PPARγ activation (Guan et al. 2002; Picard and Auwerx 2002), which stimulates glycerol transport and glyceroneogenesis (Guan et al. 2002; Tordjman et al. 2003), in addition to glucose transport and LPL expression (Figure 4.4).

Maintenance of glucose homeostasis requires optimisation of the body fat mass and distribution, as shown by animal models of obesity and lipodystrophy which are invariably insulin resistant (see e.g., Koutnikova et al. 2003). The use of PPARγ knockout mice with tissue or isoform specific deletions, together with the identification of human subjects with mutations in PPARγ, has highlighted its role as a critical regulator of adipose tissue differentiation and function, maintenance of fuel homeostasis and energy balance (Sewter and Vidal-Puig 2002). Mice harbouring dominant negative mutations of PPARγ show an altered adipose-tissue distribution (Tsai et al. 2004). As ablation or impaired function of PPARγ results in an insulin-resistant, lipodystrophic phenotype, the concept has been developed that the insulin-sensitising effects of PPARγ activation (e.g., by the TZDs) reside in its actions in adipose tissue: these include, in addition to the induction of genes within mature adipocytes that allow them to take up and esterify FA to storage TAG, the recruitment of new adipocytes allowing an increase in lipid storage capability. Abnormal accumulation of TAG and other lipid derivatives in liver and muscle can result in insulin resistance (Shulman 2000). Thus, the sequestering of lipid within the adipocyte reduces the burden of excessive FA delivery to heart, skeletal muscle, and liver, where its oxidation and/or accumulation of FA-derived lipids could cause impaired glucose utilisation via substrate competition and/or insulin desensitisation (Frayn 2002). Of interest, specific deletion of the PPARγ gene in fat and muscle causes insulin resistance (He et al. 2003; Hevener et al. 2003), but heterozygous PPARγ knockout mice have improved insulin sensitivity, and are not susceptible to insulin resistance arising from a high-fat diet (Kubota et al. 1999). Thus partial activation is required to promote normal, but not excessive, adipose tissue stores.

PPARγ activation in the mature adipocyte also alters the expression and secretion of adipocyte proteins that regulate insulin sensitivity, resulting in an insulin-sensitising adipokine profile (Rangwala and Lazar 2004). Biologically-active adipokines include leptin (insulin sensitising) and TNFα (which causes insulin resistance). Leptin is considered to be insulin sensitising through its effects to promote FA oxidation (Ceddia 2005). Increased leptin expression in PPARγ$^{+/-}$ mice has been suggested to ameliorate insulin resistance, at least in part, through its effects to promote FA oxidation in liver and muscle (Yamauchi et al., 2001). This occurs in some tissues via activation of AMPK (Minokoshi et al. 2002). In addition, leptin treatment rapidly activates PPARs in C2C12 muscle cells (Bendinelli et al. 2005). TNFα has an established role in the aetiology of insulin resistance (Hotamisligil 1999). PPARγ suppresses the production of both leptin and TNFα, the former allowing the adipocyte to continue to accumulate lipid and the latter relieving the insulin resistance that exposure to high levels of TNFα induces. Recent emphasis has been placed on the insulin-sensitising adipokine adiponectin. Adiponectin appears to mediate its effects through two distinct receptors, AdipoR1 and AdipoR2: AdipoR1 is ubiquitous (highest in skeletal muscle), while AdipoR2 expression is more restricted (most abundant in liver) (Yamauchi et al. 2003). Primary hepatocytes bind full length adiponectin which stimulates PPARα ligand activity and AMPK activation (Yamauchi et al. 2003). The insulin-sensitising effects of the TZDs are impaired in mice lacking adiponectin (Kadowaki and Yamauchi 2005).

Although PPARγ is primarily involved in the regulation of lipid storage, it may also function in regulating glycerol transport across the adipocyte plasma membrane (Figure 3.4). The aquaporins are a family of small (approx 30kDa/monomer), integral membrane proteins. Aquaporin (AQP) 7 (also termed AQPap/7 in man), which transports glycerol as well as water, is expressed at high levels in adipose tissue (Kishida et al., 2000). During differentiation, 3T3-L1 adipocytes increase adrenaline-stimulated release of glycerol in parallel with the induction of AQP7 mRNA (Kishida et al. 2000). Knockdown of AQP7 in 3T3-L1 adipocytes by introducing small interfering mRNA suppresses adrenaline-stimulated glycerol release by approximately 50% (Maeda et al, 2004). AQP7 is a direct PPARγ target gene in adipocytes (Guan et al. 2002; Kishida et al. 2001). Differentiated mouse and human adipocytes respond to the TZDs (ciglitazone and rosiglitazone) in culture with increased adipocyte AQP7 mRNA expression (Patsouris et al. 2004). The PPRE site in the human AQP7 promoter is important for high AQP7 mRNA expression in differentiated adipose cells and is responsible for the induction of AQP7 transcription by TZDs (Kondo et al. 2002). Contrasting with liver, adipose tissue normally contains very little glycerol kinase and glycerol liberated from adipocyte TAG is not recycled intracellularly to any great extent. However, it has been suggested (Guan et al. 2002), although it remains controversial (Tan et al. 2003a), that in concert with its effects to enhance adipose tissue AQP7 expression during adipocyte maturation, PPARγ activation induces adipocyte glycerol kinase mRNA and activity, allowing glycerol to serve as a direct source of glycerol 3-phosphate within the adipocyte. Under physiological conditions, this could be potentially important for expansion of the adipose-tissue mass e.g., during puberty in females to promote the secretion of biologically active adipokines such as leptin, which are essential for reproductive function (Chehab et al. 1996; Chehab et al. 1997; reviewed by Chehab 1997).

Interactions between PPARα and PPARγ in the control of tissue development are not restricted to adipose tissue. The PPARα specific ligand WY14643 and the PPARγ specific ligand GW0742 both induce increased expression of enzymes involved in mitochondrial FA β-oxidation in cell cultures of neonatal rat cardiomyocytes (Cheng et al. 2004). The highest increment in PPARα and PPARγ mRNA expression in the rat heart occurs from postnatal day 0 to day 7, whereas mRNA expression of MCAD and LCAD was observed between days 7 and 21 (Steinmetz et al. 2005). Interestingly, expression of PDK4, a PPARα/ PPARδ target gene in the heart, also increases over this period (Sugden et al. 2000). Although the delay in the rise in MCAD and LCAD expression may reflect the absence of a necessary coactivator, the mRNA expression of the RXR receptors, PGC-1α and β do not parallel the rise in MCAD and LCAD. Other PPARγ target genes in muscle include genes involved in the uptake and metabolism of glucose (e.g. c-Cbl associated protein and GLUT4) (Baumann et al. 2000).

3.8. The Role of PPARδ in the Regulation of Oxidative Metabolism

Relatively little was known about the function of PPARδ until relatively recently, when the availability of PPARδ-specific agonists and animal models allowed definition of the metabolic phenotype. PPARδ is expressed early during embryogenesis and ubiquitously in adulthood (Braissant and Wahli 1998). Most PPARδ-deficient mice die early in embryonic development due to a placental defect; those that do survive are characterised by reduced body fat mass, skin defects and altered myelinisation (Peters et al. 2000). PPARδ is implicated in the regulation of proliferation of preadipocytes (Grimaldi 2001), keratinocytes (Tan et al. 2003b), enterocytes (Poirier et al. 2001) and placenta (Lim et al. 1999). Early studies also suggested that PPARδ is involved in cholesterol metabolism (Oliver, Jr. et al. 2001) and adiposity (Barak et al. 2002). PPARα and PPARδ are similarly abundant in heart (Muoio et al., 2002); it appears that in the heart many processes are dependent on functional PPARα (Muoio et al. 2002). Nevertheless, exposure of cardiac myocytes to either PPARα- or PPARδ-specific agonists leads to significant induction of PPAR target genes involved in FA uptake and oxidation and both PPARs can be activated by FAs within this system.

Skeletal muscle expresses all three PPAR isoforms, with high levels of PPARδ expressed in both human and rodent skeletal muscle (Escher et al. 2001; Jones et al. 1995; Muoio et al. 2002). In this tissue, PPARδ promotes oxidative metabolism and enhances aerobic capacity. Slow-twitch oxidative skeletal muscle is a major site of FA catabolism, particularly during starvation and long-term exercise, when circulating lipid delivery increases. However, most evidence suggests that signalling via PPARα impacts predominantly on the regulation of lipid gene expression and fat oxidation in fast-twitch, not slow-twitch, skeletal muscle, even though fast-twitch muscle does not normally oxidise FA as avidly as slow oxidative muscle (Dulloo et al. 2001; Holness et al. 2002a), and PPARδ is predominantly expressed in mitochondrial-rich slow oxidative muscle (reviewed by Smith and Muscat 2005). Whereas PPARα ablation in the mouse results in abnormally high accumulation of neutral lipid in heart and liver in response to interventions increasing FA delivery, such as starvation, starvation of PPARα deficient mice leads to only minor abnormalities of skeletal-muscle FA metabolism (Muoio et al. 2002). In addition, skeletal muscles from PPARα null mice exhibit only minor changes in FA homeostasis and neither constitutive nor inducible mRNA expression of known PPARα target genes, including PDK 4 (Holness et al. 2002a) and uncoupling protein (UCP) 3 are negatively affected by the absence of PPARα (Muoio et al. 2002). PPARδ may therefore assume greater importance in the modulation of lipid metabolism than PPARα in skeletal muscle, a conclusion supported by findings that PPARδ mRNA expression levels in muscle are dramatically increased on fasting, under conditions where FA oxidation by skeletal muscle is increased. The failure of PPARα null mice to exhibit severe defects in skeletal muscle FA oxidation in response to starvation and exercise may reflect the fact that activation of PPARδ using GW742 also increases FA oxidation and the mRNA levels of several classical PPARα target genes, including PDK4, m-CPT1, MCD and UCP3 (Muoio et al. 2002). Similarly, PPARδ activation induces mRNA expression of UCP2 and UCP3 in human and L6 myotubes. This has led to the suggestion that significant overlap exists in the functions of PPARα and PPARδ in mediating lipid-induced regulation of β-oxidation pathways in skeletal muscle (Muoio et al. 2002). Gain- and loss-of-function experiments demonstrated that PPARδ regulated the expression of genes encoding proteins involved in FA uptake, handling and β-oxidation in differentiated mouse C2C12 myotubes (Holst et al. 2003). Similar observations were obtained using Affymetrix microarrays of rat L6 myotubes (Tanaka et al. 2003). Generation of a Cre/Lox transgenic mouse allowing skeletal muscle-specific overexpression of PPARδ resulted in increased oxidative (fast-oxidative 2a) myofibers due to an increase in total myofiber number (Luquet et al. 2003). This response resulted in an increase in oxidative capabilities, characterised by up-regulation of succinate dehydrogenase, citrate synthase, heart-FA binding protein and UCP2. In addition, these mice exhibited net reduction of body fat mass related to reduced adipocyte size in fat depots.

Subsequent studies demonstrated that overexpression of a constitutively-active PPARδ in skeletal muscle resulted in protection against the development of diet-induced obesity (Wang et al. 2004). There is also evidence that PPARδ is implicated in the response to physical exercise. The phenotypes elicited by muscle-specific overexpression of PPARδ are similar to those resulting from endurance training in rodents (Allen et al. 2001) and humans (McCall et al. 1996) and overexpression of PPARδ increases running endurance and resistance to fatigue (Wang et al. 2004). Repeated moderate exercise in mice also promotes up-regulation of PPARδ in muscle (Luquet et al. 2003).

Studies with synthetic PPARδ ligands suggest that they may improve many factors associated with dyslipidemia, obesity and insulin resistance. Treatment of insulin-resistant obese monkeys with the potent PPARδ agonist, GW1516, for 4 weeks normalises fasting plasma insulin and TAG concentrations, increases HDL-cholesterol and lowers LDL-cholesterol (Oliver, Jr. et al. 2001). Administration of GW1516 to obese mice reduced adiposity and improved insulin responsiveness (Tanaka et al. 2003; Wang et al, 2003). Mice expressing an activated mutant form of PPARδ in brown and white adipocytes are characterised by up-regulation of genes involved in FA catabolism and energy uncoupling (Wang et al. 2003). These mice are also protected against diet-induced obesity (Wang et al, 2003). These studies suggest that therapeutic activation of PPARδ could be used to decrease diet-induced obesity. Selective PPARδ agonists (e.g. GW501516) are currently in clinical trials for the treatment of dyslipidemia, insulin resistance and obesity (Smith and Muscat 2005).

3.9. PGC-1α: an Enhancer of Mitochondrial Function and Biogenesis

Many metabolic programmes can be triggered by the expression of the transcriptional coactivator PGC-1α. In addition to activating the PPARs, PGC-1α associates with other regulatory elements including cAMP response element (CREBP)-binding protein, the forkhead transcription factor FOXO1 and hepatocyte nuclear factor (HNF)-4α, which have been implicated in the development of diabetes (Gupta et al. 2005; Nakae et al. 2002; Puigserver et al. 2003; Yamagata et al., 1996; Yoon et al. 2001). PGC-1α exhibits a tissue-enriched expression pattern (Knutti and Kralli 2001; Puigserver et al. 1998; Puigserver and Spiegelman 2003). PGC-1α is robustly expressed in a range of oxidative tissues, including heart and skeletal muscle, and is rapidly induced by a range of physiological conditions (cold exposure, acute exercise and fasting) that increase the demand for mitochondrial ATP or heat production (Baar et al. 2002; Goto et al. 2000; Irrcher et al. 2003; Lehman et al. 2000; Pilegaard et al. 2003; Puigserver et al. 1998; Terada et al. 2002; Terada and Tabata 2004; Wu et al. 1999). Its expression is also increased in the postnatal heart coincident with an increase in mitochondrial FA oxidation (Lehman et al. 2000).

Recently, the PGC-1α/ERRα pathway has been demonstrated to regulate FA oxidation, oxidative phosphorylation and mitochondrial biogenesis in heart and skeletal muscle (Huss et al. 2004; Mootha et al. 2004; Schreiber et al. 2004; Willy et al. 2004). Evidence that PGC-1α regulates mitochondrial biogenesis in mammals includes studies of forced expression showing that PGC-1α induces expression of mitochondrial biogenic factors (e.g. nuclear respiratory factor 1 and 2 [NRF-1, NRF2] and the transcription factor of activated mitochondria [Tfam]) (Wu et al. 1999). In heart, PGC-1α induces mitochondrial biogenesis (Lehman et al. 2000) and, importantly, cooperates with PPARα to increase the expression of genes encoding mitochondrial FA oxidation enzymes, thereby increasing the capacity for FA oxidation and intracellular lipid clearance (Vega et al. 2000). In addition, forced expression of PGC-1α in skeletal muscle and heart triggers mitochondrial proliferation (Lehman et al. 2000; Lin et al. 2002; Russell et al. 2004).

In parallel with the increase in PPARα signalling, fasting also elevates PGC-1α mRNA expression in liver (Herzig et al. 2001; Yoon et al. 2001). Importantly, PGC-1α mRNA expression is also increased in three different rodent models of increased hepatic gluconeogenesis (Herzig et al. 2001; Yoon et al. 2001). Increased expression of PGC-1α in liver via adenovirus vector enhances hepatic glucose production (Herzig et al. 2001) (reviewed in Vidal-Puig and O'Rahilly 2001). At least two clinical studies have identified a correlation between mutations of the *Ppargc1a* (previously known as the PGC-1α gene) and insulin resistance or diabetes (Ek et al. 2001; Hara et al. 2002). The effect of PGC-1α to induce gluconeogenesis appears to occur via an interaction with the FOXO1 pathway (Puigserver et al. 2003). In liver, FOXO1 and the glucocorticoid receptor, an additional PGC-1α partner, directly regulate PDK4 expression through consensus binding sites in the PDK4 gene promoter (Kwon et al. 2004). This effect both restricts glucose oxidation and also the provision of carbon for malonyl-CoA synthesis, thereby facilitating hepatic FA oxidation by relieving malonyl-CoA-induced inhibition of CPT 1 (Sugden and Holness 1994). Thus, while the biological role of PGC-1α to enhance glucose production in starvation is of major physiological importance, abnormally high expression/activity of this coactivator will also induce the phenotype of hepatic insulin resistance.

PGC-1α expression is increased in heart after starvation (Lehman et al. 2000) and in skeletal muscle following exercise (Baar et al. 2002; Goto et al. 2000; Pilegaard et al. 2003). Skeletal muscle specific overexpression of PGC-1α results in a fibre-type switch from fast-twitch to slow-twitch fibres (Lin et al. 2002). The latter have an increased mitochondrial density compared with fast-twitch muscles, and thus an increased capacity for FA oxidation. Suppression of glucose utilisation (uptake/phosphorylation) makes a major contribution to glucose sparing during prolonged fasting (Holness and Sugden 1990) and, very recently, Kelly and colleagues have shown that PGC-1α directly activates PDK4 gene transcription in skeletal myotubes, leading to a reduction in glucose oxidation rates (Wende et al. 2005). This observation is important in view of our finding that signalling via PPARα is not obligatory for enhanced expression of PDK4 in skeletal muscle after starvation (Holness et al. 2002a).

3.10. Dysregulation of PPARs and PGC-1 in Disease States

Dysregulation of the PPARα axis and its control of cardiac FA utilisation have been detected in several cardiomyopathic disease states. Possibly as an adaptation to limited oxygen availability, in cardiac hypertrophy and heart failure the adult heart returns to a more "fetal" metabolic state, with greater reliance on glucose metabolism, (Calvani et al. 2000; van Bilsen et al. 2004). Oxygen is essential for ATP production from FAs, but glucose oxidation needs less oxygen per mole than FA oxidation and glucose utilisation can proceed, albeit less efficiently, in the absence of oxygen. However, less ATP is generated per mole of glucose oxidised, and when excess dietary lipid can be ingested, the potential exists for myocardial lipid accumulation and lipotoxicity. In animal models of heart failure related to pressure overload or ischemia, the expression of PPARα, PGC-1α and downstream target genes encoding FAO enzymes is diminished (Barger et al. 2000; Depre et al. 1998; Sack et al. 1996). Similar observations have been made in human patients with heart failure (Razeghi et al. 2002; Sack et al. 1996). It is implied that deactivation of the PPARα/PGC-1α axis is involved in the metabolic switch away from FA catabolism in the failing heart, with reversion to the fetal pattern of substrate use. Nevertheless, it remains controversial as to whether changes in cardiac PPARα signalling are adaptive or causally related to myocardial pathology. Although causality between the reciprocal changes in lipid and glucose utilisation has not been established, this metabolic switch may serve to preserve ventricular function within the context of pressure overload. The expression and activity of both PPARα and RXR are also suppressed in the hypoxic cardiac myocyte (Huss et al. 2001).

Epidemiological studies reveal that people with diabetes mellitus are at high risk for the development of cardiovascular disease. The heart in uncontrolled diabetes is constrained from switching to glucose oxidation, and relies almost exclusively on lipid as an energy source (Belke et al. 2000; Lopaschuk 2002). In the diabetic heart the PPARα/PGC-1α axis is activated (Finck et al. 2002), suggesting that dysregulation of the PPARα/PGC-1α axis in diabetes is likely to be involved in the metabolic remodelling that occurs in diabetic heart to initially sustain the requirement for ATP. However, the precise role of these metabolic "switches" in the development of heart failure has not been clearly defined. Insight has been provided by supraphysiological activation of PPARα. Cardiac-specific over-expression of PPARα (myosin heavy chain [MHC- PPARα] mice), in the absence of alterations in lipid-fuel delivery, show increased expression of genes encoding FA oxidation enzymes, together with increased rates of myocardial lipid oxidation and decreased rates of glucose uptake and oxidation (Finck et al. 2002; Finck et al. 2003). This, however, is maladaptive, since these mice develop a cardiomyopathy with enhanced sensitivity to ischemic injury and recapitulate many of the metabolic abnormalities of the diabetic heart (Lopaschuk 1996; Rodrigues and McNeill 1992), including decreased expression of genes involved in cardiac glucose utilisation, accumulation of intracellular lipid, ventricular hypertrophy and systolic dysfunction. In this situation, the profile of myocardial TAG species is similar to the lipid species in the circulation, suggesting that the expanded lipid reservoir reflects increased uptake incompletely balanced by a corresponding increase in oxidation. When mice with cardiac-specific PPARα overexpression are made diabetic, they develop a more severe cardiomyopathy than wild-type controls (Finck et al. 2002). In contrast, PPARα null mice are protected from the development of a cardiomyopathy (Campbell et al. 2002).

Effects of the TZDs to ameliorate hyperglycemia and hyperlipidemia in fatless mice demonstrate that, although predominantly expressed in adipose tissue, PPARγ ligands may exert their effects via PPARγ in tissues in addition to adipose tissue. Thus there is evidence that PPARγ is expressed in non-adipose tissue under situations where other PPAR isoforms are functionally absent. Consistent with its function as an activator of lipogenesis and glycerolipid synthesis, elevated PPARγ levels have been implicated as an early response in the progression of cardiac lipotoxicity (Muoio et al. 2002; Unger and Orci 2001). The PPARγ agonist troglitazone prevents angiotensin II-induced hypertrophy in cultured rat cardiomyocytes (Asakawa et al. 2002) and similar results are obtained in endothelin 1-induced hypertrophy (Sakai et al. 2002). In humans with ischemic heart disease, an increase in PPARγ expression has been observed when compared to patients suffering from dilated cardiomyopathy (Sakai et al. 2002). Although the underlying mechanism is not known, NFκB may be involved in the signalling pathway (Takano et al. 2000; Zingarelli et al. 2003). The only cardiac gene that exhibits a clear change in expression in response to troglitazone treatment is acyl-CoA oxidase, which is involved in peroxisomal FA oxidation (Cabrero et al. 2003). However, care should be taken when interpreting these reports as the expression levels of PPARs may vary between the main cardiac cell types (cardiac myocytes, fibroblasts and endothelial cells) and the potential direct roles of PPARγ in cardiac function are not yet clear (Kelly 2003). Of interest, the heart is closely associated with adipose tissue in adult man, and epicardial and pericardial depots can represent a significant fraction of the total mass of the heart, suggesting a possible role as a local FA source. With ageing and obesity, adipose tissue closely associated with the heart increases in mass. An intriguing possibility is that activation of PPARγ in the cardiac-associated adipose tissue may act as a potential local sink for FA, sequestering excess FA when they are delivered in excess of the cardiac ATP requirement or when they cannot be cleared through oxidation. The heart expresses LPL and, interestingly, LPL activity in isolated cardiac myocytes can be inhibited by PPARα activation (Carroll and Severson 2001). This effect may potentially help protect against a potentially toxic oversupply of FAs to the myocardium. PPARγ agonists have been reported to be cardioprotective against ischemic insult and to modify the cardiac hypertrophic growth response to pressure overload (Asakawa et al. 2002; Sivarajah et al. 2005). However, most cardiac effects are likely to be indirect because of the low level of expression of PPARγ in the adult cardiac myocytes themselves, although the adipose tissue surrounding the heart itself may be targeted directly and act as a lipid buffer.

Constitutive cardiac-specific overexpression of PGC-1 is lethal, the mice exhibiting high numbers of large mitochondria, developing a dilated cardiomyopathy (including depressed contractile function and massive edema) and ultimately dying by 6 weeks of age (Russell et al. 2005). As a consequence, subsequent studies employed the reverse tetracycline transactivator and a tetracycline responsive promoter to drive inducible cardiac-specific overexpression of PGC-1α (Russell et al. 2004). This model revealed developmental stage-specific responses to acute PGC-1α overexpression. Overexpression of PGC-1α during neonatal life causes dramatic proliferation of mitochondria, together with overt effects on cardiac function. Conversely, cardiac-specific overexpression of PGC-1α in adulthood resulted in only modest mitochondrial proliferation; however, it did elicit accumulation of aggregates of abnormal mitochondria (varied abnormal ultrastructure, including vacuoles, small granules and large membrane-bound inclusions within the mitochondria). In addition, cardiac dysfunction was observed in response to cardiac-specific overexpression of PGC-1α in adulthood, including wall thinning and chamber dilation, together with the up-regulation of atrial natriuretic factor and brain natriuretic factor.

3.11. Concluding Remarks

PPARs are central to whole-body and tissue specific lipid homeostasis, and by virtue of these actions, impact glucose homeostasis and insulin action. Activation of all three PPARs appears to decrease systemic lipid availability. PPARα and PPARδ both promote mitochondrial FA oxidation, whereas PPARγ acts to direct lipid towards storage in adipose tissue and away from oxidation in non-adipose tissues. Some physiological ligands for activation of each of the PPARs are PPAR subtype specific: such ligand specificity allows specificity of cellular downstream metabolic responses to each of the individual PPARs. The PPARs have proved to be attractive targets for the development of selective therapeutic agents for pharmacological correction of abnormal lipid metabolism in disease states. The newly-discovered role of the PPAR coactivator PGC-1α in both mitochondrial biogenesis and the coordination of lipid oxidation and glucose production provides a potential springboard for further research into mechanisms that may optimise muscle performance in exercise.

Acknowledgements. The authors' work cited in this review was supported by funding from the British Heart Foundation and Diabetes UK.

References

Adams M, Reginato MJ, Shao D, Lazar MA, Chatterjee VK (1997) Transcriptional activation by peroxisome proliferator-activated receptor gamma is inhibited by phosphorylation at a consensus mitogen-activated protein kinase site. J Biol Chem 272:5128–5132

Allen DL, Harrison BC, Maass A, Bell ML, Byrnes WC, Leinwand LA (2001) Cardiac and skeletal muscle adaptations to voluntary wheel running in the mouse. J Appl Physiol 90:1900–1908

Amri EZ, Bonino F, Ailhaud G, Abumrad NA, Grimaldi PA (1995) Cloning of a protein that mediates transcriptional effects of fatty acids in preadipocytes. Homology to peroxisome proliferator-activated receptors. J Biol Chem 270: 2367–2371

Aoyama T, Peters JM, Iritani N, Nakajima T, Furihata K, Hashimoto T, Gonzalez FJ (1998) Altered constitutive expression of fatty acid-metabolizing enzymes in mice lacking the peroxisome proliferator-activated receptor alpha (PPARalpha). J Biol Chem 273:5678–5684

Asakawa M, Takano H, Nagai T, Uozumi H, Hasegawa H, Kubota N, Saito T, Masuda Y, Kadowaki T, Komuro I (2002) Peroxisome proliferator-activated receptor gamma plays a critical role in inhibition of cardiac hypertrophy in vitro and in vivo. Circulation 105:1240–1246

Auboeuf D, Rieusset J, Fajas L, Vallier P, Frering V, Riou JP, Staels, B., Auwerx J, Laville M, Vidal H (1997) Tissue distribution and quantification of the expression of mRNAs of peroxisome proliferator-activated receptors and liver X receptor-alpha in humans: no alteration in adipose tissue of obese and NIDDM patients. Diabetes 46:1319–1327

Augustus A, Yagyu H, Haemmerle G, Bensadoun A, Vikramadithyan RK, Park SY, Kim JK, Zechner R, Goldberg IJ (2004) Cardiac-specific knock-out of lipoprotein lipase alters plasma lipoprotein triglyceride metabolism and cardiac gene expression. J Biol Chem 279:25050–25057

Baar K, Wende AR, Jones TE, Marison M, Nolte LA, Chen M, Kelly DP, Holloszy JO (2002) Adaptations of skeletal muscle to exercise: rapid increase in the transcriptional coactivator PGC-1. FASEB J 16:1879–1886

Baba H, Zhang XJ, Wolfe RR (1995) Glycerol gluconeogenesis in fasting humans. Nutrition 11:149–153

Barak Y, Liao D, He W, Ong ES, Nelson MC, Olefsky JM, Boland R, Evans RM (2002) Effects of peroxisome proliferator-activated receptor delta on placentation, adiposity, and colorectal cancer. Proc Natl Acad Sci USA 99:303–308

Barak Y, Nelson MC, Ong ES, Jones YZ, Ruiz-Lozano P, Chien KR, Koder A, Evans RM (1999) PPAR gamma is required for placental, cardiac, and adipose tissue development. Mol Cell 4: 585–595

Barger PM, Brandt JM, Leone TC, Weinheimer CJ, Kelly DP (2000) Deactivation of peroxisome proliferator-activated receptor-alpha during cardiac hypertrophic growth. J Clin Invest 105:1723–1730

Barger PM, Browning AC, Garner AN, Kelly DP (2001) p38 mitogen-activated protein kinase activates peroxisome proliferator-activated receptor alpha: a potential role in the cardiac metabolic stress response. J Biol Chem 276:44495–44501

Bastien J, Rochette-Egly C (2004) Nuclear retinoid receptors and the transcription of retinoid-target genes. Gene 328:1–16

Baumann CA, Chokshi N, Saltiel AR, Ribon V (2000) Cloning and characterization of a functional peroxisome proliferator activator receptor-gamma-responsive element in the promoter of the CAP gene. J Biol Chem 275:9131–9135

Belke DD, Larsen TS, Gibbs EM, Severson DL (2000) Altered metabolism causes cardiac dysfunction in perfused hearts from diabetic (db/db) mice. Am J Physiol Endocrinol Metab 279:E1104–E1113

Bendinelli P, Piccoletti R, Maroni P (2005) Leptin rapidly activates PPARs in C2C12 muscle cells. Biochem Biophys Res Commun 332:719–725

Berger J, Moller DE (2002) The mechanisms of action of PPARs. Annu Rev Med 53:409–435

Berkenstam A, Gustafsson JA (2005) Nuclear receptors and their relevance to diseases related to lipid metabolism. Curr Opin Pharmacol 5:171–176

Bodkin NL, Pill J, Meyer K, Hansen BC (2003) The effects of K-111, a new insulin-sensitizer, on metabolic syndrome in obese prediabetic rhesus monkeys. Horm Metab Res 35:617–624

Bowker-Kinley MM, Davis WI, Wu P, Harris RA, Popov KM (1998) Evidence for existence of tissue-specific regulation of the mammalian pyruvate dehydrogenase complex. Biochem J 329:191–196

Braissant O, Foufelle F, Scotto C, Dauca M, Wahli W (1996) Differential expression of peroxisome proliferator-activated receptors (PPARs): tissue distribution of PPAR-alpha, -beta, and -gamma in the adult rat. Endocrinology 137:354–366

Braissant O, Wahli W (1998) Differential expression of peroxisome proliferator-activated receptor-alpha, -beta, and -gamma during rat embryonic development. Endocrinology 139:2748–2754

Cabrero A, Jove M, Planavila A, Merlos M, Laguna JC, Vazquez-Carrera M (2003) Down-regulation of acyl-CoA oxidase gene expression in heart of troglitazone-treated mice through a mechanism involving chicken ovalbumin upstream promoter transcription factor II. Mol Pharmacol 64:764–772

Calvani M, Reda E, Arrigoni-Martelli E (2000) Regulation by carnitine of myocardial fatty acid and carbohydrate metabolism under normal and pathological conditions. Basic Res Cardiol 95:75–83

Camp HS, Tafuri SR, Leff T (1999) c-Jun N-terminal kinase phosphorylates peroxisome proliferator-activated receptor-gamma1 and negatively regulates its transcriptional activity. Endocrinology 140:392–397

Campbell FM, Kozak R, Wagner A, Altarejos JY, Dyck JR, Belke DD, Severson DL, Kelly DP, Lopaschuk GD (2002) A role for peroxisome proliferator-activated receptor alpha (PPARalpha) in the control of cardiac malonyl-CoA levels: reduced fatty acid oxidation rates and increased glucose oxidation rates in the hearts of mice lacking PPARalpha are associated with higher concentrations of malonyl-CoA and reduced expression of malonyl-CoA decarboxylase. J Biol Chem 277:4098–4103

Carroll R, Severson DL (2001) Peroxisome proliferator-activated receptor-alpha ligands inhibit cardiac lipoprotein lipase activity. Am J Physiol. Heart Circ Physiol 281: H888–H894

Ceddia RB (2005) Direct metabolic regulation in skeletal muscle and fat tissue by leptin: implications for glucose and fatty acids homeostasis. Int J Obes 29:1175–1183

Chakravarthy MV, Pan Z, Zhu Y, Tordjman K, Schneider JG, Coleman T, Turk J, Semenkovich CF (2005) "New" hepatic fat activates PPARalpha to maintain glucose, lipid, and cholesterol homeostasis. Cell Metab 1:309–322

Chehab FF (1997) The reproductive side of leptin. Nat Med 3:952–953

Chehab FF, Lim ME, Lu R (1996) Correction of the sterility defect in homozygous obese female mice by treatment with the human recombinant leptin. Nat Genet 12: 318–320

Chehab FF, Mounzih K, Lu R, Lim ME (1997) Early onset of reproductive function in normal female mice treated with leptin. Science 275:88–90

Cheng L, Ding G, Qin Q, Xiao Y, Woods D, Chen YE, Yang Q (2004) Peroxisome proliferator-activated receptor delta activates fatty acid oxidation in cultured neonatal and adult cardiomyocytes. Biochem Biophys Res Commun 313:277–286

Cowart LA, Wei S, Hsu MH, Johnson EF, Krishna MU, Falck JR, Capdevila JH (2002) The CYP4A isoforms hydroxylate epoxyeicosatrienoic acids to form high affinity peroxisome proliferator-activated receptor ligands. J Biol Chem 277:35105–35112

Depre C, Shipley GL, Chen W, Han Q, Doenst T, Moore ML, Stepkowski S, Davies PJ, Taegtmeyer H (1998) Unloaded heart in vivo replicates fetal gene expression of cardiac hypertrophy. Nat Med 4:1269–1275

Desvergne B, Wahli W (1999) Peroxisome proliferator-activated receptors: nuclear control of metabolism. Endocr Rev 20:649–688

Djouadi F, Weinheimer CJ, Saffitz JE, Pitchford C, Bastin J, Gonzalez FJ, Kelly DP (1998) A gender-related defect in lipid metabolism and glucose homeostasis in peroxisome proliferator- activated receptor alpha- deficient mice. J Clin Invest 102: 1083–1091

Dubois M, Pattou F, Kerr-Conte J, Gmyr V, Vandewalle B, Desreumaux P, Auwerx J, Schoonjans K, Lefebvre J (2000) Expression of peroxisome proliferator-activated receptor gamma (PPARgamma) in normal human pancreatic islet cells. Diabetologia 43:1165–1169

Dulloo AG, Samec S, Seydoux J (2001) Uncoupling protein 3 and fatty acid metabolism. Biochem Soc Trans 29:785–791

Duval C, Chinetti G, Trottein F, Fruchart JC, Staels B (2002) The role of PPARs in atherosclerosis. Trends Mol Med 8:422–430

Ek J, Andersen G, Urhammer SA, Gaede PH, Drivsholm T, Borch-Johnsen K, Hansen T, Pedersen O (2001) Mutation analysis of peroxisome proliferator-activated receptor-gamma coactivator-1 (PGC-1) and relationships of identified amino acid polymorphisms to Type II diabetes mellitus. Diabetologia 44:2220–2226

Escher P, Braissant O, Basu-Modak S, Michalik L, Wahli W, Desvergne B (2001) Rat PPARs: quantitative analysis in adult rat tissues and regulation in fasting and refeeding. Endocrinology 142:4195–4202

Fajas L, Auboeuf D, Raspe E, Schoonjans K, Lefebvre AM, Saladin R, Najib J, Laville M, Fruchart JC, Deeb S, Vidal-Puig A, Flier J, Briggs MR, Staels B, Vidal H, Auwerx J (1997) The organization, promoter analysis, and expression of the human PPARgamma gene. J Biol Chem 272:18779–18789

Fajas L, Fruchart J C, Auwerx J (1998) PPARgamma3 mRNA: a distinct PPARgamma mRNA subtype transcribed from an independent promoter. FEBS Lett 438:55–60

Finck BN, Bernal-Mizrachi C, Han DH, Coleman T, Sambandam N, LaRiviere LL, Holloszy JO, Semenkovich CF, Kelly DP (2005) A potential link between muscle peroxisome proliferator- activated receptor-alpha signaling and obesity-related diabetes. Cell Metab 1: 133–144

Finck BN, Han X, Courtois M, Aimond F, Nerbonne JM, Kovacs A, Gross RW, Kelly DP (2003) A critical role for PPARalpha-mediated lipotoxicity in the pathogenesis of diabetic cardiomyopathy: modulation by dietary fat content. Proc Natl Acad Sci USA 100:1226–1231

Finck BN, Lehman JJ, Leone TC, Welch MJ, Bennett MJ, Kovacs A, Han X, Gross RW, Kozak R, Lopaschuk GD, Kelly DP (2002) The cardiac phenotype induced by PPARalpha overexpression mimics that caused by diabetes mellitus. J Clin Invest 109:121–130

Forman BM, Chen J, Evans RM (1997) Hypolipidemic drugs, polyunsaturated fatty acids, and eicosanoids are ligands for peroxisome proliferator-activated receptors alpha and delta. Proc Natl Acad Sci USA 94:4312–4317

Frayn KN (2002) Adipose tissue as a buffer for daily lipid flux. Diabetologia 45:1201–1210

Fredenrich A, Grimaldi PA (2005) PPAR delta: an uncompletely known nuclear receptor. Diabetes Metab 31:23–27

Fu J, Gaetani S, Oveisi F, Lo VJ, Serrano A, Rodriguez DF, Rosengarth A, Luecke H, Di Giacomo B, Tarzia G, Piomelli, D (2003) Oleylethanolamide regulates feeding and body weight through activation of the nuclear receptor PPAR-alpha. Nature 425: 90–93

Gelman L, Zhou G, Fajas L, Raspe E, Fruchart J C, Auwerx J (1999) p300 interacts with the N- and C-terminal part of PPARgamma2 in a ligand-independent and -dependent manner, respectively. J Biol Chem 274:7681–7688

Gillies PJ (2003) Nutrigenomics: the Rubicon of molecular nutrition. J Am Diet Assoc 103:S50–S55

Goto M, Terada S, Kato M, Katoh M, Yokozeki T, Tabata I, Shimokawa T (2000) cDNA Cloning and mRNA analysis of PGC-1 in epitrochlearis muscle in swimming-exercised rats. Biochem Biophys Res Commun 274:350–354

Gray SL, Dalla NE, Vidal-Puig AJ (2005) Mouse models of PPAR-gamma deficiency: dissecting PPAR-gamma's role in metabolic homoeostasis. Biochem Soc Trans 33:1053–1058

Grimaldi PA (2001) The roles of PPARs in adipocyte differentiation. Prog Lipid Res 40:269–281

Grimaldi PA (2005) Regulatory role of peroxisome proliferator-activated receptor delta (PPAR delta) in muscle metabolism. A new target for metabolic syndrome treatment? Biochimie 87:5–8

Guan HP, Li Y, Jensen MV, Newgard CB, Steppan CM, Lazar MA (2002) A futile metabolic cycle activated in adipocytes by antidiabetic agents. Nat Med 8:1122–1128

Gupta RK, Vatamaniuk MZ, Lee CS, Flaschen RC, Fulmer JT, Matschinsky FM, Duncan SA, Kaestner KH (2005) The MODY1 gene HNF-4alpha regulates selected genes involved in insulin secretion. J Clin Invest 115:1006–1015

Hara K, Tobe K, Okada T, Kadowaki H, Akanuma Y, Ito C, Kimura S, Kadowaki T (2002) A genetic variation in the PGC-1 gene could confer insulin resistance and susceptibility to Type II diabetes. Diabetologia 45:740–743

He W, Barak Y, Hevener A, Olson P, Liao D, Le J, Nelson M, Ong E, Olefsky JM, Evans RM (2003) Adipose-specific peroxisome proliferator-activated receptor gamma knockout causes insulin resistance in fat and liver but not in muscle. Proc Natl Acad Sci USA 100:15712–15717

Herzig S, Long F, Jhala US, Hedrick S, Quinn R, Bauer A, Rudolph D, Schutz G, Yoon C, Puigserver P, Spiegelman B, Montminy M (2001) CREB regulates hepatic gluconeogenesis through the coactivator PGC-1. Nature 413:179–183

Hevener AL, He W, Barak Y, Le J, Bandyopadhyay G, Olson P, Wilkes J, Evans R M, Olefsky J (2003) Muscle-specific Pparg deletion causes insulin resistance. Nat Med 9:1491–1497

Holness MJ, Bulmer K, Gibbons GF, Sugden MC (2002a) Up-regulation of pyruvate dehydrogenase kinase isoform 4 (PDK4) protein expression in oxidative skeletal muscle does not require the obligatory participation of peroxisome-proliferator-activated receptor alpha (PPARalpha). Biochem J 366:839–846

Holness MJ, Smith ND, Bulmer K, Hopkins T, Gibbons GF, Sugden MC (2002b) Evaluation of the role of peroxisome-proliferator-activated receptor alpha in the regulation of cardiac pyruvate dehydrogenase kinase 4 protein expression in response to starvation, high-fat feeding and hyperthyroidism. Biochem J 364: 687–694

Holness MJ, Sugden MC (1990) Glucose utilization in heart, diaphragm and skeletal muscle during the fed-to-starved transition. Biochem J 270:245–249

Holst D, Luquet S, Nogueira V, Kristiansen, K, Leverve X, Grimaldi PA (2003) Nutritional regulation and role of peroxisome proliferator-activated receptor delta in fatty acid catabolism in skeletal muscle. Biochim Biophys Acta 1633:43–50

Hotamisligil GS (1999) The role of TNFalpha and TNF receptors in obesity and insulin resistance. J Intern Med 245:621–625

Hu E, Kim JB, Sarraf P, Spiegelman BM (1996) Inhibition of adipogenesis through MAP kinase-mediated phosphorylation of PPARgamma. Science 274:2100–2103

Huss JM, Levy FH, Kelly DP (2001) Hypoxia inhibits the peroxisome proliferator-activated receptor alpha/retinoid X receptor gene regulatory pathway in cardiac myocytes: a mechanism for O2-dependent modulation of mitochondrial fatty acid oxidation. J Biol Chem 276:27605–27612

Huss JM, Torra IP, Staels B, Giguere V, Kelly DP (2004) Estrogen-related receptor alpha directs peroxisome proliferator-activated receptor alpha signaling in the transcriptional control of energy metabolism in cardiac and skeletal muscle. Mol Cell Biol 24:9079–9091

Irrcher I, Adhihetty PJ, Sheehan T, Joseph AM, Hood DA (2003) PPARgamma coactivator-1alpha expression during thyroid hormone- and contractile activity-induced mitochondrial adaptations. Am J Physiol Cell Physiol 284:C1669–C1677

Israelian-Konaraki Z, Reaven PD (2005) Peroxisome proliferator-activated receptor-alpha and atherosclerosis: from basic mechanisms to clinical implications. Cardiol Rev 13:240–246

Issemann I, Green S (1990) Activation of a member of the steroid hormone receptor superfamily by peroxisome proliferators. Nature 347:645–650

Jensen MD, Chandramouli V, Schumann WC, Ekberg K, Previs SF, Gupta S, Landau BR (2001) Sources of blood glycerol during fasting. Am J Physiol Endocrinol Metab 281:E998–1004

Jones PS, Savory R, Barratt P, Bell AR, Gray TJ, Jenkins NA, Gilbert DJ, Copeland NG, Bell DR (1995) Chromosomal localisation, inducibility, tissue-specific expression and strain differences in three murine peroxisome-proliferator-activated-receptor genes. Eur J Biochem 233:219–226

Kadowaki T, Yamauchi T (2005) Adiponectin and adiponectin receptors. Endocr Rev 26:439–451

Kantor PF, Dyck JR, Lopaschuk GD (1999) Fatty acid oxidation in the reperfused ischemic heart. Am J Med Sci 318:3–14

Kelly DP (2003) PPARs of the heart: three is a crowd. Circ Res 92:482–484

Kersten S, Seydoux J, Peters JM, Gonzalez FJ, Desvergne B, Wahli W (1999) Peroxisome proliferator-activated receptor alpha mediates the adaptive response to fasting. J Clin Invest 103:1489–1498

Kishida K, Kuriyama H, Funahashi T, Shimomura I, Kihara S, Ouchi N, Nishida M, Nishizawa H, Matsuda M, Takahashi M, Hotta K, Nakamura T, Yamashita S, Tochino Y, Matsuzawa Y (2000) Aquaporin adipose, a putative glycerol channel in adipocytes. J Biol Chem 275:20896–20902

Kishida K, Shimomura I, Nishizawa H, Maeda N, Kuriyama H, Kondo H, Matsuda M, Nagaretani H, Ouchi N, Hotta K, Kihara S, Kadowaki T, Funahashi T, Matsuzawa Y (2001) Enhancement of the aquaporin adipose gene expression by a peroxisome proliferator-activated receptor gamma. J Biol Chem 276:48572–48579

Kliewer SA, Umesono K, Noonan DJ, Heyman RA, Evans RM (1992) Convergence of 9-cis retinoic acid and peroxisome proliferator signalling pathways through heterodimer formation of their receptors. Nature 358:771–774

Kliewer SA, Xu HE, Lambert MH, Willson TM (2001) Peroxisome proliferator-activated receptors: from genes to physiology. Recent Prog Horm Res 56:239–263

Knutti D, Kralli A (2001) PGC-1, a versatile coactivator. Trends Endocrinol Metab 12:360–365

Kondo H, Shimomura I, Kishida K, Kuriyama H, Makino Y, Nishizawa H, Matsuda M, Maeda N, Nagaretani H, Kihara S, Kurachi Y, Nakamura T, Funahashi T, Matsuzawa Y (2002) Human aquaporin adipose (AQPap) gene. Genomic structure, promoter analysis and functional mutation. Eur J Biochem 269:1814–1826

Koutnikova H, Cock TA, Watanabe M, Houten SM, Champy MF, Dierich A, Auwerx J (2003) Compensation by the muscle limits the metabolic consequences of lipodystrophy in PPAR gamma hypomorphic mice. Proc Natl Acad Sci USA 100:14457–14462

Kubota N, Terauchi Y, Miki H, Tamemoto H, Yamauchi T, Komeda K, Satoh S, Nakano R, Ishii C, Sugiyama T, Eto K, Tsubamoto Y, Okuno A, Murakami K, Sekihara H, Hasegawa G, Naito M, Toyoshima Y, Tanaka S, Shiota K, Kitamura T, Fujita T, Ezaki O, Aizawa S, Nagai R, Tobe K, Kimura S, Kadowaki T (1999) PPAR gamma mediates high-fat diet-induced adipocyte hypertrophy and insulin resistance. Mol Cell 4:597–609

Kwon HS, Huang B, Unterman TG, Harris RA (2004) Protein kinase B-alpha inhibits human pyruvate dehydrogenase kinase-4 gene induction by dexamethasone through inactivation of FOXO transcription factors. Diabetes 53:899–910

Lazennec G, Canaple L, Saugy D, Wahli W (2000) Activation of peroxisome proliferator-activated receptors (PPARs) by their ligands and protein kinase A activators. Mol Endocrinol 14:1962–1975

Le May C, Pineau T, Bigot, K, Kohl C, Girard J, Pegorier JP (2000) Reduced hepatic fatty acid oxidation in fasting PPARalpha null mice is due to impaired mitochondrial hydroxymethylglutaryl-CoA synthase gene expression. FEBS Lett 475:163–166

Lee CH, Olson P, Evans RM (2003) Minireview: lipid metabolism, metabolic diseases, and peroxisome proliferator-activated receptors. Endocrinology 144:2201–2207

Lee KJ, Kim HA, Kim PH, Lee HS, Ma KR, Park JH, Kim DJ, Hahn JH (2004) Ox-LDL suppresses PMA-induced MMP-9 expression and activity through CD36-mediated activation of PPAR-g. Exp Mol Med 36:534–544

Lehman JJ, Barger PM, Kovacs A, Saffitz JE, Medeiros DM, Kelly DP (2000) Peroxisome proliferator-activated receptor gamma coactivator-1 promotes cardiac mitochondrial biogenesis. J Clin Invest 106:847–856

Leone TC, Weinheimer CJ, Kelly DP (1999) A critical role for the peroxisome proliferator-activated receptor alpha (PPARalpha) in the cellular fasting response: the PPARalpha-null mouse as a model of fatty acid oxidation disorders. Proc Natl Acad Sci USA 96:7473–7478

Lim H, Gupta RA, Ma WG, Paria BC, Moller DE, Morrow JD, DuBois RN, Trzaskos JM, Dey SK (1999) Cyclo-oxygenase-2-derived prostacyclin mediates embryo implantation in the mouse via PPARdelta. Genes Dev 13:1561–1574

Lin J, Wu H, Tarr PT, Zhang CY, Wu Z, Boss O, Michael LF, Puigserver P, Isotani E, Olson EN, Lowell BB, Bassel-Duby R, Spiegelman BM (2002) Transcriptional co-activator PGC-1 alpha drives the formation of slow-twitch muscle fibres. Nature 418:797–801

Lin Q, Ruuska SE, Shaw NS, Dong D, Noy N (1999) Ligand selectivity of the peroxisome proliferator-activated receptor alpha. Biochemistry 38:185–190

Lopaschuk GD (1996) Abnormal mechanical function in diabetes: relationship to altered myocardial carbohydrate/lipid metabolism. Coron Artery Dis 7:116–123

Lopaschuk GD (2002) Metabolic abnormalities in the diabetic heart. Heart Fail Rev 7:149–159

Lupi R, Del Guerra S, Marselli L, Bugliani M, Boggi U, Mosca F, Marchetti P, Del Prato S (2004) Rosiglitazone prevents the impairment of human islet function induced by fatty acids: evidence for a role of PPARgamma2 in the modulation of insulin secretion. Am J Physiol Endocrinol Metab 286:E560–E567

Luquet S, Gaudel C, Holst D, Lopez-Soriano J, Jehl-Pietri C, Fredenrich A, Grimaldi PA (2005) Roles of PPAR delta in lipid absorption and metabolism: a new target for the treatment of type 2 diabetes. Biochim Biophys Acta 1740:313–317

Luquet S, Lopez-Soriano J, Holst D, Fredenrich A, Melki J, Rassoulzadegan M, Grimaldi PA (2003) Peroxisome proliferator-activated receptor delta controls muscle development and oxidative capability. FASEB J 17:2299–2301

Madsen L, Guerre-Millo M, Flindt EN, Berge K, Tronstad KJ, Bergene E, Sebokova E, Rustan AC, Jensen J, Mandrup S, Kristiansen K, Klimes I, Staels B, Berge RK (2002) Tetradecylthioacetic acid prevents high fat diet induced adiposity and insulin resistance. J Lipid Res 43:742–750

Maeda N, Funahashi T, Hibuse T, Nagasawa A, Kishida K, Kuriyama H, Nakamura T, Kihara S, Shimomura I, Matsuzawa Y (2004) Adaptation to fasting by glycerol transport through aquaporin 7 in adipose tissue. Proc Natl Acad Sci USA 101:17801–17806

Makinde AO, Kantor PF, Lopaschuk GD (1998) Maturation of fatty acid and carbohydrate metabolism in the newborn heart. Mol Cell Biochem 188:49–56

Mangelsdorf DJ, Borgmeyer U, Heyman RA, Zhou JY, Ong ES, Oro AE, Kakizuka A, Evans RM (1992) Characterization of three RXR genes that mediate the action of 9-cis retinoic acid. Genes Dev 6:329–344

McCall GE, Byrnes WC, Dickinson A, Pattany PM, Fleck SJ (1996) Muscle fiber hypertrophy, hyperplasia, and capillary density in college men after resistance training. J Appl Physiol 81:2004–2012

Minokoshi Y, Kim YB, Peroni OD, Fryer LG, Muller C, Carling D, Kahn BB (2002) Leptin stimulates fatty-acid oxidation by activating AMP-activated protein kinase. Nature 415:339–343

Mootha VK, Handschin C, Arlow D, Xie X, St Pierre J, Sihag S, Yang W, Altshuler D, Puigserver P, Patterson N, Willy PJ, Schulman IG, Heyman RA, Lander ES, Spiegelman BM (2004) Erralpha and Gabpa/b specify PGC-1alpha-dependent oxidative phosphorylation gene expression that is altered in diabetic muscle. Proc Natl Acad Sci USA 101:6570–6575

Mueller E, Drori S, Aiyer A, Yie J, Sarraf P, Chen H, Hauser S, Rosen ED, Ge K, Roeder RG, Spiegelman BM (2002) Genetic analysis of adipogenesis through peroxisome proliferator-activated receptor gamma isoforms. J Biol Chem 277:41925–41930

Muoio DM, MacLean PS, Lang DB, Li S, Houmard JA, Way JM, Winegar DA, Corton JC, Dohm GL, Kraus WE (2002) Fatty acid homeostasis and induction of lipid regulatory genes in skeletal muscles of peroxisome proliferator-activated receptor (PPAR) alpha knock-out mice. Evidence for compensatory regulation by PPAR delta. J Biol Chem 277:26089–26097

Mutch DM, Wahli W, Williamson G (2005) Nutrigenomics and nutrigenetics: the emerging faces of nutrition. FASEB J 19:1602–1616

Nakae J, Biggs WH III, Kitamura T, Cavenee WK, Wright CV, Arden KC, Accili D (2002) Regulation of insulin action and pancreatic beta-cell function by mutated alleles of the gene encoding forkhead transcription factor Foxo1. Nat Genet 32:245–253

Oliver WR Jr, Shenk JL, Snaith MR, Russell CS, Plunket KD, Bodkin NL, Lewis MC, Winegar DA, Sznaidman ML, Lambert MH, Xu HE, Sternbach DD, Kliewer SA, Hansen BC, Willson TM (2001) A selective peroxisome proliferator-activated receptor delta agonist promotes reverse cholesterol transport. Proc Natl Acad Sci USA 98:5306–5311

Patsouris D, Mandard S, Voshol PJ, Escher P, Tan NS, Havekes LM, Koenig W, Marz W, Tafuri S, Wahli W, Muller M, Kersten S (2004) PPARalpha governs glycerol metabolism. J Clin Invest 114:94–103

Peroni O, Large V, Beylot M (1995) Measuring gluconeogenesis with [2-^{13}C]glycerol and mass isotopomer distribution analysis of glucose. Am J Physiol Endocrinol Metab 269:E516–E523

Peters JM, Lee SS, Li W, Ward JM, Gavrilova O, Everett C, Reitman ML, Hudson LD, Gonzalez FJ (2000) Growth, adipose, brain, and skin alterations resulting from targeted disruption of the mouse peroxisome proliferator-activated receptor beta(delta). Mol Cell Biol 20:5119–5128

Picard F, Auwerx J (2002) PPAR(gamma) and glucose homeostasis. Annu Rev Nutr 22:167–197

Pilegaard H, Saltin B, Neufer PD (2003) Exercise induces transient transcriptional activation of the PGC-1alpha gene in human skeletal muscle. J Physiol 546:851–858

Poirier H, Niot I, Monnot MC, Braissant O, Meunier-Durmort C, Costet P, Pineau T, Wahli W, Willson TM, Besnard P (2001) Differential involvement of peroxisome-proliferator-activated receptors alpha and delta in fibrate and fatty-acid-mediated inductions of the gene encoding liver fatty-acid-binding protein in the liver and the small intestine. Biochem J 355:481–488

Puigserver P, Rhee J, Donovan J, Walkey CJ, Yoon JC, Oriente F, Kitamura Y, Altomonte J, Dong H, Accili D, Spiegelman BM (2003) Insulin-regulated hepatic gluconeogenesis through FOXO1-PGC-1alpha interaction. Nature 423:550–555

Puigserver P, Spiegelman BM (2003) Peroxisome proliferator-activated receptor-gamma coactivator 1 alpha (PGC-1 alpha): transcriptional coactivator and metabolic regulator. Endocr Rev 24:78–90

Puigserver P, Wu Z, Park CW, Graves R, Wright M, Spiegelman BM (1998) A cold-inducible coactivator of nuclear receptors linked to adaptive thermogenesis. Cell 92:829–839

Randle PJ (1998) Regulatory interactions between lipids and carbohydrates: the glucose fatty acid cycle after 35 years. Diabetes Metab Rev 14:263–283

Randle PJ, Garland PB, Hales CN, Newsholme EA (1963) The glucose fatty-acid cycle. Its role in insulin sensitivity and the metabolic disturbances of diabetes mellitus. Lancet 1:785–789

Rangwala SM, Lazar MA (2004) Peroxisome proliferator-activated receptor gamma in diabetes and metabolism. Trends Pharmacol Sci 25:331–336

Razeghi P, Young ME, Cockrill TC, Frazier OH, Taegtmeyer H (2002) Downregulation of myocardial myocyte enhancer factor 2C and myocyte enhancer factor 2C-regulated gene expression in diabetic patients with nonischemic heart failure. Circulation 106:407–411

Roden M, Price TB, Perseghin G, Petersen KF, Rothman DL, Cline GW, Shulman GI (1996) Mechanism of free fatty acid-induced insulin resistance in humans. J Clin Invest 97:2859–2865

Rodrigues B, McNeill JH (1992) The diabetic heart: metabolic causes for the development of a cardiomyopathy. Cardiovasc Res 26:913–922

Rosen ED, Sarraf P, Troy AE, Bradwin G, Moore K, Milstone DS, Spiegelman BM, Mortensen RM (1999) PPAR gamma is required for the differentiation of adipose tissue in vivo and in vitro. Mol Cell 4:611–617

Russell LK, Finck BN, Kelly DP (2005) Mouse models of mitochondrial dysfunction and heart failure. J Mol Cell Cardiol 38:81–91

Russell LK, Mansfield CM, Lehman JJ, Kovacs A, Courtois M, Saffitz JE, Medeiros DM, Valencik ML, McDonald JA, Kelly DP (2004) Cardiac-specific induction of the transcriptional coactivator peroxisome proliferator-activated receptor gamma coactivator-1alpha promotes mitochondrial biogenesis and reversible cardiomyopathy in a developmental stage-dependent manner. Circ Res 94:525–533

Sack MN, Rader TA, Park S, Bastin J, McCune SA, Kelly DP (1996) Fatty acid oxidation enzyme gene expression is downregulated in the failing heart. Circulation 94:2837–2842

Saddik M, Lopaschuk GD (1991) Myocardial triglyceride turnover and contribution to energy substrate utilization in isolated working rat hearts. J Biol Chem 266: 8162–8170

Saddik M, Lopaschuk GD (1992) Myocardial triglyceride turnover during reperfusion of isolated rat hearts subjected to a transient period of global ischemia. J Biol Chem 267:3825–3831

Sakai S, Miyauchi T, Irukayama-Tomobe Y, Ogata T, Goto K, Yamaguchi I (2002) Peroxisome proliferator-activated receptor-gamma activators inhibit endothelin-1-related cardiac hypertrophy in rats. Clin Sci (Lond) 103 Suppl 48:16S–20S

Schreiber SN, Emter R, Hock MB, Knutti D, Cardenas J, Podvinec M, Oakeley EJ, Kralli A (2004) The estrogen-related receptor alpha (ERRalpha) functions in PPARgamma coactivator 1alpha (PGC-1alpha)-induced mitochondrial biogenesis. Proc Natl Acad Sci USA 101:6472–6477

Sewter C, Vidal-Puig A (2002) PPARgamma and the thiazolidinediones: molecular basis for a treatment of 'Syndrome X'? Diabetes Obes Metab 4:239–248

Shao D, Rangwala SM, Bailey ST, Krakow SL, Reginato MJ, Lazar MA (1998) Interdomain communication regulating ligand binding by PPAR-gamma. Nature 396:377–380

Shi GQ, Dropinski JF, Zhang Y, Santini C, Sahoo SP, Berger JP, Macnaul KL, Zhou G, Agrawal A, Alvaro R, Cai TQ, Hernandez M, Wright SD, Moller DE, Heck JV, Meinke PT (2005) Novel 2,3-dihydrobenzofuran-2-carboxylic acids: highly potent and subtype-selective PPARalpha agonists with potent hypolipidemic activity. J Med Chem 48:5589–5599

Shulman GI (2000) Cellular mechanisms of insulin resistance. J Clin Invest 106:171–176

Sivarajah A, McDonald MC, Thiemermann C (2005) The cardioprotective effects of preconditioning with endotoxin, but not ischemia, are abolished by a peroxisome proliferator-activated receptor-gamma antagonist. J Pharmacol Exp Ther 313:896–901

Smith AG, Muscat GE (2005) Skeletal muscle and nuclear hormone receptors: implications for cardiovascular and metabolic disease. Int J Biochem Cell Biol 37: 2047–2063

Staels B (2005) PPARgamma and atherosclerosis. Curr Med Res Opin 21 Suppl 1:S13–S20

Staels B, Fruchart JC (2005) Therapeutic roles of peroxisome proliferator-activated receptor agonists. Diabetes 54:2460–2470

Stanley WC, Chandler MP (2002) Energy metabolism in the normal and failing heart: potential for therapeutic interventions. Heart Fail Rev 7:115–130

Steinmetz M, Quentin T, Poppe A, Paul T, Jux C (2005) Changes in expression levels of genes involved in fatty acid metabolism: up-regulation of all three members of the PPAR family (alpha, gamma, delta) and the newly described adiponectin receptor 2, but not adiponectin receptor 1 during neonatal cardiac development of the rat. Basic Res Cardiol 100:263–269

Sugden MC, Bulmer K, Augustine D, Holness MJ (2001) Selective modification of pyruvate dehydrogenase kinase isoform expression in rat pancreatic islets elicited by starvation and activation of peroxisome proliferator-activated receptor-alpha: implications for glucose-stimulated insulin secretion. Diabetes 50:2729–2736

Sugden MC, Bulmer K, Gibbons GF, Knight BL, Holness MJ (2002) Peroxisome-proliferator-activated receptor-alpha (PPARalpha) deficiency leads to dysregulation of hepatic lipid and carbohydrate metabolism by fatty acids and insulin. Biochem J 364:361–368

Sugden MC, Holness MJ (1994) Interactive regulation of the pyruvate dehydrogenase complex and the carnitine palmitoyltransferase system. FASEB J 8:54–61

Sugden MC, Holness, MJ (2003) Recent advances in mechanisms regulating glucose oxidation at the level of the pyruvate dehydrogenase complex by PDKs. Am J Physiol Endocrinol Metab 284:E855–E862

Sugden MC, Langdown ML, Harris RA, Holness MJ (2000) Expression and regulation of pyruvate dehydrogenase kinase isoforms in the developing rat heart and in adulthood: role of thyroid hormone status and lipid supply. Biochem J 352:731–738

Sznaidman ML, Haffner CD, Maloney PR, Fivush A, Chao E, Goreham D, Sierra ML, LeGrumelec C, Xu HE, Montana VG, Lambert MH, Willson TM, Oliver WR Jr, Sternbach DD (2003) Novel selective small molecule agonists for peroxisome proliferator-activated receptor delta (PPARdelta)–synthesis and biological activity. Bioorg Med Chem Lett 13:1517–1521

Takano H, Nagai T, Asakawa M, Toyozaki T, Oka T, Komuro I, Saito T, Masuda Y (2000) Peroxisome proliferator-activated receptor activators inhibit lipopolysaccharide-induced tumor necrosis factor-alpha expression in neonatal rat cardiac myocytes. Circ Res 87:596–602

Tan GD, Debard C, Tiraby C, Humphreys SM, Frayn KN, Langin D, Vidal H, Karpe F (2003a) A "futile cycle" induced by thiazolidinediones in human adipose tissue? Nat Med 9:811–812

Tan NS, Michalik L, Desvergne B, Wahli W (2003b) Peroxisome proliferator-activated receptor (PPAR)-beta as a target for wound healing drugs: what is possible? Am J Clin Dermatol 4:523–530

Tanaka T, Yamamoto J, Iwasaki S, Asaba H, Hamura H, Ikeda Y, Watanabe M, Magoori K, Ioka R X, Tachibana K, Watanabe Y, Uchiyama Y, Sumi K, Iguchi H, Ito S, Doi T, Hamakubo T, Naito M, Auwerx J, Yanagisawa M, Kodama T, Sakai J (2003) Activation of peroxisome proliferator-activated receptor delta induces fatty acid beta-oxidation in skeletal muscle and attenuates metabolic syndrome. Proc Natl Acad Sci USA 100:15924–15929

Terada S, Goto M, Kato M, Kawanaka K, Shimokawa T, Tabata I (2002) Effects of low-intensity prolonged exercise on PGC-1 mRNA expression in rat epitrochlearis muscle. Biochem Biophys Res Commun 296:350–354

Terada S, Tabata I (2004) Effects of acute bouts of running and swimming exercise on PGC-1alpha protein expression in rat epitrochlearis and soleus muscle. Am J Physiol Endocrinol Metab 286:E208–E216

Tordjman J, Khazen W, Antoine B, Chauvet G, Quette J, Fouque F, Beale EG, Benelli C, Forest C (2003) Regulation of glyceroneogenesis and phosphoenolpyruvate carboxykinase by fatty acids, retinoic acids and thiazolidinediones: potential relevance to type 2 diabetes. Biochimie 85:1213–1218

Tsai YS, Kim HJ, Takahashi N, Kim HS, Hagaman JR, Kim JK, Maeda N (2004) Hypertension and abnormal fat distribution but not insulin resistance in mice with P465L PPARgamma. J Clin Invest 114:240–249

Unger RH, Orci L (2001) Diseases of liporegulation: new perspective on obesity and related disorders. FASEB J 15:312–321

van Bilsen M, Smeets PJ, Gilde AJ, van der Vusse G (2004) Metabolic remodelling of the failing heart: the cardiac burn-out syndrome? Cardiovasc Res 61:218–226

Vega RB, Huss JM, Kelly DP (2000) The coactivator PGC-1 cooperates with peroxisome proliferator-activated receptor alpha in transcriptional control of nuclear genes encoding mitochondrial fatty acid oxidation enzymes. Mol Cell Biol 20:1868–1876

Vidal-Puig A, O'Rahilly S (2001) Metabolism. Controlling the glucose factory. Nature 413:125–126

Wang YX, Lee CH, Tiep S, Yu RT, Ham J, Kang H, Evans RM (2003) Peroxisome-proliferator-activated receptor delta activates fat metabolism to prevent obesity. Cell 113:159–170

Wang YX, Zhang CL, Yu R , Cho HK, Nelson MC, Bayuga-Ocampo CR, Ham J, Kang H, Evans RM (2004) Regulation of muscle fiber type and running endurance by PPARdelta. PLoS Biol 2:e294

Watanabe K, Fujii H, Takahashi T, Kodama M, Aizawa Y, Ohta Y, Ono T, Hasegawa G, Naito M, Nakajima T, Kamijo Y, Gonzalez FJ, Aoyama T (2000) Constitutive regulation of cardiac fatty acid metabolism through peroxisome proliferator-activated receptor alpha associated with age-dependent cardiac toxicity. J Biol Chem 275:22293–22299

Wende AR, Huss JM, Schaeffer PJ, Giguere V, Kelly DP (2005) PGC-1alpha coactivates PDK4 gene expression via the orphan nuclear receptor ERRalpha: a mechanism for transcriptional control of muscle glucose metabolism. Mol Cell Biol 25, 10684–10694

Willy PJ, Murray IR, Qian J, Busch BB, Stevens WC Jr, Martin R, Mohan R, Zhou S, Ordentlich P, Wei P, Sapp DW, Horlick RA, Heyman RA, Schulman IG (2004) Regulation of PPARgamma coactivator 1alpha (PGC-1alpha) signaling by an estrogen-related receptor alpha (ERRalpha) ligand. Proc Natl Acad Sci USA 101: 8912–8917

Wolfrum C, Borrmann C M, Borchers T, Spener F (2001) Fatty acids and hypolipidemic drugs regulate peroxisome proliferator-activated receptors alpha - and gamma-mediated gene expression via liver fatty acid binding protein: a signaling path to the nucleus. Proc Natl Acad Sci USA 98:2323–2328

Wu P, Peters JM, Harris RA (2001) Adaptive increase in pyruvate dehydrogenase kinase 4 during starvation is mediated by peroxisome proliferator-activated receptor alpha. Biochem Biophys Res Commun 287:391–396

Wu Z, Puigserver P, Andersson U, Zhang C, Adelmant G, Mootha V, Troy A, Cinti S, Lowell B, Scarpulla RC, Spiegelman BM (1999) Mechanisms controlling mitochondrial biogenesis and respiration through the thermogenic coactivator PGC-1. Cell 98:115–124

Yamagata K, Furuta H, Oda N, Kaisaki PJ, Menzel S, Cox NJ, Fajans SS, Signorini S, Stoffel M, Bell GI (1996) Mutations in the hepatocyte nuclear factor-4alpha gene in maturity-onset diabetes of the young (MODY1). Nature 384:458–460

Yamauchi T, Kamon J, Ito Y, Tsuchida A, Yokomizo T, Kita S, Sugiyama T, Miyagishi M, Hara K, Tsunoda M, Murakami K, Ohteki T, Uchida S, Takekawa S, Waki H, Tsuno NH, Shibata Y, Terauchi Y, Froguel P, Tobe K, Koyasu S, Taira K, Kitamura T, Shimizu T, Nagai R, Kadowaki T (2003) Cloning of adiponectin receptors that mediate antidiabetic metabolic effects. Nature 423:762–769

Yamauchi T, Kamon J, Waki H, Murakami K, Motojima K, Komeda K, Ide T, Kubota N, Terauchi Y, Tobe K, Miki H, Tsuchida A, Akanuma Y, Nagai R, Kimura S, Kadowaki T (2001) The mechanisms by which both heterozygous peroxisome proliferator-activated receptor gamma (PPARgamma) deficiency and PPARgamma agonist improve insulin resistance. J Biol Chem 276:41245–41254

Yoon JC, Puigserver P, Chen G, Donovan J, Wu Z, Rhee J, Adelmant G, Stafford J, Kahn CR, Granner DK, Newgard CB, Spiegelman BM (2001) Control of hepatic gluconeogenesis through the transcriptional coactivator PGC-1. Nature 413:131–138

Zhang B, Berger J, Zhou G, Elbrecht A, Biswas S, White-Carrington S, Szalkowski D, Moller DE (1996) Insulin- and mitogen-activated protein kinase-mediated phosphorylation and activation of peroxisome proliferator-activated receptor gamma. J Biol Chem 271:31771–31774

Zhou YT, Shimabukuro M, Wang MY, Lee Y, Higa M, Milburn JL, Newgard CB, Unger RH (1998) Role of peroxisome proliferator-activated receptor alpha in disease of pancreatic beta cells. Proc Natl Acad Sci USA 95:8898–8903

Zhu Y, Qi C, Calandra C, Rao M S, Reddy JK (1996) Cloning and identification of mouse steroid receptor coactivator-1 (mSRC-1), as a coactivator of peroxisome proliferator-activated receptor gamma. Gene Expr 6:185–195

Zingarelli B, Sheehan M, Hake PW, O'Connor M, Denenberg A, Cook JA (2003) Peroxisome proliferator activator receptor-gamma ligands, 15-deoxy-Delta(12,14)-prostaglandin J2 and ciglitazone, reduce systemic inflammation in polymicrobial sepsis by modulation of signal transduction pathways. J Immunol 171:6827–6837

Ziouzenkova O, Perrey S, Asatryan L, Hwang J, Macnaul KL, Moller DE, Rader DJ, Sevanian A, Zechner R, Hoefler G, Plutzky J (2003) Lipolysis of triglyceride-rich lipoproteins generates PPAR ligands: evidence for an antiinflammatory role for lipoprotein lipase. Proc Natl Acad Sci USA 100:2730–2735

Due to the importance of the CTP, our laboratory has extensively studied the CTP's structure/function relationships in order to understand its molecular mechanism. Thus the transporter has been purified in reconstitutively active form (Kaplan et al. 1990a), cloned (Kaplan et al. 1993), and overexpressed (Kaplan et al. 1995; Xu et al. 1995). Due to the high specific activity following reconstitution of the overexpressed *yeast* mitochondrial CTP (Kaplan et al. 1995), most of our recent investigative efforts have focused on this carrier source. Accordingly, we have constructed a Cys-less yeast CTP which retains wild-type functional properties (Xu et al. 2000) and have shown that in detergent micelles both the wild-type and the cys-less variants exist as homodimers (Kotaria et al. 1999). Each monomer is thought to contain six transmembrane domains (TMDs) (Kaplan et al. 1993; 1995). We then used the cys-less CTP as a template upon which to reintroduce single cysteines at desired locations which were subsequently probed with biochemical and biophysical probes (Kaplan et al. 2000a; Kaplan et al. 2000b; Ma et al. 2004; Ma et al. 2005). Utilizing this approach we have identified the secondary structure, as well as the water-accessible and -inaccessible surfaces of TMDs III and IV. This data, in combination with examination of the ability of citrate (i.e., substrate) to protect against MTSES reagent mediated inhibition of CTP function (Ma et al. 2005) has enabled identification of portions of the CTP translocation pathway. Recently we have constructed a homology model of the CTP based on the crystal structure of the mitochondrial ADP/ATP carrier (Walters and Kaplan 2004). Superposition of our functional data onto the homology-modeled structure has permitted us to glean considerable additional insight into the CTP's structure-based function. In this chapter, we will summarize our current state of knowledge regarding the structure/function relationships within the CTP (as of January 2006), and will delineate our assessment of where future efforts ought to focus.

4.1.2. Delineation of Essential Molecular Mechanistic Issues to Address in Understanding Transporter Function

We begin with a comment regarding the nature of the issues that are important to resolve in order to attain a comprehensive understanding of the functioning of a metabolite transporter. These would include a complete definition of the substrate translocation pathway, the substrate binding site(s), the gates that control accessibility from one side of the membrane *versus* the other, and the detailed 3-dimensional conformational change(s) that occur during, and in fact represent the essence of, the transport process (West 1997). Portions of this information have been obtained with the CTP, whereas other aspects are currently under investigation. We now proceed with a detailed discussion of our current state of knowledge regarding the CTP molecular mechanism. For information regarding other aspects of this transporter the reader is referred to earlier reviews that are available (Kramer and Palmieri 1992; Kaplan and Mayor 1993; Kaplan 1996; Kaplan 2001).

Yamauchi T, Kamon J, Ito Y, Tsuchida A, Yokomizo T, Kita S, Sugiyama T, Miyagishi M, Hara K, Tsunoda M, Murakami K, Ohteki T, Uchida S, Takekawa S, Waki H, Tsuno NH, Shibata Y, Terauchi Y, Froguel P, Tobe K, Koyasu S, Taira K, Kitamura T, Shimizu T, Nagai R, Kadowaki T (2003) Cloning of adiponectin receptors that mediate antidiabetic metabolic effects. Nature 423:762–769

Yamauchi T, Kamon J, Waki H, Murakami K, Motojima K, Komeda K, Ide T, Kubota N, Terauchi Y, Tobe K, Miki H, Tsuchida A, Akanuma Y, Nagai R, Kimura S, Kadowaki T (2001) The mechanisms by which both heterozygous peroxisome proliferator-activated receptor gamma (PPARgamma) deficiency and PPARgamma agonist improve insulin resistance. J Biol Chem 276:41245–41254

Yoon JC, Puigserver P, Chen G, Donovan J, Wu Z, Rhee J, Adelmant G, Stafford J, Kahn CR, Granner DK, Newgard CB, Spiegelman BM (2001) Control of hepatic gluconeogenesis through the transcriptional coactivator PGC-1. Nature 413:131–138

Zhang B, Berger J, Zhou G, Elbrecht A, Biswas S, White-Carrington S, Szalkowski D, Moller DE (1996) Insulin- and mitogen-activated protein kinase-mediated phosphorylation and activation of peroxisome proliferator-activated receptor gamma. J Biol Chem 271:31771–31774

Zhou YT, Shimabukuro M, Wang MY, Lee Y, Higa M, Milburn JL, Newgard CB, Unger RH (1998) Role of peroxisome proliferator-activated receptor alpha in disease of pancreatic beta cells. Proc Natl Acad Sci USA 95:8898–8903

Zhu Y, Qi C, Calandra C, Rao M S, Reddy JK (1996) Cloning and identification of mouse steroid receptor coactivator-1 (mSRC-1), as a coactivator of peroxisome proliferator-activated receptor gamma. Gene Expr 6:185–195

Zingarelli B, Sheehan M, Hake PW, O'Connor M, Denenberg A, Cook JA (2003) Peroxisome proliferator activator receptor-gamma ligands, 15-deoxy-Delta(12,14)-prostaglandin J2 and ciglitazone, reduce systemic inflammation in polymicrobial sepsis by modulation of signal transduction pathways. J Immunol 171:6827–6837

Ziouzenkova O, Perrey S, Asatryan L, Hwang J, Macnaul KL, Moller DE, Rader DJ, Sevanian A, Zechner R, Hoefler G, Plutzky J (2003) Lipolysis of triglyceride-rich lipoproteins generates PPAR ligands: evidence for an antiinflammatory role for lipoprotein lipase. Proc Natl Acad Sci USA 100:2730–2735

4
Molecular Structure of the Mitochondrial Citrate Transport Protein

Ronald S. Kaplan and June A. Mayor

4.1. Introduction

4.1.1. *Overview of the Mitochondrial Citrate Transport Protein*

The mitochondrial citrate transport protein (i.e., CTP) is located within the inner mitochondrial membrane and in higher eukaryotes catalyzes an obligatory 1:1 exchange of the dibasic form of a tricarboxylic acid (e.g., citrate, isocitrate, or cis-aconitate) for either another tricarboxylate, a dicarboxylate, or phosphoenolpyruvate (Palmieri et al. 1972; Bisaccia et al.1993; Palmieri 2004). The CTP occupies a prominent position within intermediary metabolism since following the efflux of citrate from the mitochondrial matrix and subsequent diffusion across the outer mitochondrial membrane, the resulting cytoplasmic citrate serves as the main carbon source fueling the fatty acid, triacylglycerol, and cholesterol biosynthetic pathways (Watson and Lowenstein 1970; Brunengraber and Lowenstein 1973; Endemann et al. 1982; Conover 1987). Additionally, via the sequential action of citrate lyase and malate dehydrogenase, cytoplasmic citrate enables the production of NAD^+, a cofactor that is required for glycolysis. Not only is CTP function essential to the bioenergetics of the physiological state, but altered functioning of this carrier in certain disease states such as diabetes (Kaplan et al. 1990b) and cancer (Kaplan et al. 1982) likely plays an important role in maintaining the aberrant bioenergetics that characterizes these pathologies.

The abbreviations used are: CTP, citrate transport protein; MTS, methanethiosulfonate; MTSES, sodium (2-sulfonatoethyl) methanethiosulfo-nate; MTSET, (2-(trimethylammonium)ethyl) methanethiosulfonate bromide; PAGE, polyacrylamide gel electrophoresis; S.E., standard error; and TMD, transmembrane domain.

Due to the importance of the CTP, our laboratory has extensively studied the CTP's structure/function relationships in order to understand its molecular mechanism. Thus the transporter has been purified in reconstitutively active form (Kaplan et al. 1990a), cloned (Kaplan et al. 1993), and overexpressed (Kaplan et al. 1995; Xu et al. 1995). Due to the high specific activity following reconstitution of the overexpressed *yeast* mitochondrial CTP (Kaplan et al. 1995), most of our recent investigative efforts have focused on this carrier source. Accordingly, we have constructed a Cys-less yeast CTP which retains wild-type functional properties (Xu et al. 2000) and have shown that in detergent micelles both the wild-type and the cys-less variants exist as homodimers (Kotaria et al. 1999). Each monomer is thought to contain six transmembrane domains (TMDs) (Kaplan et al. 1993; 1995). We then used the cys-less CTP as a template upon which to reintroduce single cysteines at desired locations which were subsequently probed with biochemical and biophysical probes (Kaplan et al. 2000a; Kaplan et al. 2000b; Ma et al. 2004; Ma et al. 2005). Utilizing this approach we have identified the secondary structure, as well as the water-accessible and -inaccessible surfaces of TMDs III and IV. This data, in combination with examination of the ability of citrate (i.e., substrate) to protect against MTSES reagent mediated inhibition of CTP function (Ma et al. 2005) has enabled identification of portions of the CTP translocation pathway. Recently we have constructed a homology model of the CTP based on the crystal structure of the mitochondrial ADP/ATP carrier (Walters and Kaplan 2004). Superposition of our functional data onto the homology-modeled structure has permitted us to glean considerable additional insight into the CTP's structure-based function. In this chapter, we will summarize our current state of knowledge regarding the structure/function relationships within the CTP (as of January 2006), and will delineate our assessment of where future efforts ought to focus.

4.1.2. *Delineation of Essential Molecular Mechanistic Issues to Address in Understanding Transporter Function*

We begin with a comment regarding the nature of the issues that are important to resolve in order to attain a comprehensive understanding of the functioning of a metabolite transporter. These would include a complete definition of the substrate translocation pathway, the substrate binding site(s), the gates that control accessibility from one side of the membrane *versus* the other, and the detailed 3-dimensional conformational change(s) that occur during, and in fact represent the essence of, the transport process (West 1997). Portions of this information have been obtained with the CTP, whereas other aspects are currently under investigation. We now proceed with a detailed discussion of our current state of knowledge regarding the CTP molecular mechanism. For information regarding other aspects of this transporter the reader is referred to earlier reviews that are available (Kramer and Palmieri 1992; Kaplan and Mayor 1993; Kaplan 1996; Kaplan 2001).

4.2. Identification of Residues that Comprise the Citrate Translocation Pathway

4.2.1. Preliminary Identification Based on MTS Reagent Accessibility

We initiated identification of those residues that comprise the citrate translocation pathway via determination of the water-accessible surfaces of transmembrane α-helices (Kaplan et al. 2000a; Ma et al. 2004). For these studies, starting with a Cys-less CTP, which displays native functional properties as the template (Xu et al. 2000), we engineered single cysteine residues at sequential positions within a given transmembrane domain. Following overexpression and functional reconstitution in liposomes of each single-Cys CTP variant, we examined the ability of two methanethiosulfonate (i.e., MTS) reagents to inhibit function. The ethylsulfonate and trimethylamine derivatives (i.e. MTSES and MTSET, respectively) are absolutely specific for cysteine, will not permeate a lipid bilayer in the absence of a translocation pathway, and react 5×10^9 more rapidly with the thiolate anion (that forms only when a cysteine sulfhydryl group is exposed to water) as compared to the unionized thiol group (Akabas et al. 1992; Holmgren et al. 1996; Karlin and Akabas 1998). Furthermore, as previously discussed, the molecular dimensions of these reagents are fairly similar to those of citrate (Xu et al. 2000), the latter also being a highly hydrophilic molecule. Thus, we believe that the MTS reagents and citrate will access similar aqueous domains within the CTP. Consequently, the theoretical underpinning for these studies consists of the idea that an accessible aqueous pathway through the lipid bilayer is likely to represent the substrate translocation pathway. This point will be further discussed below.

Utilizing the above approach we have measured the accessibility of residues in TMDs III and IV to MTS reagents. It is noteworthy that, as depicted in Figures 4.1 (TMD IV) and 4.2 (TMD III) the rate constants for inactivation vary by greater than 7 and 5 orders of magnitude in TMDs IV and III, respectively. In certain cases, the rate constants for inactivation of adjacent engineered cysteines varied by 4–5 order of magnitude, thereby indicating a dramatic difference in accessibility to the aqueous environment. Moreover, the periodicity of the rate constant data strongly suggests that substantial portions of these TMDs exist in an α-helical secondary structure. Furthermore, the rate constant data define a highly accessible surface of each helix, which likely comprises portions of the substrate translocation pathway, and a highly inaccessible surface of each helix which likely faces either the lipid bilayer or is tightly packed against other protein domains within the CTP. The results of computer modeling of these TMDs with the rate constant data superimposed on the models are depicted in Figure 4.2. As depicted in this Figure, each TMD displays a water-accessible face (depicted in red) and a water-inaccessible face depicted in blue. Moreover, TMDIII contains a series of moderately accessible residues depicted in magenta (see Figure 4.3, Panels C and D).

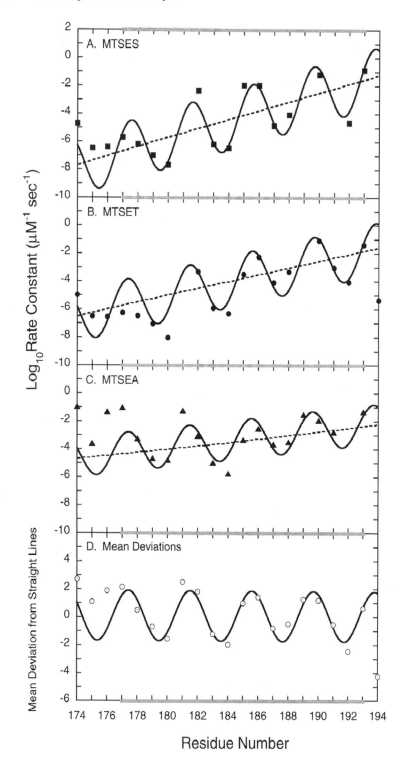

4.2.2. Evaluation of Whether Citrate Protects Against the MTS-mediated Inhibition of CTP Function

In order to test the concept that the water-accessible residues do in fact represent elements of the substrate translocation pathway through the CTP, we evaluated the ability of citrate to protect against the MTSES-mediated inhibition of CTP function (Ma et al. 2005). Thus, in the presence of a concentration of MTSES that yields 50–70% inhibition, we examined the ability of a spectrum of citrate concentrations to protect against the inhibition observed with 20 single-Cys TMDIII mutants. As depicted in Table 4.1, we observed protection at 8 sites with EC_{50} values that ranged 1.18–53.68 mM.

Figure 4.4 depicts the TMDIII protectable residues in red and the non-protectable residues in blue. Importantly, for the most part the protectable residues reside on one face of the α-helix whereas the unprotectable residues reside on the opposite face.

It is important to note that, with the exception of the S123C mutant, the other substrate-protectable residues display EC_{50} values that are significantly greater than their K_m values. In general, the condition where substrate protection is observed with a given mutant, but where the measured $EC_{50} \gg K_m$ value is diagnostic of a residue that lines the transport pathway, but is unlikely to have direct involvement in substrate binding (Fu et al. 2001; Ye and

FIGURE 4.1. Rate constants for inactivation of citrate transport by MTS reagents versus location of engineered cysteine in TMD IV. Time course data for inactivation (activation) of a given single-Cys CTP variant by the MTS reagents were fitted to a simple exponential function by unweighted Marquardt nonlinear least squares: $r_t = (r_0 - r_\infty) \bullet e^{-m.c.t} + r_\infty$ where r_t is the observed activity remaining at time t (seconds), r_0 is the initial activity, r_∞ is the asymptotic activity at $t = \infty$, c is reagent concentration (μM^{-1}) and m is the rate constant for inactivation ($\mu M^{-1} sec^{-1}$). Estimates of m (which range over 7 orders of magnitude) are reliable as judged from replicate experiments producing standard errors less than 20% of each estimate. The rate constant for inactivation, m, can be interpreted as a measure of a residue's accessibility to the MTS reagent. The broad range of estimates of m dictated that further analyses be confined to Log_{10} transformed values. For each data set (MTSES, MTSET and MTSEA) a Fourier analysis, as implemented in Mathematica, revealed a regular periodicity of 4 between residues 177 and 193 with a noticeable loss of signal when flanking sites were included. Therefore, further analyses were restricted to residues 177 through 193. For each MTS reagent, estimates of Log_{10} m were fitted to the periodic function $Log_{10} m = sn - i + a \bullet Sin((n-x) \bullet 2\pi/p)$ using unweighted Marquardt nonlinear least squares. The magnitude of the rate constant for inactivation (Log_{10} m) is a function of position (residue number, n) with the straight line ((sn - i)) describing the slope (s) and intercept (i) of the overall trend in a data set and the trigonometric function ($a \bullet Sin((n-x) \bullet 2\pi/p)$) describing the periodicity either side of the line (p is the number of residues per 2π radians, a is amplitude and x a constant setting the register with respect to position, n). Reproduced from Kaplan et al., 2000a with permission from J. Biol. Chem.

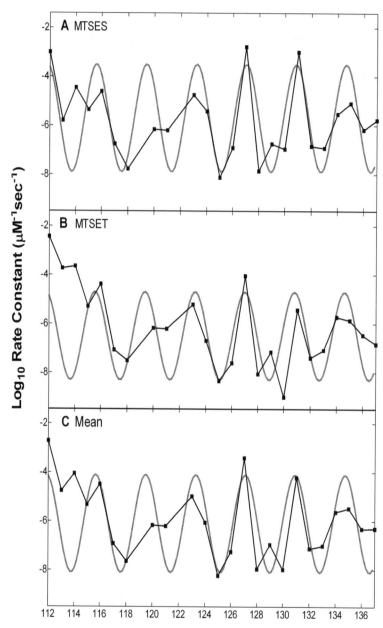

FIGURE 4.2. Rate constants for inactivation of citrate transport by MTS reagents versus location of engineered cysteine in TMD III. Time course data for inactivation of a given single-Cys CTP variant by either MTSES or MTSET were fitted to a simple exponential function by unweighted Marquardt nonlinear least squares as described in Figure 4.1. Estimates of m ranged over 5 orders of magnitude. The rate constant for inactivation, m, can be interpreted as a measure of a residue's accessibility to the MTS reagent. The broad range of estimates of m dictated that further analyses be confined to Log_{10} transformed values.

4. Molecular Structure of the Mitochondrial Citrate Transport Protein   103

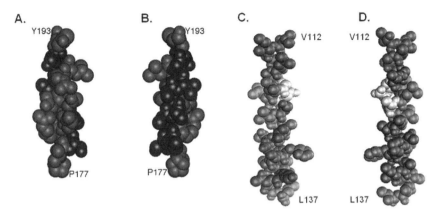

FIGURE 4.3. Computer modeling of residues 177–193 within TMDIV and 112–137 within TMDIII colored on the basis of reactivity toward the MTS reagents. TMD IV residues 177–193 (i.e., Panels A and B) were colored based on the mean of the deviations from the straight lines as depicted in Figure 4.1, Panel D. Residues with values less than zero are colored blue, while those with values greater than zero (i.e., more reactive and therefore more accessible) are colored red. The sequence was modeled initially as a standard α-helix, with N- and C-terminal residues capped as amides. The structure was lightly minimized using the CHARMm force field (Brooks et al., 1983) to allow adjustment of the backbone around Pro177. **Panel A**: Space-filling representation, showing the face of the helix containing the more reactive residues (i.e., water-accessible). **Panel B**: Space-filling representation, showing the face of the helix containing the less reactive residues (i.e., water-inaccessible). TMD III residues 112–137 (Panels C and D) were colored based on the mean MTS rate constant data depicted in Figure 4.2, Panel C. The most reactive, and thus most accessible residues (i.e., $\log_{10}$ rate constant values ranging from −2.72 −5.62 $\mu M^{-1} sec^{-1}$) are denoted in red. Residues of intermediate accessibility to the MTS reagents (i.e., $\log_{10}$ rate constant values ranging from −6.06 −6.31 $\mu M^{-1} sec^{-1}$) are denoted in magenta. Residues with the least accessibility to the MTS reagents (i.e., $\log_{10}$ rate constant values ranging from −6.91 −8.23 $\mu M^{-1} sec^{-1}$) are denoted in blue. Mutants for which MTS rate constant data could not be obtained (i.e., Cys substitution at positions 119 and 122) are denoted in gray. **Panel C** depicts a space-filling representation showing the face of the helix containing the more reactive (i.e., water-accessible) residues. **Panel D** depicts the face of the helix containing the less reactive (i.e., water-inaccessible) residues. Reproduced from Kaplan et al., 2000a and Ma et al., 2004 in modified form with permission from J. Biol. Chem. [*See* Color Plate I].

◄─────────────────────

FIGURE 4.2. For each data set (MTSES and MTSET) a Fourier analysis, as implemented in Mathematica, revealed a regular periodicity of 4 between residues 123 and 137 with a noticeable loss of signal when sites closer to the amino terminus were included. The gray traces indicate superimposed sinusoids. **Panel A** depicts the MTSES data. **Panel B** depicts results with MTSET. **Panel C** depicts the mean of the MTSES and MTSET data sets. Reproduced from Ma et al., 2004 with permission from J. Biol. Chem.

TABLE 4.1. Kinetic Properties and EC50 Values for Those Single-Cys CTP Variants that Exhibited Substrate Protection

Mutant	V_{max} (nmol/min/mg)	K_m (mM)	V_{max}/K_m (nmol/min/mg/mM)	EC_{50} (mM)
Cys-less	1402 ± 69	0.471 ± 0.052	2977 (100%)	–
Gly-115	149 ± 5[a]	0.388 ± 0.005	384 (12.9%)	4.15
Leu-116	487 ± 23[a]	0.228 ± 0.008[b]	2136 (71.7%)	1.18
Gly-117	1448 ± 4	0.355 ± 0.027	4079 (137.0%)	2.78
Leu-121	640 ± 36[a]	0.300 ± 0.018	2133 (71.7%)	5.31
Ser-123	104 ± 15[a]	2.502 ± 0.928[a]	42 (1.4%)	6.66
Val-127	229 ± 29[a]	0.136 ± 0.006[c]	1684 (56.6%)	1.84
Glu-131	13 ± 1[a]	0.170 ± 0.014[c]	76 (2.6%)	7.18
Thr-135	267 ± 32[a]	0.118 ± 0.005[c]	2263 (76.0%)	53.68

[a] Values are $p < 0.01$ from a two-tailed Student's t test between the Cys-less CTP and individual single-Cys mutants,
[b] values are $p < 0.075$, and [c] values are $p < 0.05$. Data from Ma et al., 2005 with permission from *J. Biol. Chem.*

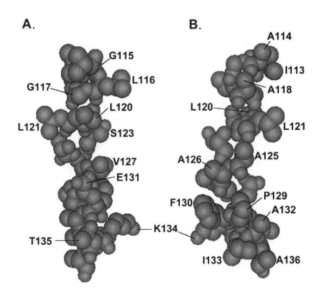

FIGURE 4.4. Homology-modeled CTP TMD III colored according to substrate protection data. Red denotes those residues that were protected by citrate against inactivation by MTSES. Blue denotes residues with which no protection was observed. **Panel A** depicts the TMD III helical face containing residues protected by citrate facing outwards. **Panel B** depicts the face of the helix containing residues not protected by citrate. The helical view depicted in panel B is rotated approx. 180° relative to that depicted in panel A. Reproduced from Ma et al., 2005 with permission from J. Biol. Chem. [*See* Color Plate II].

Maloney 2002). Furthermore, several other findings were unexpected and these will be discussed in the context of the homology modeled structure in sections 4.4.1–4.4.3.

4.2.3. Evaluation of the Temperature Dependence of Substrate Protection

It is thought that two mechanisms can account for substrate protection against modification of a given residue by a covalent reagent (Seal and Amara 1998; Chen and Rudnick 2000; Leighton et al. 2002; Henry et al. 2003; Zomot and Kanner 2003; Gasol et al. 2004). One mechanism, involves occupancy of either the transport pathway or the substrate binding site by the substrate such that the latter can directly sterically impede the approach of a reagent to the modification site. The second mechanism involves a substrate-induced change in protein conformation that results in a decreased accessibility of a given residue to the modification reagent. Since the latter mechanism would presumably require substantive movement of protein domains, it would display a significant temperature dependence. In contrast, the first mechanism, which only involves collision and competition between substrate and the modification reagent, would display little temperature dependence. With these thoughts in mind, with the eight single-Cys mutants with which substrate protection was observed, we examined the extent of this protection at 4°C *versus* 20°C (Ma et al. 2005). Similar levels of protection occurred at the two temperatures, thereby providing clear support for the concept that the citrate protection observed with these single-Cys CTP variants occurred by a direct steric blockage by substrate at the modification sites and thus citrate and MTSES both compete for sites that line the translocation pathway.

4.3. Construction of a Three-Dimensional Model of the CTP

Through the efforts of my close collaborator and colleague D. Eric Walters homology modeling was used to construct a three-dimensional model of the CTP (Walters and Kaplan 2004) based on the x-ray crystal structure of the mitochondrial ADP/ATP transporter (Pebay-Peyroula et al. 2003). As depicted in Figure 4.5, superposition of the backbone trace of the homology-modeled CTP onto the crystallographically determined structure of the ADP/ATP carrier indicates that the CTP transmembrane domains are modeled quite well with a root mean square deviation of 0.94 Å, whereas there is more variability in the modeled loops with root mean square deviation values ranging from 0.72–2.06 Å. This structure has proven to be extremely valuable in that it enables the placement and interpretation of our functional data into a structural context, as well as in the design of fine-tuned testable hypotheses regarding the molecular functioning of the CTP (see sections 4.4, 4.6, and 4.7).

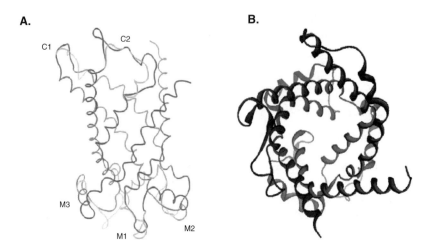

FIGURE 4.5. Homology model of the CTP. **Panel A**: Superposition of the backbone trace of the best-scoring homology-modeled CTP (red) and the crystallographically determined ADP/ATP carrier (green). The intermembrane space loops (C1 and C2) and the mitochondrial matrix loops (M1, M2, and M3) connecting the transmembrane helices are labeled. Note the nearly identical positioning of the TMDs of the two carriers, which consequently appear as one color. **Panel B**: A ribbon diagram viewing the homology-modeled CTP from the outside (i.e., the intermembrane space). **Panel A** is reproduced from Walters and Kaplan, 2004 with permission from Biophys. J. [*See* Color Plate III].

4.4. Evaluation of CTP Functional Data in the Context of the Three-Dimensional CTP Homology Model

Placing our functional data into a structural context has enabled the development of molecular explanations for several unexpected findings (Ma et al. 2005). These include the observed inability of citrate to protect against MTSES modification of cysteines substituted for either Leu120 or K134, despite the fact that both of these locations reside on the substrate-protectable face of TMDIII. In a similar vein, citrate protected against MTSES modification of the L121C mutant despite the fact that this position appears to reside on the non-protected surface of the TMDIII helix. The molecular explanations for these findings have proven to be of great interest and will now be discussed in detail.

4.4.1. Citrate Fails to Protect Against MTSES Modification of the Leu120Cys Mutant

The explanation for this unexpected finding can be derived from close examination of the entire homology-modeled CTP structure (see Figure 4.6, Panel A). Thus it becomes apparent that TMDIII intersects TMDIV at an approximate

Color Plate I

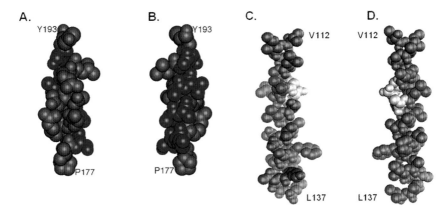

FIGURE 4.3. Computer modeling of residues 177–193 within TMDIV and 112–137 within TMDIII colored on the basis of reactivity toward the MTS reagents. TMD IV residues 177–193 (i.e., Panels A and B) were colored based on the mean of the deviations from the straight lines as depicted in Figure 4.1, Panel D. Residues with values less than zero are colored blue, while those with values greater than zero (i.e., more reactive and therefore more accessible) are colored red. The sequence was modeled initially as a standard α-helix, with N- and C-terminal residues capped as amides. The structure was lightly minimized using the CHARMm force field (Brooks et al., 1983) to allow adjustment of the backbone around Pro177. **Panel A**: Space-filling representation, showing the face of the helix containing the more reactive residues (i.e., water-accessible). **Panel B**: Space-filling representation, showing the face of the helix containing the less reactive residues (i.e., water-inaccessible). TMD III residues 112–137 (Panels C and D) were colored based on the mean MTS rate constant data depicted in Figure 4.2, Panel C. The most reactive, and thus most accessible residues (i.e., $\log_{10}$ rate constant values ranging from -2.72 -5.62 $\mu M^{-1} \text{sec}^{-1}$) are denoted in red. Residues of intermediate accessibility to the MTS reagents (i.e., $\log_{10}$ rate constant values ranging from -6.06 -6.31 $\mu M^{-1} \text{sec}^{-1}$) are denoted in magenta. Residues with the least accessibility to the MTS reagents (i.e., $\log_{10}$ rate constant values ranging from -6.91 -8.23 $\mu M^{-1} \text{sec}^{-1}$) are denoted in blue. Mutants for which MTS rate constant data could not be obtained (i.e., Cys substitution at positions 119 and 122) are denoted in gray. **Panel C** depicts a space-filling representation showing the face of the helix containing the more reactive (i.e., water-accessible) residues. **Panel D** depicts the face of the helix containing the less reactive (i.e., water-inaccessible) residues. Reproduced from Kaplan et al., 2000α and Ma et al., 2004 in modified form with permission from J. Biol. Chem. [*See* page 103].

Color Plate II

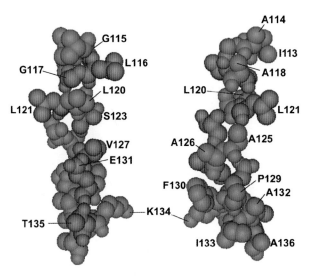

FIGURE 4.4. Homology-modeled CTP TMD III colored according to substrate protection data. Red denotes those residues that were protected by citrate against inactivation by MTSES. Blue denotes residues with which no protection was observed. **Panel A** depicts the TMD III helical face containing residues protected by citrate facing outwards. **Panel B** depicts the face of the helix containing residues not protected by citrate. The helical view depicted in panel B is rotated approx. 180° relative to that depicted in panel A. Reproduced from Ma et al., 2005 with permission from J. Biol. Chem. [*See* page 104].

Color Plate III

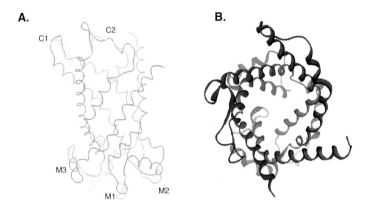

FIGURE 4.5. Homology model of the CTP. **Panel A**: Superposition of the backbone trace of the best-scoring homology-modeled CTP (red) and the crystallographically determined ADP/ATP carrier (green). The intermembrane space loops (C1 and C2) and the mitochondrial matrix loops (M1, M2, and M3) connecting the transmembrane helices are labeled. Note the nearly identical positioning of the TMDs of the two carriers, which consequently appear as one color. **Panel B**: A ribbon diagram viewing the homology-modeled CTP from the outside (i.e., the intermembrane space). **Panel A** is reproduced from Walters and Kaplan, 2004 with permission from Biophys. J. [*See* page 106].

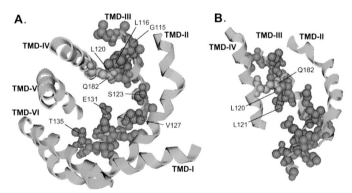

FIGURE 4.6. Depiction of TMD III substrate protection data in the context of the CTP homology-modeled structure. Protected residues are shown in red, unprotected in blue, and Q182 in TMD IV in orange. For clarity, only the TMDs are depicted. **Panel A**: View into the transport pathway from the cytosolic surface of the bilayer. **Panel B**: Cross-sectional view of the transport pathway depicting TMDs II - IV. Helices I, V, and VI have been removed for clarity. This panel emphasizes the accessibility of L121 to the transport path. Reproduced from Ma et al., 2005 with permission from J. Biol. Chem. [*See* page 107].

Color Plate IV

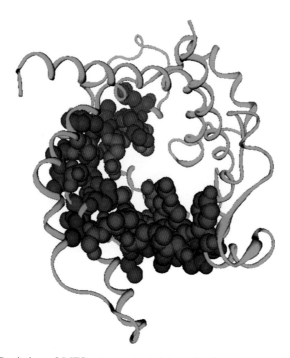

FIGURE 4.7. Depiction of MTS rate constant data and substrate protection data superimposed onto the CTP homology model. Single-Cys CTP mutants with which the rate constants for MTS reagent inactivation have been determined are depicted in space-filling representation within the CTP homology-modeled structure. The substrate-protectable residues of TMDIII and the MTS-accessible residues of TMDIV are shown in red. Residues in TMDIII that are not protected by substrate, as well as the MTS-inaccessible residues of TMDIV are depicted in blue. Finally, moderately accessible TMDIII residues are denoted in purple. The view presented is looking down into the translocation pathway from the cytosolic surface of the bilayer. [*See* page 109].

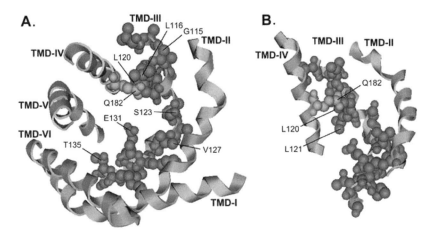

FIGURE 4.6. Depiction of TMD III substrate protection data in the context of the CTP homology-modeled structure. Protected residues are shown in red, unprotected in blue, and Q182 in TMD IV in orange. For clarity, only the TMDs are depicted. **Panel A**: View into the transport pathway from the cytosolic surface of the bilayer. **Panel B**: Cross-sectional view of the transport pathway depicting TMDs II - IV. Helices I, V, and VI have been removed for clarity. This panel emphasizes the accessibility of L121 to the transport path. Reproduced from Ma et al., 2005 with permission from J. Biol. Chem. [*See* Color Plate III].

angle of 30°. When packed in this manner, the Leu-120 side-chain folds behind the side-chain of Gln-182 in TMDIV. Thus, we believe that the Gln-182 sterically blocks the approach of citrate to Leu-120, thereby preventing protection against MTSES modification. Importantly, as depicted in Figure 4.6 (Panel A), the Gln-182 side chain is more proximal to the transport pathway, and as expected its modification by MTSES is protected by citrate (Ma et al. 2006).

4.4.2. *Citrate Protects Against MTSES Modification of the Leu121Cys Mutant*

A second unexpected finding was the observation that even though within the TMDIII helix, Leu-121 is displaced approximately 100° from the substrate protectable surface of TMDIII, and thus would appear to reside on a portion of TMDIII pointing away from the transport path, we nonetheless observed substrate protection. However, as depicted in Figure 4.6 (Panel B), a cross-sectional view of the transport pathway clearly indicates the potential accessibility of this residue's side-chain to the transport path. Additionally, the observation that both Leu-120 and Leu-121 reside near the intersection of TMDs III and IV, suggests that a low-magnitude helix-helix movement during the transport cycle may profoundly influence the accessibility of these residues to the transport pathway.

4.4.3. Citrate Does Not Protect Against Modification of the Lys134Cys Mutant

Based on the high MTS reagent accessibility of the Lys134Cys mutation (see Figure 4.2), and its location on the substrate-protectable surface of TMDIII, we were surprised to observe that citrate did not protect against MTSES-mediated inhibition. An examination of the homology-modeled TMDIII reveals that the Lys-134 side-chain resides near, but not entirely within, the transport path. We believe that our results support the idea that upon removal of the length and charge of the lysine side-chain, the resulting Cys side-chain is sufficiently distant from the transport path such that citrate does not access this sulfhydryl group with sufficient frequency to prevent covalent modification by MTSES.

An important take-home message from these studies is that MTS reagent accessibility studies, when conducted in combination with substrate protection investigations, represent a powerful set of approaches enabling identification of residues that form the substrate translocation pathway. However, if only MTS reagent inhibition studies are conducted, then one must interpret the findings with appropriate caution. Thus, the MTS approach is, on its own, very effective for identifying water-accessible and -inaccessible surfaces of a TMD, as well as the likely secondary structure of such domains based on the periodicity of the accessibility data. However, when it comes to identification of those residues that line the translocation pathway one must interpret such data with caution realizing that there will likely be a small percentage of mistaken identification of such residues. Clearly, accessibility data in combination with substrate protection data is a much better indicator of translocation pathway lining residues.

4.4.4. Superposition of TMD III and IV Functional Data onto the CTP Homology-Modeled Structure

Significant insight into CTP functioning is obtained upon superposition of our functional data onto the modeled CTP structure. For example, Figure 4.7 depicts the substrate-protectable residues of TMDIII and the MTS-accessible residues of TMDIV in red, and the residues in TMDIII that are not protected by substrate, as well as the MTS-inaccessible residues of TMDIV in blue. Finally, moderately accessible TMDIII residues are denoted in purple. This figure indicates that as one views the transport pathway from the external (i.e., cytosolic) surface of the bilayer, a substantial portion of the pathway is enclosed by TMD III and IV residues colored in red or purple. In contrast, the residues in these TMDs denoted in blue, face either other domains within the CTP or the lipid bilayer. Thus, our data clearly indicate the location of the pathway within the homology-modeled monomeric structure and offer firm support for the idea that two pathways exist per functional dimer. Also, as we have previously noted, Cys substitution mutagenesis with residues pointing into the pathway, is considerably more disruptive of CTP function than is mutagenesis of residues that do not comprise the pathway (Ma et al. 2004; Ma et al. 2006). Thus, of 14 essential

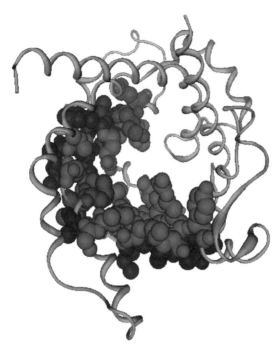

FIGURE 4.7. Depiction of MTS rate constant data and substrate protection data superimposed onto the CTP homology model. Single-Cys CTP mutants with which the rate constants for MTS reagent inactivation have been determined are depicted in space-filling representation within the CTP homology-modeled structure. The substrate-protectable residues of TMDIII and the MTS-accessible residues of TMDIV are shown in red. Residues in TMDIII that are not protected by substrate, as well as the MTS-inaccessible residues of TMDIV are depicted in blue. Finally, moderately accessible TMDIII residues are denoted in purple. The view presented is looking down into the translocation pathway from the cytosolic surface of the bilayer. [*See* Color Plate IV].

residues that we have identified within the CTP, the side-chains of all but one project into the translocation pathway (Ma et al., 2006), implying that the precisely defined 3-dimensional architecture of the pathway is critical to support CTP function.

4.5. Criteria for Identification of Residues Involved in Substrate Binding Versus those Involved in Other Aspects of the Transport Mechanism

An important issue regarding the molecular mechanism of any transport protein has to do with the identification of those residues that form the substrate binding site *versus* those residues that are important to other aspects of the translocation mechanism. We suggest that several criteria should be used in the identification of binding site residues. *First,* a residue that plays a prominent role in substrate binding should be essential for function, such that replacement of the residue via mutagenesis causes near total disruption of transporter function. The identification of an "essential" residue is relatively straightforward as it is obtained via a simple specific activity determination and typically involves residues, the mutation of which, causes a $\geq 98\%$ inactivation of transporter function. This first condition represents a necessary but not sufficient criterion. *Second,* mutation of a substrate binding site residue should result in a profound increase in the K_m value. Here we would expect that at least a 5-10-fold increase and often an even more profound increase would be obtained. Mutation of residues that are mechanistically essential but are not involved in substrate binding will often result in a large decrease in the V_{max} with little change in the K_m. *Third,* upon mutagenesis of a binding site residue to a Cys, we would expect substrate to protect against chemical modification of the engineered Cys, unless the remaining residues that comprise the substrate binding site are of sufficient distance from the engineered Cys such that citrate binding to these side-chains does not sterically block access of the MTS reagent to the cysteine sulfhydryl group. *Fourth,* the EC_{50} for protection should be similar to the K_m value (Note: it is often overly rigorous to expect the EC_{50} to be exactly the same as the K_m value, since an inhibitor that forms a covalent bond will ultimately overwhelm a substrate's ability to competitively protect against this inhibition). *Fifth,* the substrate protection should be temperature-independent, thereby ruling out the possibility that protection arises indirectly as a consequence of a protein conformational change, but rather is due to a direct steric blockage by the substrate of the approach of the inhibitor to the site of modification. *Sixth,* if a non-transportable competitive inhibitor exists for the transporter, one would expect its K_d for binding to be greatly increased (≥ 5–10-fold) upon mutation of a binding site residue to Cys. As mentioned above, with the CTP we have identified 14 essential residues (Ma et al. 2006). In a very recent development, we have been able to modify our transport assay such that we are now able to accurately measure K_m values that are increased up to 100-fold over the native value. Current studies indicate that some of these essential residues display profound K_m shifts. We believe that completion of these studies should enable an accurate, data-based modeling of the citrate binding site within the CTP during the next year.

4.6. The Location of the Monomer-Monomer Interface in Homodimeric Mitochondrial Transporters

4.6.1. Predictions With the CTP

To date, extensive work from many laboratories utilizing a battery of approaches indicates that mitochondrial transporters exist in detergent micelles as homodimers and that this likely represents the functional form of these carriers (see Kotaria et al. 1999 for a review of this literature). For example with the CTP, utilizing both a gel filtration approach and a charge-shift native PAGE (polyacrylamide gel electrophoresis) approach, we observed that both the wild-type and the Cys-less CTP exist as homodimers (Kotaria et al. 1999). The question arises as to the role that a dimeric structure plays in the CTP mechanism. Insight into this issue can be gleaned from kinetic data indicating that the CTP operates by a sequential type mechanism which includes the formation of a ternary complex of two substrate molecules with the transporter prior to the translocation step (Bisaccia et al. 1993). This kinetic mechanism has important structural consequences in that two substrate binding sites must exist on opposite sides of the membrane and these sites must be exposed and occupied prior to the translocation event (Kramer and Palmieri 1992). Since superposition of our functional data onto the CTP homology-modeled structure indicates that each monomer contains a translocation pathway (Figure 4.7), a working model posits that the obligatory exchange catalyzed by the CTP occurs when a binding site within each monomer is occupied by substrate at opposite sides of the membrane. The two monomers in the homodimer must be out-of-phase (Pebay-Peyroula et al. 2003) and must closely communicate such that a conformational change leading to transport in one direction through one monomer is linked to a conformational change leading to transport in the opposite direction in the other monomer. Thus the composition of the dimer interface is essential to the correct functioning of the homodimer.

We have proposed that TMDIII forms a substantive portion of the dimer interface (Ma et al. 2004; Walters and Kaplan 2004). Specifically we posit that G115 and G119, two residues which are highly conserved and which upon mutation to Cys considerably inactivate CTP function, form part of the interface and facilitate close packing between helices TMDIII and TMDIII' (i.e., TMDIII in the other monomer). In addition, we have inferred that Ala118 also resides at the dimer interface. This supposition was based on our findings that the Ala118Cys CTP variant displayed a 7-fold increase in the K_m, moderate accessibility to MTSES, and no substrate protection against MTSES-mediated inhibition of CTP function (Ma et al. 2004; Ma et al. 2005). These findings, in combination with our molecular modeling efforts, led to the conclusion that Ala118 may reside in close proximity to Ala118' of the other monomer and together form an essential portion of the dimer interface. Further we posit that disruption of the interface by replacement with the bulkier Cys results in a substantial conformational change in the substrate binding site, leading to an increase in the observed K_m. Interestingly, the V_{max} value of the mutant also increases suggesting that Cys substitution at this site removes an intrinsic molecular constraint upon the transport rate, possibly one that is involved in the coordination of the function of the two monomers.

A final hypothesis we have put forward regarding the composition of the dimer interface is depicted in Figure 4.8, and posits that E122 (a residue which upon mutation to Cys completely inactivates CTP function and which projects approximately 180° away from the transport pathway) of one monomer resides in close proximity to the TMDIII-TMDIV interface of the other monomer (Walters and Kaplan 2004). It is interesting to note, that the amide side-chain

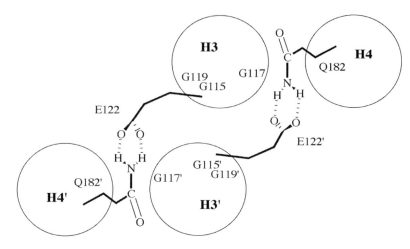

FIGURE 4.8. Schematic representation of a proposed dimer interface involving E122 of H3 and Q182' of H4'. Helices 3 and 4 (i.e., H3 and H4) are identical to TMDs III and IV. Reproduced from Walters and Kaplan, 2004 with permission from Biophys. J.

of Q182' (i.e., Q182 of the other monomer) resides at the TMDIII-TMDIV interface such that it can either face into the transport pathway or away from the pathway into the interface. Moreover, we have proposed that E122-Q182' and E122'-Q182 interactions might serve to couple the conformational changes of the two translocation pathways, enabling the CTP to function as an obligatory antiporter.

4.6.2. Comparisons with the ADP/ATP Carrier Dimer Interface

It should be noted that the above predictions regarding the dimer interface were developed by interpreting our functional data in the context of the CTP homology-modeled structure. An interesting question to consider is how these predictions compare with data available from the crystal structure of the mitochondrial ADP/ATP carrier. In the first structure, published by Pebay-Peyroula et al. (2003), only the monomeric form of the carrier was crystallized. However, in more recent studies crystals of two different dimeric forms of the transporter, which differ greatly in their orientation, were obtained (Nury et al. 2005). One of the crystal structures is compatible with existing data on the ADP/ATP carrier putative dimer interface and indicates the presence of cardiolipin at the interface. While the details of the protein-cardiolipin-protein contacts that they propose are beyond the scope of this chapter, in general they suggest that the interface involves interactions between cardiolipin, the matrix loops and TMDs II, IV, and VI, as well as possible interactions between the loops themselves. Presently, it is difficult to compare their findings with predictions that we have put forward for the CTP, since to date most of our work with the

CTP has involved the TMDs and not the extramembranous loops. Accordingly, experiments are underway in our laboratory which are aimed at testing the role of the external loops in CTP function.

4.7. Perspectives and Future Directions

We believe it accurate to say that substantial progress has been made in our understanding of the functioning of the CTP at the molecular level during the last 5 years, and yet in many ways we are now poised to address the most interesting questions. Thus, we believe we are close to attaining a detailed delineation of the citrate binding site(s). The precise detail of this site will be established via molecular modeling that will be guided by our extensive array of functional data. A second question that we are beginning to address has to do with the composition of the dimer interface. Our homology-model has permitted the development of a number of hypotheses concerning this matter, which we plan to test using a combination of functional assay and measurement of the distance between engineered cysteines utilizing both disulfide crosslinking and EPR. A third unresolved issue has to do with the conformational changes that occur during the opening and closing of the gates that modulate movement through the permeation pathway during the transport cycle. This can probably be best addressed via X-ray crystallographic analysis of trapped conformations representing different steps of the transport cycle. Towards this end, we are in the midst of extensive crystallization trials utilizing Fab fragments generated from conformation-specific anti-CTP monoclonal antibodies, various CTP inhibitors, and inactive CTP mutants that may be locked in a given conformation. Through a combination of the above approaches we believe that we will achieve a complete understanding of the molecular basis for CTP functioning during the next 5–10 years.

Acknowledgements. The authors would like to acknowledge the efforts of a talented list of colleagues whom we have had the good fortune of working with and who have made substantial contributions towards conducting much of the work described in this chapter. This list includes Chunlong Ma, Sreevidya Remani, as well as Drs. Rusudan Kotaria, Yan Xu, and D. Eric Walters. This work was supported by National Institutes of Health Grant GM-054642 to R.S.K.

References

Akabas MH, Stauffer DA, Xu M, Karlin A (1992) Acetylcholine receptor channel structure probed in cysteine-substitution mutants. Science 258:307–310

Bisaccia F, De Palma A, Dierks T, Kramer R, Palmieri F (1993) Reaction mechanism of the reconstituted tricarboxylate carrier from rat liver mitochondria. Biochim. Biophys. Acta 1142:139–145

Brooks BR, Bruccoleri RE, Olafson BD, States DJ, Swaminathan S, Karplus M (1983) CHARMM: a program for macromolecular energy, minimization, and dynamics calculations. J Comp Chem 4:187–217

Brunengraber H, Lowenstein JM (1973) Effect of (-)-hydroxycitrate on ethanol metabolism. FEBS Lett 36:130–132

Chen J-G, Rudnick G (2000) Permeation and gating residues in serotonin transporter. Proc Nat Acad Sci. USA 97: 1044–1049

Conover TE (1987) Does citrate transport supply both acetyl groups and NADPH for cytoplasmic fatty acid synthesis? Trends Biochem Sci. 12:88–89

Endemann G, Goetz PG, Edmond J, Brunengraber H (1982) Lipogenesis from ketone bodies in the isolated perfused rat liver. Evidence for the cytosolic activation of acetoacetate. J Biol Chem 257:3434–3440

Fu D, Sarker RI, Abe K, Bolton E, Maloney PC (2001) Structure/function relationships in OxlT, the oxalate-formate transporter of oxalobacter formigenes. Assignment of transmembrane helix 11 to the translocation pathway. J Biol Chem 276: 8753–8760

Gasol E, Jimenez-Vidal M, Chillaron J, Zorzano A, Palacin M (2004) Membrane topology of system xc-light subunit reveals a reentrant loop with substrate-restricted accessibility. J Biol Chem 279:31228–31236

Henry LK, Adkins EM, Han Q, Blakely RD (2003) Serotonin and cocaine-sensitive inactivation of human serotonin transporters by methanethiosulfonates targeted to transmembrane domain I. J Biol Chem 278: 37052–37063

Holmgren M, Liu Y, Xu Y, Yellen G (1996) On the use of thiol-modifying agents to determine channel topology. Neuropharmacology 35:797–804

Kaplan RS (1996) Mitochondrial transport processes. In: Schultz SG, Andreoli T, Brown A, Fambrough D, Hoffman J, Welsh J (eds) Molecular Biology of Membrane Transport Disorders. Plenum Press, New York, pp. 277–302

Kaplan RS (2001) Structure and function of mitochondrial anion transport proteins. J Membrane Biol 179:165–183

Kaplan RS, Morris HP, Coleman PS (1982) Kinetic characteristics of citrate influx and efflux with mitochondria from Morris hepatomas 3924A and 16. Cancer Res 42:4399–4407

Kaplan RS, Mayor JA, Johnston N, Oliveira DL (1990a) Purification and characterization of the reconstitutively active tricarboxylate transporter from rat liver mitochondria. J Biol Chem 265:13379-13385

Kaplan RS, Oliveira DL, Wilson GL (1990b) Streptozotocin-induced alterations in the levels of functional mitochondrial anion transport proteins. Arch Biochem Biophys 280:181-191

Kaplan RS, Mayor JA, Wood DO (1993) The mitochondrial tricarboxylate transport protein. cDNA cloning, primary structure, and comparison with other mitochondrial transport proteins. J Biol Chem 268:13682-13690

Kaplan RS, Mayor JA, Gremse DA, Wood DO (1995). High level expression and characterization of the mitochondrial citrate transport protein from the yeast *Saccharomyces cerevisiae*. J Biol Chem 270:4108–4114

Kaplan RS, Mayor JA, Brauer D, Kotaria R, Walters DE, Dean AM (2000a) The yeast mitochondrial citrate transport protein: probing the secondary structure of transmembrane domain IV and identification of residues that likely comprise a portion of the citrate translocation pathway. J Biol Chem 275:12009–12016

Kaplan RS, Mayor JA, Kotaria R, Walters DE, Mchaourab HS (2000b) The yeast mitochondrial citrate transport protein: Determination of secondary structure and solvent accessibility of transmembrane domain IV using site-directed spin labeling. Biochemistry 39:9157–9163

Karlin A, Akabas MH (1998) Substituted-cysteine accessibility method. Methods Enzymol 293:123–145

Kotaria R, Mayor JA, Walters DE, Kaplan RS (1999) Oligomeric state of wild-type and cysteine-less yeast mitochondrial citrate transport proteins. J Bioenerg Biomemb 31:543–549

Kramer R, Palmieri F (1992) Metabolite carriers in mitochondria. In: Ernster L (ed) Molecular Mechanisms in Bioenergetics. Elsevier, New York, pp. 359–384

Leighton BH, Seal RP, Shimamoto K, Amara SG (2002) A hydrophobic domain in glutamate transporters forms an extracellular helix associated with the permeation pathway for substrates. J Biol Chem 277: 29847–29855

Ma C, Kotaria R, Mayor JA, Eriks LR, Dean AM, Walters DE, Kaplan RS (2004) The mitochondrial citrate transport protein: probing the secondary structure of transmembrane domain III, identification of residues that likely comprise a portion of the citrate transport pathway, and development of a model for the putative TMDIII-TMDIII' interface. J Biol Chem 279:1533–1540

Ma C, Kotaria R, Mayor JA, Remani S, Walters DE, Kaplan RS (2005) The yeast mitochondrial citrate transport protein: characterization of transmembrane domain III residue involvement in substrate translocation. J Biol Chem 280:2331–2340

Ma C, Remani S, Kotaria R, Mayor JA, Walters DE, Kaplan RS (2006) The mitochondrial citrate transport protein: evidence for a steric interaction between glutamine 182 and leucine 120 and its relationship to the substrate translocation pathway and identification of other mechanistically essential residues. Biochim Biophys Acta, Manuscript Submitted

Nury H, Dahout-Gonzalez C, Treqeguet V, Lauguin G, Brandolin G, Pebay-Peyroula E (2005) Structural basis for lipid-mediated interactions between mitochondrial ADP/ATP carrier monomers. FEBS Lett 579:6031–6036

Palmieri F (2004) The mitochondrial transporter family (SLC25): physiological and pathological implications. Pflugers Arch - Eur J Physiol 447:689–709

Palmieri F, Stipani I, Quagliariello E, Klingenberg M (1972) Kinetic study of the tricarboxylate carrier in rat liver mitochondria. Eur J Biochem 26:587–594

Pebay-Peyroula E, Dahout-Gonzalez C, Kahn R, Trezeguet V, Lauquin GJ-M, Brandolin G (2003) Structure of mitochondrial ADP/ATP carrier in complex with carboxytractyloside. Nature 426:39–44

Seal RP, Amara SG (1998) A reentrant loop domain in the glutamate carrier EAAT1 participates in substrate binding and translocation. Neuron 21: 1487–1498

Walters DE, Kaplan RS (2004) Homology-modeled structure of the yeast mitochondrial citrate transport protein. Biophys J 87:907–911

Watson JA, Lowenstein JM (1970) Citrate and the conversion of carbohydrate into fat. Fatty acid synthesis by a combination of cytoplasm and mitochondria, J Biol Chem 245:5993–6002

West IC (1997) Ligand conduction and the gated-port mechanism of transmembrane transport. Biochim Biophys Acta 1331:213–234

Xu Y, Mayor JA, Gremse D, Wood DO, Kaplan RS (1995) High-Yield bacterial expression, purification, and functional reconstitution of the tricarboxylate transport protein from rat liver mitochondria. Biochem Biophys Res Comm 207: 783–789

Xu Y, Kakhniashvili DA, Gremse DA, Wood DO, Mayor JA, Walters DE, Kaplan RS (2000) The yeast mitochondrial citrate transport protein: probing the roles of cysteines, Arg181, and Arg189 in transporter function. J Biol Chem 275:7117–7124

Ye L, Maloney PC (2002) Structure/function relationships in OxlT, the oxalate/formate antiporter of *Oxalobacter formigenes*: assignment of transmembrane helix 2 to the translocation pathway. J Biol Chem 277: 20372–20378

Zomot E, Kanner BI (2003) The interaction of the gamma-aminobutyric acid transporter GAT-1 with the neurotransmitter is selectively impaired by sulfhydryl modification of a conformationally sensitive cysteine residue engineered into extracellular loop IV. J Biol Chem 278: 42950–42958

5
Regulation of Pyruvate and Amino Acid Metabolism

Thomas C. Vary, Wiley W. Souba and Christopher J. Lynch

5.1. Introduction

Amino acids serve multiple functions in body ranging from the basic building blocks of proteins to precursors for glucose and urea synthesis. The circulating amino acids are generally categorized with respect to either their function (gluconeogenic or ketogenic) or availability (essential or nonessential). The metabolism of amino acids is directly controlled through mitochondrial function. In addition, there is growing evidence that amino acids participate in processes that control cellular functions. Amino acids are not only utilized as metabolic substrates, increasingly it has become evident they also participate in the regulation of various cellular processes including protein synthesis, glycogen synthesis and autophagy. The regulatory processes affected by amino acids may involve mitochondria. This chapter will provide an overview of the current understanding of the role of the mitochondrial function in controlling amino acid metabolism.

5.2. Metabolism of Amino Acids via Pyruvate Dehydrogenase

The complexity of the processes involved in utilization and synthesis of glucose represent one of the fundamental metabolic processes of the human body. The overriding reason for the complexity of glucose homeostasis is the central nervous system's requirement for glucose to sustain energy production. The liver and kidney represent the major organ systems capable of gluconeogenesis in man. Skeletal muscle, by virtue of its mass, represents a major source of gluconeogenic substrates. As such there is an intricate balance in the release of amino acids from protein stores in muscle and the utilization of amino acids for gluconeogenesis in the liver. Thus, interorgan transfer of carbon skeletons derived from amino acids contributes significantly to the flux of nutrients delivered to the brain. In the most simplistic scenario, when glucose becomes unavailable via ingestion of food, amino acid availability is increased by net protein degradation in skeletal muscle. There are several fates of the amino acids released by proteolysis in muscle (Figure 5.1). First, a certain amount of amino acids can be reincorporated into proteins via protein synthesis. Second, amino acids can undergo oxidation in mitochondria producing in the process energy. At least six amino acids (alanine, aspartate, glutamate, leucine, isoleucine and valine) can be oxidized by skeletal muscle. Third, amino acids can be released into the venous blood system. All amino acids are readily released from skeletal muscle via specialized transport systems. Fourth, skeletal muscle has the ability to interconvert the nitrogen from amino acids to alanine and glutamine, thereby promoting de novo synthesis of these two amino acids. Mitochondrial regulation of amino acid and carbohy-

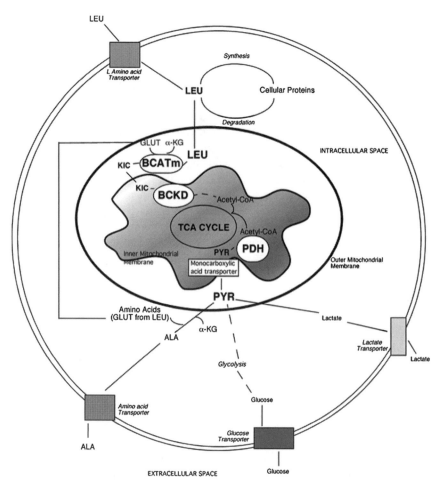

FIGURE 5.1. Interrelationship between mitochondrial amino acid and pyruvate oxidation in skeletal muscle. LEU – leucine; BCATm – Mitochondrial branched chain amino transferase; α-KG – α-ketoglutarate; KIC – α-ketoisocaproic acid; BCKD – branched chain amino acid dehydrogenase; TCA – tricarboxylic acid cycle; PDH – pyruvate dehydrogenase complex; PYR – pyruvate; ALA – alanine; GLUT – glutamate.

drate oxidation plays a vital role in the extent of de novo alanine and glutamine synthesis.

Indeed, the majority of amino acids released by skeletal muscle during starvation are alanine and glutamine (Felig 1975). The total content of alanine plus glutamine in skeletal muscle proteins is slightly more than 10%, whereas they comprise as much as 60% of amino acids released from muscle. Hence, these two amino acids are released in amounts greater than can be accounted for by proteolysis of skeletal muscle proteins. Hence, they undergo "de novo" synthesis (Figures 5.2 and 5.3).

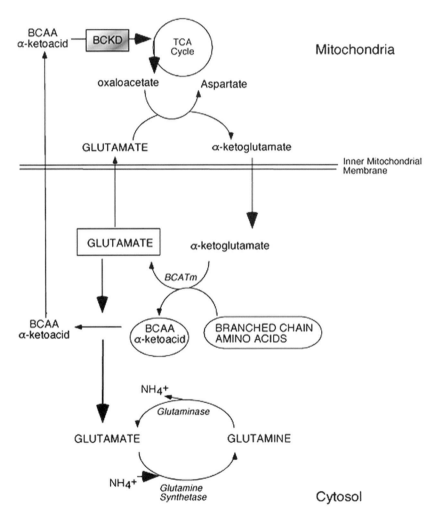

FIGURE 5.2. Relationship between branched chain amino acid metabolism and glutamine synthesis in skeletal muscle. BCKD – branched chain amino acid dehydrogenase; TCA – tricarboxylic acid cycle.

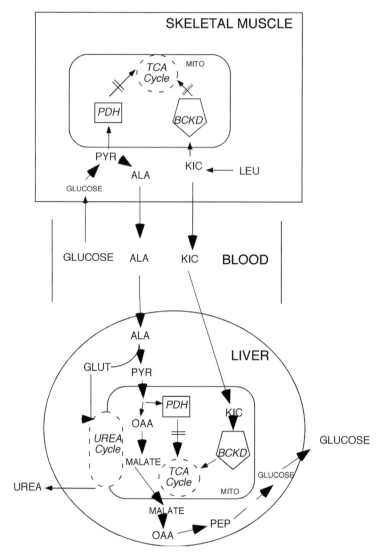

FIGURE 5.3. Mitochondrial regulation of interorgan amino acid fluxes between skeletal muscle and liver. LEU – leucine; BCATm – mitochondrial branched chain amino transferase; ; α-KG – α-ketoglutarate; KIC – α-ketoisocaproic acid; BCKD – branched chain amino acid dehydrogenase; TCA – tricarboxylic acid cycle; PDH – pyruvate dehydrogenase complex; PYR – pyruvate; ALA – alanine; GLUT – glutamate; OAA – oxaloacetate; PEP – phosphoenolpyruvate; MITO – mitochondria.

Amino acids are released from the skeletal muscle and are taken up by liver and converted into glucose with the release of the nitrogen group being detoxified via the formation of urea (Figure 5.3). Urea is then excreted from the body via normal kidney function. The physiological significance of amino acid oxidation for energy is not thought to be as important as the ability of skeletal muscle to convert amino acids to alanine and glutamine that are released in quantities that greatly exceed their concentration in muscle proteins.

5.2.1. Alanine Cycle

The pathway by which alanine is formed and transported back to liver has been referred to as the glucose/alanine cycle. About 30% of the alanine released by skeletal muscle is derived from net protein degradation, while the remaining 70% is synthesized de novo. Glucose carbon is metabolized in skeletal muscle to pyruvate via the glycolytic pathway. Once pyruvate is formed, there are essentially three pathways that it can follow in muscle tissues. First, it can be transported into the mitochondria, and oxidized via the pyruvate dehydrogenase complex to acetyl-CoA and then subsequently completely oxidized via the tricarboxylic acid cycle. Second, it can be reduced to lactate with the consumption of NADH. Lastly, pyruvate can be transaminated to alanine in a reaction catalyzed by the alanine aminotransaminase. The pyruvate dehydrogenase (PDC) complex catalyzes the first irreversible reaction in the mitochondrial oxidation of glucose carbon and as such is the primary determinant of glucose oxidation (Randle et al. 1984; Randle et al. 1988; Randle 1998; Sugden et al. 1989). The regulation of this complex is important to glucose homeostasis. Oxidation of pyruvate results in the depletion of body glucose carbon sources because glucose cannot be synthesized from acetyl-CoA and the PDH reaction is physiologically irreversible (Randle et al. 1984; Randle et al. 1988; Randle 1998; Sugden et al. 1989). Thus, the activity of the PDH complex determines whether pyruvate is oxidized via the Krebs cycle or converted to lactate via lactate dehydrogenase or converted to alanine via alanine transaminase.

The source of the nitrogen for de novo alanine synthesis occurs via transaminations with other amino acids. As much as 60% of the nitrogen for de novo alanine synthesis is derived from branched chain amino acids (valine, isoleucine, and leucine) (Hayman and Miles 1982). For the branched chain amino acids, transamination is catalyzed by the branched chain aminotransferase. There are two aminotransferases; one that has a preference for valine (a deficiency of which gives rise to hypervalenemia) and one that has a preference for leucine or isoleucine (a deficiency of which gives rise to leucine-isoleucinaemia). The resulting α-keto acids (α-ketoisocaproic; α-keto-3-methylvalarate; α-ketoisovalate) are available for oxidative decarboxylation via the mitochondrial branched chain amino acid dehydrogernase (BCKD) an enzyme located in the inner mitochondrial membrane (Figure 5.1) and eventually metabolized to either acetyl-CoA (leucine) or succinyl CoA (isoleucine and valine) which enter the citric acid cycle. A deficiency in BCKD leads to the accumulation of the α-keto acids in the blood and subsequent elimination from the body (see section on Maple syrup urine disease).

5.2.2. Glutamine Cycle

Glutamine serves various metabolic functions in different tissues and under different physiologic states. Its functions within the cell are characterized by its

(a) utility in nitrogen transport and ammonia detoxification (b) importance in maintaining the cellular redox state (c) position as a metabolic intermediate and (d) and ability to serve as an energy source. Glutamine like alanine must arise from amino acid metabolism in skeletal muscle. Glutamine serves as the primary ammonia shuttle between tissues. This "nitrogen shuttle" provides a non-toxic means of transporting ammonia from the periphery to the viscera. After reaching its target organ ammonia can be regenerated for excretion or ureagenesis.

Glutamine is synthesized from intramuscular glutamate and ammonia via glutamine synthetase (Figures 5.2 and 5.4).

The glutamine synthetase (GS) pathway accounts for as much as 50% of the glutamine released with elevated rates of glutamine release (Vary et al. 1989a; Vary and Murphy 1989). Expression of GS is increased in muscle during a number of catabolic states, with induction of GS mRNA levels by 400–700% (Abcouwer et al. 1996; Abcouwer et al. 1997; Ardawi and Majzoub 1991; Labow et al. 2001). Induction of GS gene expression in skeletal muscle is largely dependent on adrenal derived hormones (Abcouwer et al. 1995; Chandrasekher et al. 1999; Lukaszewicz et al. 1997).

Glutamate is most likely formed from 2-oxoglutarate via transamination from other amino acids. The 2-oxoglutarate arises from mitochondrial citrate and transport of the carbon out of the mitochondria via malate/aspartate shuttle. The ammonia arises from the reactions catalyzed by the glutamate dehydrogenase

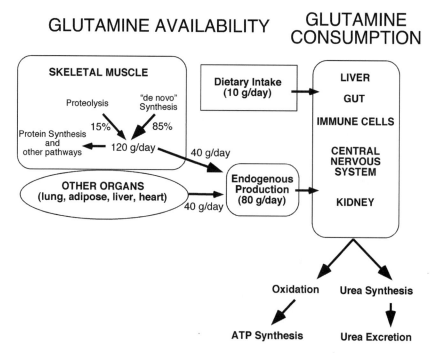

FIGURE 5.4. Glutamine balance showing relationship between availability and consumption.

reaction. Activation of PDH complex activity through dichloroacetate leads to increased glutamine production presumably by lowering the availability of pyruvate for transamination with glutamate to form alanine (Vary et al. 1988b). Hence, the supply of glutamate would be elevated and available for de novo glutamine synthesis. The liver serves as the major regulatory tissue responsible for whole-body ammonia detoxification and glutamine homeostasis (Haussinger et al. 1992; Meijer et al. 1990). Transport of glutamine across the plasma membrane of liver cells may represent a rate-limiting step in its metabolism, especially when intracellular catabolism is accelerated (Haussinger et al. 1985; Low et al. 1993). In normal rat and human hepatocytes, glutamine transport is predominantly mediated by a Na^+-dependent transporter with narrow substrate specificity (glutamine, histidine, and asparagine) termed system N (Bode et al. 1995; Kilberg et al. 1980).

In the mitochondria, ammonia is derived from glutamine by hydrolysis of glutamine to glutamate by the enzyme glutaminase. Hydrolysis of glutamate to α-ketoglutarate by the enzyme glutamate dehydrogenase can also generate ammonia. In the liver, ammonia can combine with CO_2 to form carbamoyl phosphate that subsequently is detoxified by the urea cycle. The amino group on glutamate and glutamine may also enter the urea cycle directly through a reaction involving oxaloacetate to generate aspartate. One of the major functions of the liver is the maintenance of ammonia homeostasis within the body. Glutamine metabolism is distributed heterogeneously throughout the liver with glutamine consumption via glutaminase concentrated in the periportal hepatocytes while glutamine synthesis via glutamine synthetase is localized within the perivenous hepatocytes. This distribution provides the liver with an elegant mechanism to detoxify the blood of ammonia while contributing glutamine to the systemic supply.

Maintenance of acid-base homeostasis is an essential function of the kidney, particularly under conditions associated with metabolic acidosis. During metabolic acidosis, aminogenesis is higher in kidney, allowing for the excretion of protons in the form of ammonium ions in urine. The use of glutamine as a nitrogen shuttle is also important in the excretion of nitrogenous wastes and the maintenance of acid-base homeostasis. Circulating glutamine contributes 80–90% of the ammonia produced during acidosis. Under normal conditions, renal uptake of glutamine from the blood is minimal. During chronic metabolic acidosis, however, renal extraction of glutamine increases markedly. Enhanced renal glutamine uptake and metabolism results from a coordinated series of events triggered by the disruption of normal acid-base balance. The homeostatic response involves alterations in interorgan glutamine flux, such that glutamine release from skeletal muscle is doubled, the splanchnic bed shifts from net glutamine uptake to glutamine release, and the kidney becomes the major site of glutamine consumption.

Three Na^+-dependent amino acid transport systems (system A, system ASC (ATB0), and system N) transport glutamine across cell membranes. Almost all the glutamine filtered at the glomerulus is reclaimed in the convoluted proximal tubule, leaving little margin for increased glutamine uptake from the tubular lumen during acidosis. The major portion of the increased renal glutamine taken up enters the tubular epithelium across the basolateral membrane from the blood via system N1 (SN1) amino acid transporter. Acidosis causes a 10-fold increase in whole kidney SN1 mRNA level and a 100-fold increase in the cortex (Karinch et al. 2002). Acidosis increases Na^+-dependent glutamine uptake (SN1) into basolateral and brush-border membrane vesicles isolated from rat cortex. The SN1 carrier appears to be the most important cell membrane transporter of circulating glutamine into the renal tubular cell during acidosis.

Once circulating glutamine is transported into the renal tubular cell, it is acted upon by a kidney-specific isozyme of glutaminase; hydrolyzing glutamine and generating ammonia for excretion in the urine (Deferrari et al. 1994; Dejong et al. 1993). The catabolism of glutamine provides the majority of nitrogen excreted in the urine. Furthermore, the availability of ammonia also facilitates the excretion of acid loads by conjugating H^+ ions and generating ammonium ions for excretion. The metabolism of glutamine produces bicarbonate ions that can function to further neutralize hydrogen ions in the blood. The adaptive changes in proximal renal tubules include 1) escalating glutamine transport both in brush border and basolateral membranes increases the availability of glutamine for ammoniagenesis and simultaneously augments secretion of hydrogen ions into urine and blood, 2) not surprisingly, glutamine consumption via phosphate-dependent glutaminase, glutamate dehydrogenase, and phosphoenol carboxykinase become accelerated within the kidney during acidosis (Deferrari et al. 1994; Deferrari et al. 1997), and 3) upsurges in activity of Na^+/H^+ antiporter and Na^+/HCO_3^- co-transporter increases H^+ secretion into urine and bicarbonate ion reabsorption into the blood.

Aside from ureagenesis, glutamine plays a key role in regulating glutathione synthesis. Glutathione is a tripeptide composed of glutamate, cysteine and glycine and represents the major source of cellular reducing equivalents, protecting the cell against oxidative injury. As a source of intracellular glutamate, glutamine provides one of the constituents of glutathione, but glutamine can also contribute to glutathione synthesis indirectly by exchanging intracellular glutamate (derived from glutamine) across the cell membrane for extracellular cysteine, another component of glutathione. The importance of glutathione resides in its ability to catalyze the reduction of organic hydroperoxides and hydrogen peroxide to limit reactive oxygen species damage. Glutathione peroxide lowers oxygen reactive oxygen species through several mechanisms. First, it converts hydrogen peroxide to water. Second, it converts peroxidated fatty acids to hydroxyl fatty acids. Hydroxyl radicals tend to interact with unsaturated fatty acids in phospholipids and other lipid components of membranes forming lipid hydroperoxides ultimately resulting in increased hydrophilicity of membranes and inhibition of several enzymes.

The flow of glutamine into and out of the liver can vary widely. Physiological concentrations of ammonia within the portal circulation can produce a 'feed-forward stimulation' of hepatic glutaminase (Haussinger 1990), allowing the liver to increase glutamine consumption when glutamine is abundant (Welbourne and Oshi 1990). Similarly, the effects of starvation, starvation in conjunction with sepsis (Fischer et al., 1996), proinflammatory cytokines (Pacitti et al. 1993), burn injury (Lohmann et al. 1998), advanced malignant disease (Inoue et al. 1995), and glucocorticoid excess can all increase glutamine transport and utilization within the liver (Bobe et al. 2002; Bode et al. 1995; Low et al. 1992). However, during metabolic acidosis glutamine flow is directed away from the liver and is significantly increased in the kidney (Welbourne and Oshi 1990). Under these conditions, hepatic ureagenesis is decreased, but renal ammoniagenesis is increased to facilitate H^+ excretion.

5.3. Regulation of Pyruvate Dehydrogenase Complex

The determining factor regulating formation of alanine and lactate from pyruvate is the activity of the pyruvate dehydrogenase complex. The pyruvate dehydrogenase (PDH) complex catalyzes the first irreversible reaction in the mitochondrial oxidation of glucose (Harris et al. 2002; Holness and Sugden 2003; Randle et al. 1984; Randle 1998; Tsai et al. 1973). Pyruvate undergoes oxidative decarboxylation by the PDH complex in the presence of NAD^+ and CoA to form acetyl-CoA, NADH and CO_2 (Figure 5.5). As such the PDH complex is the primary determinant of glucose oxidation in animal cells (Harris et al. 2002; Holness and Sugden 2003; Randle et al. 1984; Randle 1998; Tsai et al. 1973). The regulation of this complex is important to glucose homeostasis as it links glycolysis to tricarboxylic acid cycle and ultimately to ATP production by the mitochondria. Oxidation of pyruvate results in the depletion of body glucose carbon sources because glucose cannot be synthesized from acetyl-CoA and the PDH reaction is physiologically irreversible (Harris et al. 2002; Holness and Sugden 2003; Randle et al. 1984; Randle 1998; Tsai et al. 1973). Thus, the activity of the PDH complex determines whether pyruvate is oxidized to CO_2 and H_2O or converted to lactate via lactate dehydrogenase or to alanine via alanine transaminase.

Flux through the pyruvate dehydrogenase complex is tightly controlled through several mechanisms (Figure 5.6). The activity of the PDH complex is controlled both by end-product inhibition and by reversible phosphorylation (Harris et al. 2002; Holness and Sugden 2003; Randle et al. 1984; Randle 1998; Tsai et al. 1973). *In vivo*, the cycle of phosphorylation and dephosphorylation is accepted as the major form of regulation of the PDH complex, and the predominant mechanism controlling glucose oxidation in other conditions such as starvation or diabetes (Harris et al. 2002; Holness and Sugden 2003; Randle et al. 1984; Randle 1998; Tsai et al. 1973). The PDH complex is inactivated by phosphorylation, catalyzed by a PDH kinase (PDK), while a PDH phosphatase dephosphorylates the PDH complex and reactivates the enzyme complex (Harris et al. 2002; Holness and Sugden 2003; Randle et al. 1984; Randle 1998; Tsai et al. 1973). The PDH kinase and the PDH phosphatase are both active, and the relative activities of each of these two competing enzymes determines the proportion of PDH in the active, dephosphorylated form (PDH_a). Decreased flux through the

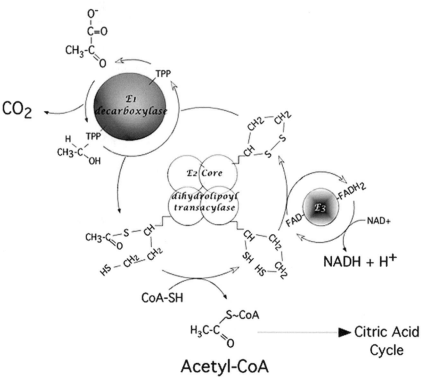

FIGURE 5.5. Individual reactions catalyzed by subunits of pyruvate dehydrogenase complex.

PDH complex results from an increased proportion of inactive (phosphorylated) PDH complex (no change in total PDH activity) (Berger et al. 1976; Randle et al. 1984; Randle et al. 1988; Sugden et al. 1989). Inactivation of PDH complex limits oxidation of pyruvate derived from glycolysis and, hence, promotes increased lactate and alanine production.

Accelerated oxidation of fatty acids raises the mitochondrial [acetyl-CoA]/[CoA] and [NADH]/[NAD$^+$] concentration ratios, thereby stimulating the PDH kinase reaction (Hutson and Randle 1978; Kerbey et al. 1976; Vary et al. 1986). In other conditions, the PDH$_a$ may be increased *in vitro* by decreasing the fatty acid concentration in fed animals (Randle et al. 1988; Sugden et al. 1989), or in hearts by inhibiting fatty acid oxidation with 2-tetradecylglycidate (Caterson et al. 1982) or ranolazine (Clark et al. 1993; Clark et al. 1996). This mechanism represents a rapid onset/offset mechanism for short-term adjustments to the availability of fatty acids (Randle et al. 1988; Sugden et al. 1989).

In starvation and diabetes, the short-term effects of oxidation of fatty acids to decrease the PDH$_a$ are supplemented by a more long-term mechanism involving

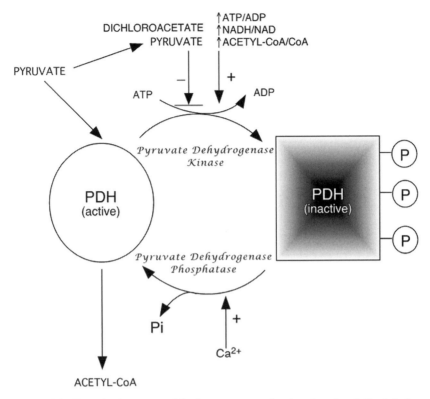

FIGURE 5.6. Control of pyruvate dehydrogenase complex by phosphorylation/ dephosphorylation mechanism.

a stable increase in the activity of the PDH kinase (PDK). The activation of the PDK is observed under *in vitro* conditions where fatty acid oxidation does not occur. PDK activity is enhanced under these conditions by two mechanisms during starvation and diabetes. First, the specific activity of PDH kinase intrinsic to the complex is increased (Kerbey et al. 1979; Kerbey et al. 1984). Second, the abundance of the kinase is also increased (Jones and Yeaman 1991; Kerbey and Randle 1982).

Indeed, PDK exists as genetically distinct isozymes (PDK1, PDK2, PDK3 and PDK4) (Bowker-Kinley et al., 1998; Gudi et al., 1995; Harris et al., 1995; Popov et al., 1993; Popov et al., 1994; Rowles et al., 1996). All members of this family contain conserved motifs that presumably form the kinase domain. The expression of these PDK isoenzymes occurs in a tissue-specific manner (Bowker-Kinley et al., 1998; Wu et al., 1988). The mRNA for PDK1 is localized almost exclusively in heart and pancreatic islets. PDK3 is limited to kidney, brain and testis. The mRNA for PDK2 was present in all tissues tested but its abundance was low in spleen and lungs. The mRNA for PDK4 was predominantly expressed in skeletal muscle and heart. The specific activities of the isoenzymes varies 25-fold from 50 nmol/min/mg for PDK2 to 1250 nmol/min/mg for PDK3

(PDK1 = 650 nmol/min/mg; PDK4 = 400 nmol/min/mg). The isoenzymes also vary in the ability of dichloroacetate (DCA) to inhibit their activities (PDK2 > PDK4 > PDK1 > PDK3; from most sensitive to DCA to least sensitive to DCA), in the ability to be activated by NADH (PDK4 > PDK1 = PDK2 > PDK3), and in their ability to be activated by acetyl-CoA (PDK2 > PDK1 > PDK4 = PDK3). In vivo, overexpression of PDK2 inactivates PDC in cardiac muscle (Vary et al unpublished data).

PDK4 is upregulated in most tissues during starvation, hormonal imbalances (diabetes, hyperthyroidism) and insulin resistance (Harris et al. 2002; Holness et al. 2002b; Holness and Sugden 2003; Randle 1998; Wu et al. 1999), but not aging (Moreau et al. 2004). The factors that appear to regulate PDK4 include fatty acid oxidation and adequate insulin signaling. The signaling pathways responsible for the changes in PDK4 expression remain unresolved. However, PDK4 expression is either directly or indirectly regulated by peroxisome proliferators-activated receptor alpha (PPAR-α). PPAR-α deficiencies limit the upregulation of PDK4 in liver and kidney during starvation. Administration of WY-14,643, a selective agonist for the peroxisome proliferator-activated receptor-alpha (PPAR-α), induced large increases in pyruvate dehydrogenase kinase activity, PDK4 protein, and PDK4 mRNA in gastrocnemius muscle from wild–type but not in PPAR-α-null mice (Wu et al. 1999; Wu et al. 2001). Because long-chain fatty acids activate PPAR-α endogenously, increased levels of these compounds in starvation and diabetes may signal increased expression of PDK4 in skeletal muscle.

However, there is some controversy regarding the role of PPAR-alpha in regulating PDK4 expression in skeletal muscle and heart (Holness et al. 2002a; Holness et al. 2002b; Spriet et al. 2004). Whereas, activation of PPAR-alpha in gastrocnemius, a muscle composed primarily of fast, glycolytic fibers, increases PDK4 expression, the same effect is not observed in oxidative skeletal muscles. Likewise, skeletal muscle PDK-4 mRNA was progressively augmented over the course of a 40 h fast in humans, whereas PPAR-alpha protein and forkhead homolog in rhabdomyosarcoma (FKHR) mRNA abundance were unaffected by the fast (Spriet et al. 2004). The changes in PDK-4 expression and PDH activity did not coincide with increases in the transcriptional activators PPAR-alpha and FKHR. PDK4 protein expression in oxidative skeletal muscle and cardiac muscle appears regulated by a fatty acid-dependent mechanism that is not obligatorily dependent on signaling via PPAR-alpha, but may involve declines in plasma insulin concentrations. In this regard, PDK activity and the abundance of the PDK isoform 4 protein and mRNA become elevated prior to overt diabetes, but concomitant with rises in plasma fatty acids in animals that become diabetic spontaneously (Bajotto et al. 2004). Indeed, effects of insulin on PDK activity have shown that there are two phases to the insulin effect on PDC complex (Feldhoff et al. 1993). Acute activation of the PDH complex does not appear to result from lower PDK activity but rather correlates with an insulin-mediated lowering of plasma free fatty acids. However, with daily insulin therapy in diabetes, activation of PDH results from a decreased PDH kinase activity. In cardiac muscle, PPAR-alpha activation lowers glucose oxidation mainly by decreasing the flux of pyruvate through PDC due to negative feedback inhibition of PDC by fatty acid oxidation reaction products rather than by the phosphorylation of the PDC complex.

5.3.1. PDC in Regulation of Gluconeogenesis

Maintenance of plasma glucose concentrations is achieved through glycogenolysis and by gluconeogenesis. Gluconeogenesis is the metabolic process by which glucose is formed from non-carbohydrate precursors. The process converts three carbon fragments derived primarily from lactate, glycerol and amino acids. The carbon source for gluconeogenesis arises from several sources including lactate, alanine, cysteine, and serine; substances released from peripheral organs in response to inhibition of PDC and accelerated proteolysis (Figure 5.3) (Vary 1999).

The PDC lies at the crossroads of utilization of pyruvate for energy production and gluconeogenesis. Inactivation of PDC secondary to increased PDK activity promotes gluconeogenesis in starvation or diabetes by limiting oxidation of pyruvate thereby conserving three-carbon substrates (Vary 1999). This helps maintain glucose levels during starvation, but is detrimental in diabetes by inducing sustained hyperglycemia.

The hallmark responsible for the transition of the liver and kidney to gluconeogenesis is lowering of plasma insulin concentrations or development of insulin resistance resulting in a sustained elevation in plasma lipids. Lipids act to lower PDC activity through increased acetyl-CoA/CoA ratios and expression of PDK4. The lipid-activated transcription factor, peroxisome proliferator-activated receptor-alpha (PPAR-α), plays a pivotal role in the cellular metabolic response to fatty acids. In liver, decreased concentrations of insulin and increased concentrations of fatty acids and glucocorticoids promote PDK4 gene expression in starvation and diabetes through PPAR-α (Huang et al. 2002). The decreased level of insulin is likely responsible for the increase in PDK2 mRNA levels in starvation and diabetes. Renal protein expression of PDK4 was only marginally induced by fasting in PPAR alpha null mice (Sugden et al. 2001).

Specific up-regulation of PDK4 protein expression in starvation, by maintaining PDC activity relatively low, facilitates pyruvate carboxylation to oxaloacetate and therefore entry of acetyl-CoA derived from fatty acid beta-oxidation into the TCA cycle, allowing adequate ATP production for brisk rates of gluconeogenesis. Indeed, activation of PDH by dichloroacetate (DCA) decreases peripheral release of alanine and lactate, thereby interrupting the Cori and alanine cycles and reducing the availability of three-carbon precursors for gluconeogenesis. DCA inhibits hepatic triglyceride and cholesterol biosynthesis. Administration of DCA in patients with non-insulin-dependent diabetes markedly reduces circulating very-low-density lipoprotein cholesterol and triglyceride concentrations (Stacpoole and Greene 1992). Unfortunately, DCA is metabolized to glyoxylate, which is converted to oxalate and, in the presence of adequate thiamine levels, to other metabolites which are toxic to humans (Stacpoole et al.

1990). Hence, chronic use of DCA for diabetes mellitus or hyperlipoproteinemias is limited by its neurologic and other forms of toxicity.

5.3.2. PDH Phosphatase

Activation of the PDH complex occurs through dephosphorylation catalyzed by two genetically different isozymes of pyruvate dehydrogenase phosphatase, PDP1c and PDP2c. PDP1c and PDP2c display marked biochemical and cellular distribution differences (Huang et al. 2003; Karpova et al. 2003). PDP1c is highly expressed in rat heart, brain, and testis and is detectable but less abundant in rat muscle, lung, kidney, liver, and spleen, whereas PDP2 is abundant in rat kidney, liver, heart, and brain and is detectable in spleen and lung (Huang et al. 2003). The catalytic subunit of pyruvate dehydrogenase phosphatase 1 (PDP1c) is a magnesium-dependent protein phosphatase heterodimer that regulates the activity of mammalian pyruvate dehydrogenase complex. The activity of PDP1c strongly depends upon the simultaneous presence of calcium ions and the E2 component of PDC. The stimulatory effect of E2 on PDP1c can be partially mimicked by a monomeric construct consisting of the inner lipoyl-bearing domain and the E1-binding domain of E2 component (Chen et al. 1996). This strongly suggests that the E2-mediated activation of PDP1c largely reflects the effects of co-localization and mutual orientation of PDP1c and E1 component facilitated by their binding to E2. Ca^{2+} appears to aid in the association of PDP1c with E2 bound, phosphorylated PDH (Chen et al. 1996). Site directed mutagenesis reveals that residues Asp54, and Asp347 contribute to the binuclear metal-binding center, and Asn49 contributes to the phosphate-binding sites (Karpova et al. 2004). The regulatory subunit of PDP (PDPr) lowers the sensitivity of the catalytic subunit of PDP1c to Mg^{2+} (Yan et al. 1996).

In contrast, the activity of PDP2c displays little, if any, dependence upon either calcium ions or E2 (Karpova et al. 2003). Instead, PDP2c is sensitive to the biological polyamine spermine, which, in turn, has no effect on the enzymatic activity of PDP1 (Huang et al. 1998). Like PDP1, PDP2 is a Mg^{2+}-dependent enzyme, but its sensitivity to Mg^{2+} ions is almost 10-fold lower than that of PDP1. Spermine increases the sensitivity of PDP2c to Mg^{2+}, apparently by interacting with PDPr (Yan et al. 1996). Furthermore, PDP2c does not appreciably bind to PDC under the conditions when PDP1c exists predominantly in the PDC-bound state. Both PDP1c and PDP2c can efficiently dephosphorylate all three phosphorylation sites located on the alpha chain of the E1 component. For PDC phosphorylated at a single site, the relative rates of dephosphorylation of individual sites are: site 2 > site 3 > site 1. Phosphorylation of sites 2 or 3 in addition to site 1 does not have a significant impact on the rates of dephosphorylation of individual sites by PDP1c, suggesting a random mechanism of dephosphorylation. In contrast, a significant decrease in the overall rate of dephosphorylation of pyruvate dehydrogenase by PDP2c occurs under these conditions, suggesting that the mechanism of dephosphorylation of PDC phosphorylated at multiple sites by PDP2c is not purely random. Starvation and diabetes lower PDP2 mRNA abundance, PDP2 protein content, and PDP activity in rat heart and kidney. Refeeding and insulin treatment effectively reversed these effects of starvation and diabetes, respectively on PCP (Huang et al. 2003). These marked differences in the site-specificity displayed by PDP1c and PDP2c may be important under conditions (e.g. starvation and diabetes) associated with increases in phosphorylation of sites 2 and 3 of pyruvate dehydrogenase and provide a plausible mechanism for the observed changes in phosphorylation of different sites of the PDH complex upon reactivation (Randle et al. 1984).

5.3.3. Role of Pyruvate Dehydrogenase Complex in Sepsis-Induced Hypermetabolism

Severe trauma and sepsis initiate a pattern of physiologic and metabolic adaptations characterized by an increase in oxygen consumption and alterations in carbohydrate, fat, and protein metabolism. Virtually every organ system of the body becomes affected displaying recognizable metabolic adjustments. The host's metabolic response to sepsis is associated with the rapid breakdown of the body's reserves of protein and carbohydrates, and an increased dependence upon fatty acids. Sepsis shifts interorgan substrate fluxes causing an exaggerated response in the normal physiologic mechanisms responsible for maintenance of whole-body metabolic homeostasis. The degree of metabolic dysfunction is ultimately indicative of the extent of organ dysfunction induced by the invading organism or traumatic injury. Two of the earliest and most recognizable manifestations of organ dysfunction in sepsis are a massive excretion of urea resulting in a large negative nitrogen balance and alterations in glucose kinetics resulting in increased rates of glucose appearance (150–200%) and metabolic clearance. As such, sepsis represents the most common causes of lactic acidosis in the presence (59%) or absence (49%) of shock (Stacpoole et al. 1994).

Sepsis increases the rate of glucose appearance by stimulating hepatic glucose production (Keller et al. 1982; Lang et al. 1987; Vary et al. 1988b). Of the gluconeogenic precursors, lactate accounts for 60–70% of the glucose carbon used for gluconeogenesis. Net lactate extraction by the liver is increased 2–3 fold following severe infection (Dahn et al. 1987; McGuinness et al. 1995; Wilmore et al. 1980). An increased rate of glucose production indicates that hepatic lactate utilization is accelerated in sepsis. Thus, a constant supply of gluconeogenic precursors must be maintained for the enhanced and persistent rates of gluconeogenesis during sepsis.

The release of lactate and alanine from skeletal muscle is increased, indicating both the Cori cycle and Felig cycle are providing adequate precursors for gluconeogenesis in sepsis. Skeletal muscle also represents a major source for the production of gluconeogenic substrates. In contrast to protein catabolism, altering substrate delivery to the liver by modulating skeletal muscle glucose metabolism by treatment with dichloroacetate to activate PDC short-circuited the characteristic derangements in glucose kinetics during sepsis. Thus, the accelerated glucose appearance and glucose clearance appears dependent upon increased lactate and alanine production by skeletal muscle during sepsis. Elevated plasma lactate concentrations are prognostic indicators of illness severity and/or death in septic patients (Cerra 1989; Clark et al. 1996; Siegel et al. 1990; Vary et al. 1988a).

Skeletal muscle, by virtue of its mass, represents the major source of gluconeogenic substrates. Sepsis dramatically enhances the peripheral utilization of glucose, such that the glucose metabolic clearance rate is increased over 2-fold compared with non-septic animals (Lang et al. 1990; Vary et al. 1988b). Uptake of glucose by skeletal muscle is enhanced by systemic infection. Stimulation of glucose uptake is not associated with increased deposition of glycogen (Vary et al. 1995) but instead there is an acceleration of glycolysis with the subsequent release lactate and alanine from skeletal muscle (Gump et al. 1975; Shaw et al. 1985; Vary et al. 1988b; Vary and Murphy 1989). The increased lactate production by skeletal muscle provides the necessary gluconeogenic precursors for maintenance of sustained rates of gluconeogenesis in sepsis. Furthermore, lactate production most likely exceeds its utilization, thereby accounting for elevated plasma lactate concentrations in sepsis. A strong correlation is obtained between arterial lactate concentration and the rate of glucose appearance or rate of recycling of glucose carbon (Lang et al. 1987) implying that circulating lactate concentrations are an important determinant of gluconeogenesis.

Lactate arises from the reduction of pyruvate. Lactate production occurs whenever the rate of pyruvate production from glycolysis or other sources [malic enzyme or pyruvate carboxylase] exceeds glucose oxidation by the mitochondria. Therefore, glucose oxidation should decrease resulting in increased gluconeogenesis, with a higher rate of diversion of pyruvate to lactate being important physiologically in the regulation of plasma lactate concentrations. In septic patients, whole-body glucose oxidation is inhibited and glucose turnover and recycling is increased (Jahoor et al. 1989; Shaw et al. 1985). Accelerated lactate release by the forearm of post-surgical patients occurs at a time when glucose oxidation is depressed, indicating such a scenario occurs in vivo during sepsis (Brandi et al. 1993).

Mitochondria isolated from skeletal muscle of septic rats have a defect in their ability to oxidize pyruvate, but not other substrates, including fatty acids (Mela-Riker et al., 1992). This implies that there is specific sepsis-induced inhibition of the PDC. The inhibition of PDH complexes is a consequence of an increased acetyl-CoA/CoA ratio, secondary to augmented fatty acid oxidation (Vary et al. 1986). The importance of inhibiting the PDH complex in controlling plasma lactate concentrations is indicated by two observations. First, a negative linear correlation exists between plasma lactate concentrations and skeletal muscle PDHC in the active dephosphorylated form (Vary 1996). Second, activation of the PDH complex with administration of dichloroacetate (DCA), an noncompetitive inhibitor of the PDH kinase, normalizes muscle and plasma lactate and alanine concentrations and inhibits skeletal muscle lactate and alanine production during sepsis (Lang et al. 1987; Vary et al. 1988b; Vary et al. 1989a; Vary et al. 1989b; Vary 1996).

Does the ability of DCA to activate the PDH complex, thereby enhancing pyruvate oxidation and subsequently lowering skeletal muscle lactate and alanine concentrations and production, modulate glucose kinetics during sepsis? After injection of DCA into septic rats, the elevated rates of glucose appearance (Lang et al. 1987) and turnover were reduced (Vary et al. 1988b). Furthermore, the sepsis-increased glucose metabolic clearance rate is normalized (Vary et al. 1988b). Hence, activation of the PDH complex successfully reversed or attenuated the sepsis-induced increases in glucose rate of appearance, glucose carbon recycling, and glucose metabolic rate. These observations indicate enhanced rates of gluconeogenesis are dependent in part upon the sustained delivery of precursors from skeletal muscle from muscle to liver. Thus, stimulated glucose uptake and glycolysis coupled with an inhibition of PDH complex in skeletal muscle and subsequent release of lactate and alanine drives the accelerated rates of gluconeogenesis during sepsis.

5.3.4. Potential Beneficial Therapeutic Effects of Pyruvate

The inhibition of the PDC and the subsequent hyperlactatemia and hyperalaninemia under a variety of conditions has prompted the notion of using inhibitors of the PDHK to modify metabolic derangements with the assumption that this would lead to an increased beneficial effect. Indeed, dichloroacetate has been touted as a useful therapeutic modality in treating hyperlactatemia (Vary et al. 1988a) and in animal models of malaria prolongs survival (Holloway et al. 1995). Activation of PDC improves heart function (Kline et al. 1997) and survival following hemorrhagic shock (Granot and Steiner 1985). Unfortunately, a multicenter double-bind randomized trial failed to show improvement in hemodynamics or survival following activation of PDC despite lowering plasma lactate concentrations and increasing plasma arterial pH (Stacpoole et al. 1992). One potential reason for such a result was that both hemodynamic and non-hemodynamic (metabolic) underlying causes were present in the patient population many of which independently predicted survival and most of who were refractory to standard care.

Alternatively, elevating the cellular content of pyruvate also inhibits the PDHK. In addition, pyruvate may act as an endogenous antioxidant and free radical scavenger. Recognition that pyruvate acts both as an effective activator of PDC and ROS scavenger has led investigators to use this compound as a potential therapeutic agent for the treatment of various pathologic conditions (Liedtke et al. 1976; Mochizuki and Neely 1980; Mongan et al. 1999; Slovin et al. 2001). The usefulness of pyruvate as a therapeutic agent is abrogated by its very poor stability in solution. Aqueous solutions of pyruvate rapidly undergo an aldol-like condensation reaction to form 2-hydroxy-2-methyl-4-ketoglutarate (parapyruvate), a compound that is a potent inhibitor of a critical step in the mitochondrial tricarboxylic acid cycle (Montgomery and Webb 1956).

Ethyl pyruvate, a stable derivative of pyruvic acid in solution, circumvents the problems relating to stability of pyruvate in solution (Sims et al. 2001).

Original reports in the literature indicated that ethyl pyruvate could attenuate deleterious effects of chemically-induced oxygen free radical damage in the lens (Varma et al. 1998). Subsequently, treatment with ethyl pyruvate under a variety of conditions ameliorates much of the structural and functional damage caused by the insult. Treatment with ethyl pyruvate solution could improve survival in rodent models of hemorrhagic shock or myocardial ischemia (Tawadrous et al. 2002; Woo et al. 2004), acute endotoxemia and bacterial peritonitis (Fink 2003a; Ulloa et al. 2002), and sepsis-induced renal failure (Miyaji et al. 2003). The protective actions may be related to metabolic (increased ATP production from pyruvate) and non-metabolic (anti-inflammatory) mechanisms (Fink 2003b) that down-regulate a number of proinflammatory genes. The anti-inflammatory effects of ethyl pyruvate may be mediated through inhibition of NF-kappaB signaling to p65 by binding ethyl pyruvate to p65 at Cys(38) (Han et al. 2005).

5.3.5. Role of Pyruvate Dehydrogenase Complex in Prostate Cancer

Prostate cancer is the most common cancer detected in American men and the second leading cause of cancer-related deaths in the United States. Most prostatic cancers are detected in asymptomatic men who are found to have focal nodules or areas of induration within the prostate at the time of digital rectal exam or more recently through screening for the prostate-specific antigen in the blood or transrectal ultrasound. The PSA that alone is not specific for cancer but is more commonly associated with benign prostate hyperplagia which occurs frequently in the same age groups as prostate cancer patients.

The prostate gland secretes a thin, milky fluid that contains calcium, citrate, phosphate, a clotting enzyme, and a profibrinolysin and eventually becomes incorporated, with secretions from other tissues, into semen. The prostate gland has the unique function of accumulating and secreting extraordinarily high levels of citrate. The prostate secretory epithelial cells synthesize citrate which, due to a limiting mitochondrial (m-) aconitase, accumulates rather than being oxidized. Thus citrate becomes essentially an end product of metabolism in prostate. For continued net citrate production, a continual source of oxaloacetate (OAA) and acetyl CoA is required. Glucose via pyruvate oxidation provides the source of acetyl-CoA.

Citrate production is regulated by testosterone and/or by prolactin (Costello et al. 2000). Both hormones selectively regulate the level and activity of pyruvate dehydrogenase E1 alpha (E1a), thereby regulating the availability of acetyl CoA for citrate synthesis. This process is under regulatory control by testosterone and prolactin, which serve to modulate the expression of the genes associated with these mitochondrial enzymes. The regulation of these genes by testosterone and prolactin is specific for prostate cells, and appears to be mediated through signaling pathways involving protein kinase C. Of interest to the pathogenesis of prostate cancer, the ability of the prostate to secrete citrate is retarded in cancer cells whereas increased citrate production is characteristic of benign prostate hyperplagia. These disease entities therefore represent a unique regulation of mitochondrial function in that there is regulation of the gene expression of key regulatory enzymes that are under hormonal control.

5.4. Branched-Chain Amino Acid Dehydrogenase

Branched chain amino acids (BCAA) (leucine, valine, isoleucine) are essential amino acids whose major function in the body may not reside with energy production but rather as a nutrient signal regulating protein metabolism. Of the branched-chain amino acids, leucine is well suited as a nutrient signal for several reasons. First, leucine is the most potent of the branched chained amino acids with regard to accelerating protein synthesis. Second, it cannot be synthesized de novo by mammals. Thus, changes in the plasma leucine concentrations must arise from nutrition, tissue breakdown or changes in the rate of oxidation or loss. Third, it is the most abundant essential amino acid in dietary protein and thus is a crucial nutrient following ingestion of a protein meal. Fourth, once taken up by the gut, leucine cannot be directly metabolized by the liver, which lacks the branched-chain amino transferase (BCATm), the mitochondrial enzyme responsible for removal of the nitrogen. This is in contrast to the liver's capacity to initiate the catabolism of other amino acids. In contrast to liver, most other peripheral tissues have excess capacity (BCATm) to catalyze this first step. This may facilitate postprandial rises in plasma leucine concentrations as leucine would not be cleared in first pass by the liver.

BCAA catabolism entails two enzymatic steps common to all BCAAs followed by subsequent steps that are dependent upon BCAA being utilized. The first step involves reversible transamination of the BCAA with α-ketoglutarate to form branched chain alpha-keto acids (BCKAs) and glutamate (Figure 5.1). The glutamate formed is then transaminated with pyruvate to form alanine and regenerate α-ketoglutarate or the glutamate is converted to glutamine. Following transamination, BCKAs undergo irreversible oxidative decarboxylation. The BCAA catabolic enzymes are distributed widely in body tissues and, with the exception of the nervous system, all reactions occur in the mitochondria of the cell. Transamination provides a mechanism for dispersing BCAA nitrogen according to the tissue's requirements for glutamate and other dispensable amino acids. The intracellular compartmentalization of the branched-chain aminotransferase isozymes (mitochondrial branched-chain aminotransferase, cytosolic branched-chain aminotransferase) impacts on intra- and interorgan exchange of BCAA metabolites, nitrogen cycling, and net nitrogen transfer.

The branched-chain ketoacid dehydrogenase (BCKDC) multienzyme complex catalyzes the rate-limiting step and is the first irreversible step in leucine oxidation. As such, the BCKDC is the most important regulatory enzyme in the catabolic pathways of the BCAAs. The reaction converts branched-chain α-keto acids [α-ketoisocaproate acid (KIC)] to CoA derivatives (such as isovaleryl CoA). The BCKDC is a multi-enzyme complex. It contains a branched-chain keto acid decarboxylase (E1, which has an $\alpha 2\beta 2$ structure with a covalently bound thiamine pyrophosphate cofactor), dihydrolipoyl transacylase (E2, which has a covalently bound lipoic acid), and the flavin-linked dihydrolipoyl dehydrogenase (E3) (for review see Ref. (Harper et al. 1985)).

Activity of the complex is controlled by covalent modification with phosphorylation of its branched-chain alpha-ketoacid dehydrogenase subunits by a specific kinase [branched-chain kinase (BDK)] causing inactivation and dephosphorylation by a specific phosphatase [branched-chain phosphatase (BDP)] causing activation. Tight control of BCKDC activity is important for conserving as well as disposing of BCAAs. Phosphorylation of the complex occurs when there is a need to conserve BCAAs for protein synthesis; dephosphorylation occurs when BCAAs are present in excess. The relative activities of BDK and BDP set the activity state of BCKDC. BDK activity is regulated by alpha-ketoisocaproate inhibition and altered level of expression.

The α-keto acid of leucine, KIC, is thought to inhibit the BCKD kinase, promoting activation of BCKD complex. BCKD kinase phosphorylates the E1α subunit of BCKD at two sites, termed sites 1 and 2 (Harris et al. 1986; Paxton et al. 1986). Phosphorylation at site 1 (Ser^{293}) inhibits BCKD activity, whereas phosphorylation at site 2 is silent (Zhao et al. 1994). The BCKDH kinase is responsible for inactivation of the complex by phosphorylation, and the activity of the kinase is inversely correlated with the activity state of the BCKDH complex, which suggests that the kinase is the primary regulator of the complex. Ligands for peroxisome proliferator-activated receptor-alpha (PPAR-α) in rats caused activation of the hepatic BCKDH complex in association with a decrease in the kinase activity, which suggests that higher rates of fatty acid oxidation upregulates the BCAA catabolism (Shimomura et al. 2004b). Less is known about BDP but a novel mitochondrial phosphatase was identified recently that may contribute to the regulation of BCKDC.

Whereas BCAA transamination occurs primarily in extrahepatic tissues, liver is thought to be the major site of BCAA oxidation. It has been shown that kinase activity and expression are affected by nutritional states imposed by low-protein diet feeding, starvation, and diabetes. Evidence has also been presented that certain hormones, particularly insulin, glucocorticoid, thyroid hormone and female sex hormones, affect the activity and expression of the kinase. The findings indicate that nutritional and hormonal control of the activity and expression of branched-chain alpha-keto acid dehydrogenase kinase provides an important means of control of the activity of the branched-chain alpha-keto acid dehydrogenase complex, with inactivation serving to conserve branched-chain amino acids for protein synthesis in some situations and activation serving to provide carbon for gluconeogenesis in others (Shimomura et al. 2001). In rats fed a standard laboratory diet ($\sim$20% protein), which is more than sufficient to meet the animal's protein requirements, the liver enzyme is essentially active (Block et al. 1990; Gillim et al. 1983; Miller et al. 1988). Only low-protein diets had a significant impact on activity state. Dietary protein restriction results in increased expression of BCKD kinase (Popov et al. 1995), and high levels of kinase were found with BCKD, leading to increased steady-state levels of phosphorylated

E1α and conservation of essential BCAA. Once dietary protein exceeded the requirement, little kinase was found with BCKD in the PEG pellet. Decreasing dietary protein lowers BCKD kinase mRNA and activity (Popov et al. 1995). Lower concentrations of E2 and E1 were also observed in the low-protein diet in addition to the changes in BCKD kinase (Lynch et al. 2003a and b). Hence, dietary intake of leucine can regulate its metabolism in the mitochondria.

5.4.1. Maple Syrup Urine Disease (Branched Chain Ketoaciduria)

Maple syrup urine disease (MSUD) is a disorder caused by a deficiency of the branched-chain alpha-ketoacid dehydrogenase complex (BCKD). The defect leads to an accumulation of both the branched chain amino acids and branched chain α-keto acids. Indeed, the molar ratio of leucine to alanine is elevated 10- to 20-fold in this disease. The disease gets its name from the maple syrup odor of the urine which is caused by the excretion of excessive amounts of α-keto-3-methylvalarate in the urine of affected patients, producing an odor similar to that of maple syrup. MSUD is characterized by psychomotor retardation, feeding problems, and a maple syrup odor of the urine initially diagnosed in newborns. Mutations in the genes encoding BCKD complex components E1α, E1β, E2 and E3 mapped to human chromosomes 19q13.1-q13.2, 6p22-p21, 1p31, and 7131-q32, respectively are the causative agents. The mode of inheritance is autosomal recessive with an incidence of MSUD of approximately 1 in 180,000 births in the United States. The incidence increases to as high as 1 in 200 births in populations that have significant inbreeding communities, such as the Mennonites living in Pennsylvania.

There are five distinct phenotypes of MSUD: classical, intermittent, intermediate, thiamine-responsive, and E3-deficient. Classical MSUD typically presents in newborns and represents the most common form of the disease. Newborns typically develop ketonuria within 48 hours of birth and present with irritability, poor feeding, vomiting, lethary and dystonia. By four days of age, neurologic abnormalities include alternating lethary and irritability, dystonia, apnea, seizures, and indications of cerebral edema. The initial symptoms may not develop until the infant is four to seven days of age, depending upon the feeding regimen and breastfeeding may delay onset of symptoms to the second week.

Dietary therapy to promote normal growth and development and prompt treatment of episodes of acute metabolic decompensation constitute the cornerstone of management of these patients. Despite careful dietary management, episodes of metabolic intoxication may occur in affected older infants or children by increased net endogenous protein catabolism induced by intercurrent illness, exercise, injury, surgery, cancer or fasting. Clinical manifestations include epigastric pain, vomiting, anorexia and muscle fatigue. Neurologic signs may include hyperactivity, sleep disturbance, stupor, decreased cognitive function, dystonia and ataxia. The best outcomes occur in newborns, where therapy is begun before they become symptomatic or are treated rapidly after symptoms develop. Acute metabolic decompensation can result in brain injury and requires prompt treatment to avoid neurologic sequelae. Even with aggressive treatment MSUD can be fatal in the newborn period or during decompensation.

The intermediate, intermittent and thiamine-responsive forms of MSUD may present at any time during infancy or childhood and appear to be brought on by episodes of stress. Intermediate MSUD is a rare disorder caused by mutations in the E1α component of BCKD but in contrast to the classical MSUD, residual BCKD activity typically is 3 to 30% of normal. Patients may become symptomatic at any age, when the flux through BCATm exceeds flux through BCKD. The intermittent form of the disease is the second most common type of MSUD. Affected patients have normal growth and development. They typically present with ketoacidosis during episodes of catabolic stress in childhood, including intercurrent illnesses such as otitis media. Signs of neurotoxicity develop, including ataxia, lethargy, seizures and coma. Death may occur without appropriate recognition and treatment. Thiamine-responsive MSUD is a rare phenotype that has been associated with mutations in the E2 component, the subunit that requires thiamine as a cofactor for complete enzymatic activity (Figure 5.7).

Clinical presentation is similar to the intermediate form. In general, symptoms in affected patients are not completely resolved with thiamine supplementation alone, necessitating dietary restriction of branched-chain amino acids to achieve metabolic control. Mutations in the gene encoding the E3 component of BCKD (Figure 5.7) result in E3-deficient MSUD. Because the E3 component of BCKD is also present in the PDH complex (Figure 5.5) and α-ketoglutarate dehydrogenase, affected patients show symptoms of combined deficiencies of BCKD and the pyruvate complexes including lactic acidosis superimposed on MSUD.

5.4.2. *Role of Mitochondria in Cell Signaling Pathways Activated by Leucine*

Leucine is a direct acting nutrient signal that regulates protein synthesis in a number of tissues. The effects on protein synthesis are brought about, at least in part, by activation of cell-signaling pathways. These pathways are regulated by adapter proteins, such as RAPTOR (Hara et al. 2002; Kim et al. 2002), and by phosphorylation/dephosphorylation cascades (e.g., mTOR to ribosomal protein S6 kinase-1 (S6K1) to S6 and mTOR to 4E-BP1 to eIF4E·eIF4G) that, in turn, modulate factors involved in initiation of protein synthesis (Figure 5.8).

In striated muscle, the mTOR signaling pathway appears to play a role in stimulation of protein synthesis by growth promoting anabolic hormones. Two well known substrates of mTOR, namely 70 kDa ribosomal protein S6 kinase (S6K1) and the translational repressor, eukaryotic initiation factor 4E binding protein 1 (4E-BP1), have been implicated in the regulation of protein synthesis. At least two signaling pathways appear to be important in leucine's acute regulation of protein synthesis, a rapamycin-sensitive pathway involving mTOR and an unknown rapamycin-insensitive pathway that may or may not involve

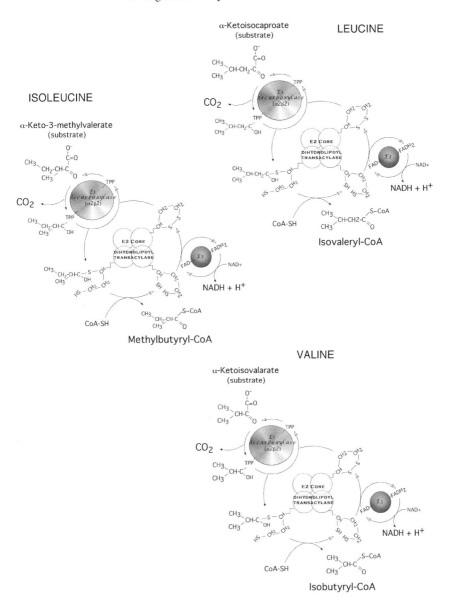

FIGURE 5.7. Individual reactions catalyzed by subunits of branched chain amino acid dehydrogenase complex.

mTOR. It is presently unclear exactly how leucine activates mTOR aside from the fact that the mechanism differs from that used by insulin. For example, no leucine receptor has been yet identified. In fact, it has not been unequivocally established whether effects of leucine are mediated by leucine or a metabolite derived

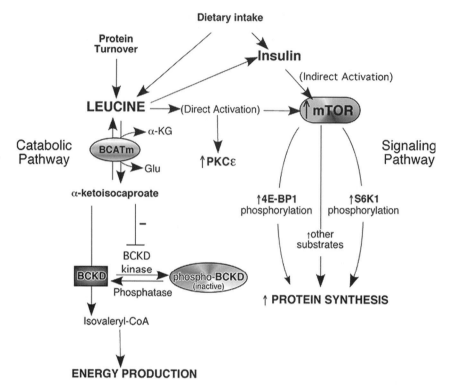

FIGURE 5.8. Dual role of leucine in contributing to energy metabolism and control of signaling pathways involved in gene expression. mTOR – mammalian target of rapamycin; Glu – glutamate; α-KG – α-ketoglutarate; PKCε – protein kinase C epsilon isoform; 4E-BP1 – eukaryotic initiation factor 4E binding protein 1; S6K1 – ribosomal protein S6 protein kinase 1; BCKD – branched chain amino acid dehydrogenase.

from leucine transamination such as α-ketoisocaproate (α-KIC) or leucine oxidation.

Leucine's effects on protein metabolism are self-limiting because leucine promotes its own disposal by an oxidative pathway, thereby terminating its positive effects on body protein accretion. A strong case can therefore be made that the proper leucine concentration in the various compartments of the body is critically important for maintaining body protein levels beyond simply the need of this essential amino acid for protein synthesis. In this regard, mitochondrial metabolism may be required for leucine activation of mTOR. Specifically, a metabolically-linked signal arising from activation of leucine metabolism in the mitochondria results in mTOR activation (McDaniel et al. 2002). A potential target cited was the branched chain ketoacid dehydrogenase (BCKD) complex. There are several compelling arguments in favor of the notion that BCKD kinase activities might provide a link between leucine signal and the activation of mTOR. The first is that α-ketoisocaproate acid is a physiological inhibitor of BCKD kinase. The second is that the order of potency of Leu > Ile > Val and importance of leucine in the activation of mTOR signaling is similar to that of their respective keto acids' ability to activate the BCKD complex via inhibition of the BCKD kinase. A third observation that supports this potential mechanism is that mTOR appears to be associated with mitochondria membrane *in situ*.

One way to delineate this effect is to simultaneously examine the dose dependent effects of orally administered leucine on acute activation of S6K1 (an mTOR substrate) and phosphorylation of BCKD (Lynch et al. 2003a). Increasing doses of leucine given orally via gavage directly correlated with elevations in plasma leucine concentration. Phosphorylation of S6K1 (Thr389, the phosphorylation site leading to activation) in adipose tissue was maximal at a dose of leucine that increased plasma leucine approximately threefold. Changes in BCKD phosphorylation state required higher plasma leucine concentrations. Hence there exists a disconnection between the concentration of leucine required for activation of S6K1 system and phosphorylation of BCKD. Only at concentrations significantly higher than those required for activation of S6K1 is the phosphorylation state of BCKD lowered, allowing for oxidation of leucine to occur. Hence as the leucine signal becomes too high, there is a stimulation of leucine oxidation in peripheral tissue thereby lowering the plasma leucine concentration. Thus, leucine metabolism may only be important in regulating the magnitude and duration of the signal for mTOR activation.

Leucine dependent stimulation of mRNA translation *in vivo* occurs in part, through a rapamycin-sensitive pathway (Anthony et al. 2000; Anthony et al. 2001). Rapamycin treatment, an inhibitor of the protein kinase mTOR, reduced the mTOR-dependent 4E-BP1 and S6K1 phosphorylation. However, leucine retained its ability to stimulate protein synthesis. Thus, inhibition of mTOR-mediated signaling was not capable of completely eliminating the leucine-dependent stimulation of protein synthesis. An additional study examining oral leucine administration in diabetic rats demonstrated enhanced levels of protein synthesis in skeletal muscle without corresponding changes in 4E-BP1 or S6K1 (Anthony et al. 2002). Similarly rates of skeletal muscle protein synthesis in perfused hindlimbs can be enhanced by elevated leucine concentrations in the absence of phosphorylation changes of 4E-BP1 and S6K1. Collectively, this evidence underscores the ability of leucine to modulate translation initiation and protein synthesis through an mTOR-independent mechanism(s).

5.4.3. *Branched Chain Amino Acid Metabolism in Liver Disease*

Hepatic encephalopathy is a neuropsychiatric syndrome occurring in patients with acute or chronic liver disease including cirrhosis. During the last 20 years there has been much interest in nutritional treatment for patients with advanced cirrhosis. Most studies have measured the potential benefit of nutritional supplements of dietary proteins, generic protein hydrolysates, or specific branched-chain amino acid (BCAA)-enriched formulas in regard to nutritional parameters and hepatic encephalopathy. Nutritional factors play a major role both in the pathogenesis as well as management of hepatic encephalopathy (HE). Physicians treating patients with chronic liver disease often restrict the intake of dietary protein to prevent a rise in blood ammonia levels. The role of protein restriction in patients with chronic hepatic encephalopathy (CHE) has been questioned recently as the efficacy of protein withdrawal in patients with CHE has never been subjected to a controlled trial. Evidence suggests that protein intake plays only a limited role in precipitating encephalopathy. In fact, measures taken to suppress endogenous protein breakdown are more effective than dietary restrictions in reducing the load of amino acids on the decompensated liver. A protein intake of less than 40 g per day, as has been indicated, contributes to a negative nitrogen balance, which along with increased endogenous protein catabolism, worsens encephalopathy. A positive nitrogen balance may have positive effects on encephalopathy. Rather, depressed plasma branched-chain amino acid (BCAA) levels, implicated in the pathogenesis of HE, also supervene in cirrhosis only when malnutrition is present as well. Therefore, the emphasis in the nutritional management of patients with CHE should not be on the reduction of protein intake (Srivastava et al. 2003). Eleven randomised trials (556 patients) assessing BCAA versus carbohydrates, neomycin/lactulose, or isonitrogenous control showed BCAA significantly increased the number of patients improving from hepatic encephalopathy at the end of treatment although there were methodological concerns (Als-Nielsen et al. 2003; Fabbri et al. 1996). Chemically-induced liver cirrhosis in rats raised the activity of hepatic BCKDH complex and decreased plasma BCAA and branched-chain α-keto acid concentrations, suggesting that the BCAA requirement is increased in liver cirrhosis. Because the effects of liver cirrhosis on the BCKDH complex in human liver are different from those in rat liver, further studies are needed to clarify the differences between rats and humans. In the valine catabolic pathway, crotonase and beta-hydroxyisobutyryl-CoA hydrolase are very important to regulate the toxic concentration of mitochondrial methacrylyl-CoA, which occurs in the middle part of valine pathway and highly reacts with free thiol compounds. Both enzyme activities in human and rat livers are very high compared to that of the BCKDH complex. It has been found that both enzyme activities in human livers were significantly reduced by liver cirrhosis and hepatocellular carcinoma, suggesting a decrease in the capability to dispose methacrylyl-CoA (Shimomura et al. 2004a). The findings described here suggest that alterations in hepatic enzyme activities in the BCAA catabolism are associated with liver failure.

Metabolic acidosis increases protein degradation resulting in muscle wasting and a negative nitrogen balance. The branched-chain amino acids serve as useful markers of these changes and their catabolism is increased in acidosis, particularly for the spontaneous acidosis associated with renal failure (Bailey 1998; Wang et al. 1997). As a result, the neutral nitrogen balance is compromised and malnutrition results. Glucocorticoids mediate these changes through the recently discovered ATP-dependent ubiquitin-proteasome pathway. Metabolic acidosis and glucocorticoids act in concert to stimulate branched-chain amino acid (BCAA) oxidation in adrenalectomized rats. In muscles of normal rats, metabolic acidosis increases the maximal activity of the rate-limiting enzyme, branched-chain alpha-ketoacid dehydrogenase (BCKAD) and a genetic response to catabolic conditions like uremia is implicated by concurrently higher levels of BCKAD subunit mRNA. Independent transactivation response elements to acidification or glucocorticoids were localized in the E2 promoter. In summary, catabolic responses to low extracellular pH and glucocorticoids include enhanced expression of genes encoding BCKAD subunits. (Price and Wang 1999).

5.5. Summary

The regulation of the PDH and BCDH was originally established nearly 30 years ago as a means to help explain the control of carbohydrate and amino acid metabolism during transitions from feeding to starvation and during conditions involved in altered carbohydrate metabolism. These types of studies also delineated the role of glutamine in amino nitrogen metabolism in the body. In recent years, there has been renewed interest in both the structure, function and interaction of the components of the PDH and BCDH to define the molecular interactions responsible for control mechanisms. More importantly, there has been a growing body of evidence suggesting that these amino acids can function as key effector molecules altering the signaling cascades responsible for regulatory several intracellular systems responsible for synthesis and expression of specific gene products. Future studies will have to integrate the mechanisms of metabolism of pyruvate, glutamine and leucine with their roles in the controlling gene expression.

Acknowledgements. This work was supported NIGMS Grant GM-39277 (TCV), NIAAA Grant AA12814 (TCV), NIDDK Grant DK053843 (CJL) and NIDDK Grant DK062880 (CJL) awarded by the National Institutes of Health.

References

Abcouwer SF, Bode BP, Souba WW (1995) Glucocorticoids regulate rat glutamine synthetase expression in a tissue specific manner. J Surg Res 59: 59–65
Abcouwer SF, Norman J, Fink G, Carter G, Lustig RJ, Souba WW (1996) Tissue-specific regulation of glutamine symthetase gene expression in acute pancreatitis is confirmed by using interleukin-1 receptor knockout mice. Surgery 120: 255–263
Abcouwer SF, Lohmann R, Bode BP, Lustig RJ, Souba WW (1997) Induction of glutamine synthetase expression after major burn injury is tissue specific and temporally variable. J Trauma 42: 421–427
Als-Nielsen B, Koretz RL, Kjaergard LL, Gluud C (2003) Branched-chain amino acids for hepatic encephalopathy. Cochrane Database Systematic Rev 2: CD001939
Anthony JC, Yoshizawa F, Anthony TG, Vary TC, Jefferson LS, Kimball SR (2000) Leucine stimulates translation initiation in skeletal muscle of postabsorptive rats via a rapamycin-sensitive pathway. J Nutr 130: 2413–2419

Anthony JC, Anthony TG, Kimball SR, Jefferson LS (2001) Signaling pathway involved in translation control of protein synthesis in skeletal muscle by leucine. J Nutr 131: 856S–860S

Anthony JC, Lang CH, Crozier SJ, Anthony TG, MacLean DA, Kimball SR, Jefferson LS (2002) Contribution of insulin to the translational control of protein synthesis in skeletal muscle by leucine. Am J Physiol Endo Metab 282: E1092–E1101

Ardawi MS, Majzoub MF (1991) Glutamine metabolism in skeletal muscle of septic rats. Metabolism 40: 155–164

Bailey JL (1998 .Metabolic acidosis and protein catabolism: mechanisms and clinical implications. 24: 13–19

Bajotto G, Murkami T, KNagasaki M, Tamura T, Tamura N, Harris RA, Shimomura Y, Sato Y (2004) Downregulation of the skeletal muscle pyruvate dehydrogenase complex in the Otsuka Long-Evans Tokushima Fatty rat both before and after the onset of diabetes mellitus. Life Sci 75: 2117–2130

Berger M, Hagg SA, Goodman MN, Ruderman NB (1976) Glucose metabolsim in perfused skeletal muscle. Biochem J 158: 191–202

Block KP, Aftring RP, Buse MG (1990) Regulation of rat liver branched-chain alpha-keto acid dehydrogenase activity by meal frequency and dietary protein. J Nutr 120: 793–799

Bobe BP, Fuchs BC, Hurley BP, Conroy HL, Suetterlin JE, Tanabe KK, Rhoads DB, Abouwer SF, Souba WW (2002) Molecular and functioanl analysis of glutamine uptake in huma hepatoma and liver-derived cells. Am J Gastrointest Liver Physiol 283: G1062–G1073

Bode BP, Kaminski DL, Souba WW, Li AP (1995) Glutamine transport in isolated hepatocytes and transformed liver cells. Hepatology 21: 511–520

Bowker-Kinley MM, Davies WI, Wu P, Harris RA, Popov KM (1998) Evidence for existence of tissue specific regulation of the mammalian pyruvate dehydrogenase complex. Biochem J 329: 191–196

Brandi LS, Santoro D, Natali A, Altomonte F, Baldi S, Frascerra S, Ferrannini E (1993) Insulin resistance of stress: sites and mechanisms. Clin Sci 85: 525–535

Caterson ID, Fuller SJ, Randle PJ (1982) Effect of fatty acid inhibitor 2-tetra-decyglycidic acid on pyruvate dehydrogenase complex activity in starved and alloxan-diabetic rats. Biochem J 208: 53–60

Cerra FB (1989) Hypermetabolism-organ failure syndrome: a metabolic response to injury. Crit Care Clin 5: 289–301

Chandrasekher S, Souba WW, Abcouver SF (1999) Indentification of glucocorticoid-responsive elements that control transcription of rat glutamine synthetase. Am J Lung Cell Mol Physiol 276: L319–L331

Chen G, Wang L, Liu S, Chuang C, Roche TE (1996) Activated function of the pyruvate dehydrogenase phosphatase through Ca2+-facilitated binding to the inner lipoyl domain of the dihydrolipoyl acetyltransferase. J Biol Chem 271: 28064–28070

Clark B, Spedding M, Patmore L, McCormack JG (1993) Protective effects of ranolazine in guinea-pig hearts during low-flow ischaemia and their association with increases in active pyruvate dehydrogenase. Br J Pharmacol 109: 748–750

Clark B, Wyatt KM, McCormack JG (1996) Ranolazine increase active pyruvate dehydrogenase in perfused normoxic rat hearts: evidence for an indirect mechanism. J Mol Cell Cardiol 28: 342–350

Costello LC, Liu Y, Zou J, Franklin RB (2000) The pyruvate dehydrogenase E1 alpha gene is testosterone and prolactin regulated in prostate epithelial cells. Endocrine Res 26: 23–39

Dahn MS, Lange P, Lobell K, Hans B, Jacobs LA, Mitchell RA (1987) Splanchnic and total body oxygen consumption differences in septic and injured patients. Surgery 101: 69–80

Deferrari G, Garibotto G, Robaudo, C, al, e. (1994) Renal ammoniagenesis and interorgan flow of glutamine in chronic metabolic acidosis. Contrib Nephrol 110: 144–149

Deferrari G, Garibotto G, Robaudo C, al, e. (1997) Protein and amino acid metabolism in splanchnic organs in metabolic acidosis. Miner Electrolyte Metab 23: 229–233

Dejong CH, Deutz NE, Soeters PB (1993) Renal ammonia and glutamine metabolism during liver insufficiency-induced hyperammonemia in the rat. J Clin Invest 92: 2834–2840

Fabbri A, Magrini N, Bianchi G, Zoli M, Marchesini G (1996) Overview of randomized clinical trials of oral branched-chain amino acid treatment in chronic hepatic encephalopathy. J Perent Enteral Nutr 20: 159–164

Feldhoff PW, Arnold J, Oesterling B, Vary TC (1993) Insulin-induced activation of pyruvate dehydrogenase complex in skeletal muscle of diabetic rats. Metabolism 42: 615–623

Felig P (1975) Amino acid metabolism in man. Ann Rev Biochem 44: 993–955

Fink MP (2003a) Ringer's ethyl pyruvate: a novel resuscitation fluid for treatment of hemorrhagic shock and sepsis. J Trauma 54: S141–S143

Fink MP (2003b) Ethyl pyruvate: a novel anti-inflammatory agent. Crit Care Clin 31: S51–S56

Fischer CP, Bode BP, Souba WW (1996) Starvation and endotoxin act independently and synergistically to coordinate hepatic glutamine transport. J Trauma 40: 688–693

Gillim SE, Paxton R, Cook GA, Harris RA (1983) Activity state of the branched chain alpha-ketoacid dehydrogenase complex in heart, liver, and kidney of normal, fasted, diabetic, and protein-starved rats. Biochem Biophys Res Commun 111: 74–81

Granot H, Steiner I (1985) Successful treatment of irreversible hemorrhagic shock in dogs with fructose-1,6-diphosphate and dichloroacetate. Circ Shock 15: 163–173

Gudi R, Bowker-Kinley MM, Kedishvili NY, Zhao Y, Popov KM (1995) Diversity of the pyruvate dehydrogenase kinase gene family in humans. J Biol Chem 270: 28989–28994

Gump FE, Long CL, Geiger JW, Kinney JH (1975) The significance of altered gluconeogenesis in surgical catabolism. J Trauma 15: 704–713

Han Y, Englert JA, Yang R, Delude RL, Fink MP (2005) Ethyl pyruvate inhibits nuclear factor-kappaB-dependent signaling by directly targeting p65. J Pharm Exp Ther 312: 1097–1105

Hara K, Maruki Y, Long X, Yoshino K, Oshiro N, Hidayat S, Tokuagna C, Avrch J, Yonezawa K (2002) Raptor, binding partner of target of rapamycin (TOR), mediates TOR action. Cell 110: 177–189

Harper AE, Miller RH, Block KP (1985) Branched-chain amino acid metabolism. Ann Rev Nutr 4: 409–454

Harris RA, Paxton R, Powell SM, Goodwin GW, Kuntz MJ, Han AC (1986) Regulation of branched-chain alpha-ketoacid dehydrogenase complex by covalent modification. Adv Enzyme Regul 25: 219–237

Harris RA, Popov KM, Zhao Y, Kedishvili NY, Shimomura Y, Crabb DW (1995) A new family of protein kinases-the mitochondrial protein kinases. Adv Emz Reg 35: 147–162

Harris RA, Bowker-Kinley MM, Huang B, P W (2002) Regulation of activity of the pyruvate dehydrogenase complex. Adv Enz Reg 42: 249–259

Haussinger D, Soboll S, Meijer AJ, Gerok W, Tager JM (1985) Role of plasma membrane in hepatic glutamine metabolism. Eur J Biochem 152: 597–603

Haussinger D (1990) Liver glutamine metabolism. J Parenter Enteral Nutr 14: 56S–62S

Haussinger D, Lamers WH, Moorman AF (1992) Hepatocyte heterogeneity in metabpolism of amino acids and ammonia. Enzyme(Basal) 46: 72–93

Hayman MW, Miles JM (1982) Branced chain amino acids as a major source of alanine nitrogen in man. Diabetes 31: 86–89

Holloway PA, Knox K, Bajaj N, Chapman D, White NJ, O'Brian R, Stacpoole PW, Krishna S (1995) Plasmodium berghei infection: dichloroacetate improves survival in rats with lactic acidosis. Exp Parasitol 80: 624–632

Holness MJ, Smith ND, Bulmer K, Hopkins T, Gibbons GF, Sugden MC (2002a) Evaluation of the role of peroxisome-proliferator-activated receptor alpha in the regulation of cardiac pyruvate dehydrogenase kinase 4 protein expression in response to starvation, high-fat feeding and hyperthyroidism. Biochem J 364: 687–694

Holness MJ, Bulmer K, Gibbons GF, Sugden MC (2002b) Up-regulation of pyruvate dehydrogenase kinase isoform 4 (PDK4) protein expression in oxidative skeletal muscle does not require the obligatory participation of peroxisome-proliferator-activated receptor alpha (PPARalpha). Biochem J 366: 839–846

Holness MJ, C SM (2003) Regulation of pyruvate dehydrogenase activity by reversible phosphorylation. Biochem Soc Trans 31: 1143–1151

Huang B, Gudi R, Wu P, Harris RA, Hamilton RA, Popov KM (1998) Isoenzymes of pyruvate dehydrogenase phosphatase. DNA-derived amino acid sequences, expression, and regulation. J Biol Chem 273: 17680–17688

Huang B, Wu P, Bowker-Kinley MM, Harris RA (2002) Regulation of pyruvate dehydrogenase kinase expression by peroxisome proliferator-activated receptor-alpha ligands, glucocorticoids, and insulin. Diabetes 51: 276–283

Huang B, Wu P, Popov KM, Harris RA (2003) Starvation and diabetes reduce the amount of pyruvate dehydrogenase phosphatase in rat heart and kidney. Diabetes 52: 1371–1371

Hutson NJ, Randle PJ (1978) Enhanced activity of pyruvate dehydrogenase kinase in rat heart mitochondria in alloxan-diabetes or starvation. FEBS Lett 92: 73–76

Inoue Y, Bode B, Copeland EM, Souba WW (1995) Tumor necrosis factor mediates tumor-induced increases in hepatic amino acid transport. Cancer Res 55: 3525–3550

Jahoor F, Shengraw RE, Myoshi H, Wallfish H, Herdon DN, Wolfe RR (1989) Role of insulin and glucose oxidation in mediating protein catabolism of burns and sepsis. Am J Physiol Endo Metab 257: E323–E331

Jones BS, Yeaman SJ (1991) Long-term regulation of pyruvate dehydrogenase complex. Evidence that kinase-activator protein (KAP) is free pyruvate dehydrogenase kinase. Biochem J 275: 781–784

Karinch A, Wolfgang CL, Lin CM, M P, Souba WW (2002) Regulation of the system N1 glutamine transporter in the kidney in chronic metabolic acidosis in rats. Am J Physiol Renal Physiol 283: F1011–F1019

Karpova T, Danchuk S, Kolobova E, Popov KM (2003) Characterization of the isozymes of pyruvate dehydrogenase phosphatase: implications for the regulation of pyruvate dehydrogenase activity. Biochim Biophys Acta 1652: 126–135

Karpova T, Danchuk S, Huang B, Popov KM (2004) Probing a putative active site of the catalytic subunit of pyruvate dehydrogenase phosphatase 1 (PDP1c) by site-directed mutagenesis. Biochim Biophys Acta 1700: 43–51

Keller DLP, Puinno PA, Fong BC, Spitzer JA (1982) Glucose and lactate kinetics in septic rats. Metabolism 31: 252–257

Kerbey AL, Randle PJ, Cooper RH, Whitehouse S, Pask HT, Denton RM (1976) Regulation of pyruvate dehydrogenase in rat heart. Biochem J 219: 653–646

Kerbey AL, Radcliffe PM, Randle PJ, Sugden PH (1979) Regulation of kinase reactions of pig heart pyruvate dehydrogenase complex. Biochem J 181: 427–433

Kerbey AL, Randle PJ (1982) Pyruvate dehydrogenase kinase/activator in rat heart mitochondria. Biochem J 206: 103–111

Kerbey AL, Richardson LJ, Randle PJ (1984) The roles of intrinsic kinase and of kinase/acvtivator protein in the enhanced phosphorylation of pyruvate dehydrogenase complex in starvation. FEBS Lett 176: 115–119

Kilberg MS, Handlogton ME, Christensen HN (1980) Characteristics of an amino acid transport system in rat liver for glutamne, asparagine, histidine, and closely related analogs. J Biol Chem 255: 4011–4019

Kim DH, Sarbassov DD, Ali SM, King JE, Latek RR, Erdjumnet-Bromage H, Tempest P, Sabatini DM (2002) mTOR interacts with raptor to form a nutrient-sensitive complex that signals to cell growth machinery. Cell 110: 163–175

Kline JA, Maiorano PC, Schoeder JD, Grattn RM, Vary TC, Watts JA (1997) Activation of pyruvate dehydrogenase improves heart function and metabolism after hemorrhagic shock. J Mol Cell Cardiol 29: 2465–2474

Labow BI, Souba WW, Abcouwer SF (2001) Mechanisms governing the expression of the enzymes of glutamine metabolism-glutaminase and glutamine synthetase. J Nutr 131: 2467S–2474S

Lang CH, Bagby GJ, Blakesley HL, Spitzer JJ (1987) Glucose kinetics and pyruvate dehydrogenase activity in septic rats treated with dichloroacetate. Circ Shock 23: 131–141

Lang CH, Dobrescu C, Meszaros K (1990) Insulin-mediated glucose uptake by individual tissues during sepsis. Metabolism (39): 1096–1107

Liedtke AJ, Nellis SH, Neely JR, Hughes HC (1976) Effects of treatment with pyruvate and tromethamine in experimental myocardial ischemia. Circ Res 39: 378–389

Lohmann RG, Souba WW, Zakrzewski K, Bode BP (1998) Burn-dependent stimulation of rat hepatic amino acid transport. Metabolism 47: 608–616

Low SY, Taylor PM, Hundal HS, Pogson CI, Rennie MJ (1992) Transport of L-glutamine and L-glutamate across sinusoidal membranes of rat liver. Effects of starvation, diabetes and corticosteroid treatment. Biochem J 284: 333–340

Low SY, Salter M, Knowles RG, Pogson CI, Rennie MJ (1993) A quantitative analysis of the control of glutamine catabolism in rat liver cells. Use of selective inhibitors. Biochem J 295: 617–624

Lukaszewicz G, Souba WW, Abcouwer SF (1997) Induction of glutamine synthetase gene expression during endotoxemia is adrenal gland dependent. Shock 7: 332–338

Lynch CJ, Halle B, Fujii H, Vary TC, Wallin R, Damuni Z, Hutson SM (2003a) Potential role of leucine metabolism in the leucine signaling pathway involving mTOR. Am J Physiol Endocrinol & Metab 285: E854–E863

Lynch CJ, Halle B, Fujii H, Vary TC, Wallin R, Damuni Z, Hutson SM (2003b) Potential role of leucine metabolism in the leucine signaling pathway involving mTOR. Am J Physiol Endo Metab 285: E854–E863

McDaniel M, Marshall C, Pappan K, Kwon G (2002) Metaboic and aurocrine regulation of the mammalian target of rapamycin by pancreatic b-cells. Diabetes 51: 2877–2885

McGuinness OP, Jacobs J, Moran C, Lacy B (1995) Impact of infection on hepatic disposal of a peripheral glucose infusion in the conscious dog. Am J Physiol Endo Metabol 269: E199–E207

Meijer AJ, Lamers WH, Chamuleau RAFM (1990) Nitrogen metabolism and ornithine cycle function. Physiol Rev 70: 710–748

Mela-Riker L, Bartos D, Vleiss AA, Widner L, Muller P, Trunkey DD (1992) Chronic hyperdynamic sepsis in the rat II. Characterization of liver and muscle energy metabolism. Circ Shock 36: 83–92

Miller RH, Eisenstein RS, Harper AE (1988) Effects of dietary protein intake on branched-chain keto acid dehydrogenase activity of the rat. Immunochemical analysis of the enzyme complex. J Biol Chem 263: 3454–3461

Miyaji T, Hu X, Yen PS, Muramatsu Y, Iyer S, Hewitt SM, Star RA (2003) Ethyl pyruvate decreases sepsi-induced acute renal failure and multiple organ damage in aged mice. Kidney Int 64: 1620–1631

Mochizuki S, Neely JR (1980) Energy metabolism during reperfusion following ischemia. J Physiologie 76: 805–812

Mongan PD, Fontana JL, Chen R, Bunger R (1999) Intravenous pyruvate prolongs survival during hemorrhagic shock in swine. Am J Physiol Heart Circ Physiol 277: H2253–H2263

Montgomery CM, Webb JL (1956) Metabolic studies on heart mitochondria: II. The inhibitory action of parapyruvate on the tricarboxylic acid cycle. J Biol Chem 221: 359–368

Moreau R, Heath SH, Doneanu CE, Harris RA, Hagen TM (2004) Age-related compensatory activation of pyruvate dehydrogenase complex in rat heart. Biochem Biophys Res Comm 325: 48–58

Pacitti AJ, Inoue Y, Souba WW (1993) Tumor necrosis factor stimulates amino acid transport in plasma membrane vesicles from rat liver. J Clin Invest 91: 474–483

Paxton R, Kuntz M, Harris RA (1986) Phosphorylation sites and inactivation of branched-chain alpha-ketoacid dehydrogenase isolated from rat heart, bovine kidney, and rabbit liver, kidney, heart, brain, and skeletal muscle. Arch Biochem Biophys 244: 187–201

Popov KM, Kedishvili NY, Zhao Y, Shimomura Y, Crabb DW, A HR (1993) Primary structure of the pyruvate dehydrogenase kinase establishes a new family of eukaryotic protein kinases. J Biol Chem 268: 26602–26606

Popov KM, Kedishvili NY, Zhao Y, Gudi R, Harris RA (1994) Molecular cloning of the p45 subunit of pyruvate dehydrogenase kinase. J Biol Chem 269: 29720–29724

Popov KM, Zhao Y, Shimomura Y, Jaskiewicz J, Kedishvili NY, Irwin J, Goodwin GW, Harris RA (1995) Dietary control and tissue specific expression of branched-chain alphaketoacid dehydrogenase kinase. Arch Biochem Biophys 316: 148–154

Price SR, Wang X (1999) Glucocorticoids and acidification independently increase transcription of branched-chain ketoacid dehydrogenase subunit genes. Miner Electrolyte Metab 25: 224–227

Randle PJ, Fuller SJ, Kerbey AL, Sale GG, Vary TC (1984) Molecular mechanisms regulating glucose oxidation in insulin deficient animals. Horm Cell Reg 8: 139–150

Randle PJ, Kerbey AL, Espinal J (1988) Mechanisms of decreasing glucose oxidation in diabetes and starvation: Role of lipid fuels and hormones. Diabetes Met Rev 4: 6230–638

Randle PJ (1998) Regulatory interactions between lipids and carbohydrates: the glucose fatty acid cycle after 35 years. Diabetes-Metabolism Rev 14: 263–283

Rowles J, Scherer SW, Xi T, Majer M, Nicle DC, Rommens JM, Popov KM, Riebow NL, Xia J, Tsui LC, Bogardus C, Prochazaka M (1996) Cloning and characterization of PDK4 on 7q21.3 encoding a fourth pyruvate dehydrogenase kinase isoenzyme in human. 271, 1996. J Biol Chem 271: 22376–22382

Shaw JHF, Klein S, Wolfe R (1985) Assessment of alanine, urea and glucose interrationships in normal subjects and subjects with sepsis with stable isotopes. Surgery 97: 557–567

Shimomura Y, Obayashi M, Murakami T, Harris RA (2001) Regulation of branched-chain amino acid catabolism: nutritional and hormonal regulation of activity and expression of the branched-chain alpha-keto acid dehydrogenase kinase. Curr Opin Clin Nutr Metabol Care 4: 419–423

Shimomura Y, Honda T, Goto H, Nonami T, Kurokawa T, Nagasaki M, Murakami T (2004a) Effects of liver failure on the enzymes in the branched-chain amino acid catabolic pathway. Biochem Biophys Res Comm 313: 381–385

Shimomura Y, Murakami T, Nakai N, Nagasaki M, Harris RA (2004b) Exercise promotes BCAA catabolism: effects of BCAA supplementation on skeletal muscle during exercise. J Nutr 134: 1583S-1587S

Siegel JH, Rivikind AI, Dalal S, Goodarzi S (1990) Early physiologic predictors of injury and death in blunt multiple trauma. Arch Surg 125: 498–508

Sims CA, Wattanasirichaigoon S, Menconi MJ, Ajami AM, Fink MP (2001) Ringer's ethyl pyruvate solution ameliorates ischemia/reperfusion-induced intestinal mucosal injury in rats. Crit Care Med 29: 1513–1518

Slovin PN, Huang CJ, Cade JR, Wood CE, Nasiroglu O, Privette M, Orbach P, Skimming JW (2001) Sodium pyruvate is better than sodium chloride as a resuscitation solution in a rodent model of profound hemorrhagic shock. Resuscitation 50: 109–115

Spriet L, Tunstall R, Watt MJ, Mehan KA, Hargreaves M, Cameron-Smith D (2004) Pyruvate dehydrogenase activation and kinase expression in human skeletal muscle during fasting. J Appl Physiol 96: 2082–2087

Srivastava N, Singh N, Joshi YK (2003) Nutrition in the management of hepatic encephalopathy. Tropical Gastroenter 24: 59–62

Stacpoole PW, Harwood HJ Jr, Cameron DF, Curry SH, Samuelson DA, Cornwell PE, Sauberlich HE (1990) Chronic toxicity of dichloroacetate: possible relation to thiamine deficiency in rats. Fund Appl Toxicol 14: 327–337

Stacpoole PW, Wright EC, Baumgartner TG, Bersin RM, Buchalter S, Curry SH, Duncan C, Harman EM, Henderson GN, Jenkinson S, et al (1992) A controlled clinical trial of dichloroacetate for treatment of lactic acidosis in adults. The Dichloroacetate-Lactic Acidosis Study Group N Eng J Med 327: 1564–1569

Stacpoole PW, Greene YJ (1992) Dichloroacetate Diabetes Care 15: 785–791

Stacpoole PW, Wright EC, Baumgartner TG, Bersin RM, Buchalter S, Curry SH, Duncan C, Harman EM, Henderson GN, Jenkinson S, al et (1994) Natural history and course of acquired lactic acidosis in adults. DCA-lactic acidosis study group. Am J of Med 97: 47–54

Sugden MC, Holness MJ, Palmer TN (1989) Fuel selection and carbon flux during the starved-to-fed transition. Biochem J 263: 313–323

Sugden MC, Bulmer K, Gibbons GF, Holness MJ (2001) Role of peroxisome proliferator-activated receptor-alpha in the mechanism underlying changes in renal pyruvate dehydrogenase kinase isoform 4 protein expression in starvation and after refeeding. Arch Biochem Biophys 395: 246–252

Tawadrous ZS, Delude RL, Fink MP (2002) Resuscitation from hemorrhagic shock with Ringer's ethyl pyruvate solution improves survival and ameliorates intestinal mucosal hyperpermeability in rats. Shock 17: 473–477

Tsai CS, Burgett MW, Reed LJ (1973. a-keto acid dehydrogenase complexes. J Biol Chem 248: 8348–8352

Ulloa L, Ochani M, Yang H, Tanovic M, Halperin D, Yang D, Czura CJ, Fink MP, Tracey KJ (2002) Ethyl pyruvate prevents liability in mice with established lethal sepsis and systemic inflammation. Proc Nat Acad Sci (USA) 99: 12351–12356

Varma SD, Devamanoharan PS, Ali AH (1998) Prevention of intracellular oxidative stress to lens by pyruvate and its ester. Free Radical Res 28: 131–135

Vary TC, Siegel JH, Nakatani T, Sato T, Aoyama H (1986) Effect of sepsis on the activity of the pyruvate dehydrogenase complex in skeletal muscle and liver. Am J Physiol Endo Metabol 250: E634–E640

Vary TC, Siegel JH, Rivikind A (1988a) Clinical and therapeutic significance of metabolic patterns of lactic acidosis. Perspect Crit Care: 85–132

Vary TC, Siegel JH, Zechnich A, Tall BD, Morris JG, Placko R, Jawoar D (1988b) Pharmacological reversal of abnormal glucose regulation, BCAA utilization and muscle catabolism in sepsis by dichloroacetate. J Trauma 28: 1301–1311

Vary TC, Placko R, Siegel JH (1989a) Pharmacological modulation of increased release of glucoenogenic precursors from extra-splanchnic organ in sepsis. Circ Shock 29: 59–76

Vary TC, Siegel JH, Placko R, Tall BD, Morris JG (1989b) Effect of dichloroacetate on plasma and hepatic amino acids in sterile inflammation and sepsis. Arch Surg 124: 1071–1077

Vary TC, Murphy JM (1989) Role of extra-splanchnic organs in the metabolic response to sepsis: effect of insulin. Circ Shock 29: 41–57

Vary TC, Drnevich D, Brennan W (1995) Mechanisms regulating glucose metabolism in sepsis. Shock 3: 403–410

Vary TC (1996) Sepsis-induced alterations in pyruvate dehydrogenase complex activity in rat skeletal muscle: Effects on plasma lactate. Shock 6: 89–94

Vary TC (1999) Inter-organ protein and carbohydrate metabolic resltionships during sepsis: necessary evils or uncanny coincidences? Curr Opin Clin Nutr Metabol Care 2: 235–242

Wang X, Jurkovitz C, Price SR (1997) Branched-chain amino acid catabolism in uremia: dual regulation of branched-chain alpha-ketoacid dehydrogenase by extracellular pH and glucocorticoids. Miner Electrolyte Metab 23: 206–209

Welbourne TC, Oshi S (1990) Interorgan glutamine metabolism during acidosis. J Parenter Enteral Nutr 14: 77S–85S

Wilmore DW, Goodwin CW, Aulick LH, Powanda MC, Mason AD, Pruitt BA (1980) Effect of injury and infection on visceral metabolism and circulation. Ann Surg 192: 491–504

Woo YJ, Taylor MD, Cohen JE, Jayasankar V, Bish LT, Burdick J, Pirolli TJ, Berry MF, Hsu V, Grand T (2004) Ethyl pyruvate preserves cardiac function and attenuates oxidative injury after prolonged myocardial ischemia. J Thorac Cardio Surg 127: 1262–1269

Wu P, Sato J, Zhau Y, Jeaskiewicz J, Popov KM, Harris RA (1988) Starvation and diabetes increase the amount of pyruvate dehydrogenase kinase isoenzyme 4 in rat heart. Biochem J 329: 191–19

Wu P, Inskeep K, Bowker-Kinley MM, Popov KM, Harris RA (1999) Mechanism responsible for inactivation of skeletal muscle pyruvate dehydrogenase complex in starvation and diabetes. Diabetes 48: 1593–1599

Wu P, Peters JM, Harris RA (2001) Adaptive increase in pyruvate dehydrogenase kinase 4 during starvation is mediated by peroxisome proliferator-activated receptor alpha. Biochem Biophys Res Comm 287: 391–396

Yan J, Lawson JE, Reed LJ (1996) Role of the regulatory subunit of bovine pyruvate dehydrogenase phosphatase. Proc Nat Acad Sci (USA) 93: 4953–4956

Zhao Y, Hawes J, Popov KM, Jaskiewicz J, Shimomura Y, Crabb DW, Harris RA (1994) Site-directed mutagenesis of phosphorylation sites of the branched chain alpha-ketoacid dehydrogenase complex. J Biol Chem 269: 18583–18587

6
Amino Acids and the Mitochondria

Nicola King

6.1. Introduction and Summary

This chapter describes some of the important physiological functions of amino acids in the mitochondria and the alterations caused by specific pathologies. To some extent all of the featured items are dependent upon the movement of amino acids across the highly selective permeability barrier that is the inner mitochondrial membrane. The performance of this transport by specific carriers is the subject of the first section. Once inside the mitochondrial matrix the amino acids become involved in a bewildering number of critical metabolic pathways. The second section elaborates on two of the most significant namely: the malate-aspartate shuttle essential for the transfer of reducing equivalents between the cytoplasm and the mitochondria; and the urea cycle, which is responsible for maintaining sub-toxic levels of ammonia. The final section covers the changes to mitochondrial amino acid metabolism that occur under different pathological conditions. In this case three examples have been chosen comprising ischemia-reperfusion in heart, myocardial hypertrophy and the special relationship that exists between glutamine and cancer cells.

6.2. Amino Acid Transport Across the Mitochondrial Inner Membrane

6.2.1. Introduction

When non-respiring mitochondria are suspended in isotonic solutions containing different amino acids they swell suggesting that these compounds can cross the mitochondrial inner membrane (Palmieri et al. 2001; Porter 2000). Amino acid flux through the mitochondrial inner membrane has also been measured directly using radiotracers via inhibitor stop techniques, oil- or Millipore- filtration (reviewed in LaNoue and Schoolwerth 1979; Passarella et al. 2003). These investigations have greatly magnified our knowledge of the transport characteristics of the various carriers (Figure 6.1), although surprisingly only a relatively small number of the genes and proteins responsible have been identified.

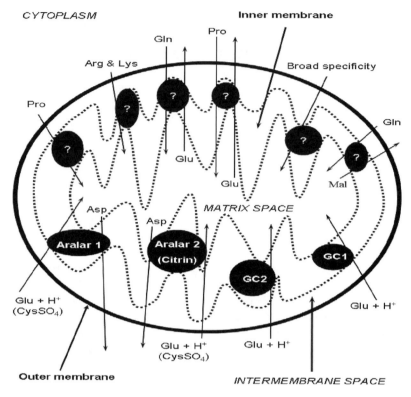

FIGURE 6.1. Amino acid carriers in the inner mitochondrial membrane. Question marks indicate carriers that have been functionally characterized where the genes and proteins responsible are unknown. For the sake of clarity only the major substrates are shown. In contrast with those carriers, which have been identified at the molecular level, the names of the proteins and all the natural amino acid substrates are given.

6.2.2. *Acidic Amino Acid Carriers*

The main acidic amino acid carriers of the mitochondrial inner membrane are the aspartate-glutamate exchangers and the glutamate carriers. Some of the most important characteristics of the aspartate-glutamate exchangers will be discussed later (section 6.3.2), because of their key role in the malate-aspartate shuttle. However, it is worth mentioning here that these carriers have narrow substrate specificities, i.e. the only other natural amino acid to be accepted is cysteine-sulfinate (LaNoue and Schoolwerth 1979). This compound, which is important for sulfur homeostasis (Townsend et al. 2004), can be involved in a straight exchange for aspartate or in an electrogenic exchange for glutamate and a proton (LaNoue and Schoolwerth 1979). There are two aspartate-glutamate exchangers that have so far been identified, which are encoded by the genes, SLC25A12 and SLC25A13 (Palmieri et al. 2001). The encoded proteins named aralar 1 and aralar 2 (also called citrin) respectively display the typical features of the mitochondrial carrier family including EF hand Ca^{2+} binding motifs in their N terminals (4 in each protein) (Palmieri et al. 2001). Aralar 1 is most prevalent in heart, skeletal muscle and brain, whereas aralar 2 is expressed in many tissues but highest levels are in liver (Palmieri et al. 2001). Mutations in the gene encoding aralar 2 cause adult type II citrullinemia, a rare disease marked by a sudden onset of disorientation, restlessness, drowsiness and coma, leading in the majority to death within a few years due to cerebral edema (Yasuda et al. 2001).

GC1 (SLC25A22) and GC2 (SLC25A23) are glutamate proton symporters in the mitochondrial inner membrane (Fiermonte et al. 2001). They have chiefly been implicated in the regulation of amino acid deamination in association with glutamate dehydrogenase (LaNoue and Schoolwert 1979; Bradford and McGivan 1973). GC1 is widely distributed in human tissues with high levels in pancreas, liver, brain and testis and lower levels in heart, kidney, lung, small intestine and spleen (Fiermonte et al., 2001). In contrast, GC2 expression is largely confined to the brain and testis (Fiermonte et al. 2001). When the carriers were reconstituted into proteoliposomes their substrate specificity was virtually confined to glutamate; they were potently inhibited by α-cyanocinnamate and bongkrekic acid; the K_m for GC1 was around 5 mmoles, whilst that for GC2 was approximately 0.25 mmoles (Fiermonte et al. 2001).

A recent report has suggested that a member of the excitatory amino acid transporter (EAAT) family, EAAT1 is expressed in cardiac mitochondria (Ralphe et al. 2004). This transporter family, which contains 5 members, constitutes the high affinity sodium dependent glutamate and aspartate transporters that are most easily recognizable for their contribution to neurotransmitter re-uptake in the central nervous system (Kanai and Hediger 2004). It is known that EAAT1-EAAT3 are present in heart (King et al. 2004; Kugler 2004), where EAAT2 and EAAT3 are believed to reside within the cardiac sarcolemma (King et al. 2001; King et al. 2004; Kugler 2004). The exact function of EAAT1 in cardiac mitochondria remains to be determined, although Ralphe et al. (2005) proposed that it may be an additional aspartate-glutamate carrier involved in the malate-aspartate shuttle.

6.2.3. Neutral Amino Acids

Early investigations of substrate flux across the mitochondrial inner membrane suggested there was a single neutral amino acid carrier with a broad specificity (reviewed in LaNoue and Schoolwerth 1979; Porter 2000). At this time, proline was noted as a likely exception since it was not a very effective agent in protecting other neutral amino acids against mersalyl inhibition (LaNoue and Schoolwerth 1979), yet proline oxidation occurs exclusively in the mitochondria (Passarella et al. 2003). Subsequently, Atlante et al. (1994) demonstrated in rat kidney mitochondria the existence of a proline uniporter and a proline-glutamate antiporter (Figure 6.1), the latter of which was dependent upon the membrane potential difference.

Another amino acid that has sparked controversy regarding its transport mechanism is glutamine. The discussions seem to have been prompted by investigations of the regulation of glutaminase activity, which catalyzes the deamination of glutamine to glutamate and ammonia in the matrix, and the extent to which this is limited by glutamine entry across the inner mitochondrial membrane (LaNoue and Schoolwerth 1979; Passarella et al. 2003). This dispute has yet to be resolved; however the purification and reconstitution into liposomes of a 41.5 kiloDalton carrier protein has helped clarify the specifics of glutamine transport (Indiveri et al. 1998). The reconstituted glutamine carrier operated either as a uniporter or as a homoexchanger (Indiveri et al. 1998). Glutamine uptake was highest when the mitochondria were solubilized at pH 6.5 and the protein was reconstituted at the same pH (Indiveri et al. 1998). The only other amino acids tested, amongst a group containing aspartate, glutamate, leucine, alanine, glycine, proline and serine, that was transported was asparagine (Indiveri et al. 1998).

6.2.4. Basic Amino Acid Carriers

A basic amino acid transporter has been described in brain mitochondria (Porter 2000; Dolinska et al. 1998). The uptake of arginine was sensitive to lysine and ornithine suggesting that these compounds were also substrates with the uptake inhibited by the nitric oxide inhibitor L-N-monomethyl arginine, but neither glutamine nor histidine nor glutamate affected transport (Dolinska et al. 1998). The K_m for arginine uptake in rat cerebral mitochondria was 0.08 mmoles (Dolinska et al. 1998).

6.3. Amino Acid Metabolism in the Mitochondria Under Normal Conditions

6.3.1. Introduction

As important intermediary metabolites, amino acids are involved in numerous major pathways many of which have a mitochondrial component. These different reactions may contribute to: energy production, including for example gluconeogenesis, ketogenesis and supplementation of citric acid cycle intermediates (Jungas et al. 1992; Hutson et al. 2005; Pisarenko 1996); cellular and organ ionic homeostasis, such as the regulation of nitrogen (Hutson et al. 2005; Medina 2001) and sulfur levels (Townsend et al. 2004); and the synthesis of many biologically active compounds e.g. purines, pyramidines, hormones, neurotransmitters and other proteins (Mathews and Van Holde 1990). A comprehensive review of all these processes would be worthy of a complete volume and as such is beyond the scope of this chapter. Nonetheless, it would be neglectful to ignore all the metabolic pathways involving amino acids that occur in the mitochondria. We will therefore restrict ourselves to discussing just two of arguably the most significant pathways namely: the malate-aspartate shuttle and the urea cycle.

6.3.2. The Malate-Aspartate Shuttle

It has long been known that the mitochondrial inner membrane is impermeable to the NADH/NAD$^+$ couple (Lehninger 1951). This presents a problem since the initial stages of energy production from glucose or lactate yield pyruvate and NADH in the cytoplasm. In order to achieve maximal ATP production this cytoplasmic pyruvate must be transferred into the mitochondria for oxidative metabolism within the citric acid cycle accompanied by an equivalent oxidation of NADH also within the mitochondria (Safer 1975).

The existence of a "shuttle" utilising the cytoplasmic aspartate aminotransferase and the mitochondrial malate dehydrogenase (Figure 6.2), which could achieve redox segregation was initially proposed in 1963 (Borst). However, Borst (1963) was not entirely confident in his scheme, because the system was symmetrical resulting in at equilibrium equal NAD$^+$/NADH ratios in the cytoplasm and the mitochondria. Previous work had shown that the NAD$^+$/NADH ratio was more oxidized in the cytoplasm than the mitochondria (Williamson et al. 1967). Two further pieces of evidence were required before the malate-aspartate shuttle as shown in Figure 6.2 came to be recognised as a vital conduit of cellular metabolism.

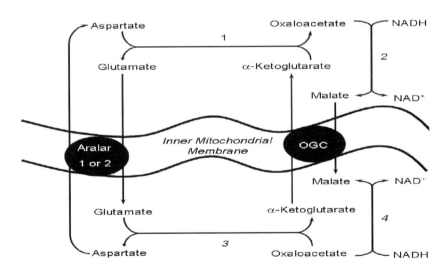

FIGURE 6.2. The malate-aspartate shuttle. Arrows indicate the direction of reactions. Enzymes are numbered: 1 Cytoplasmic glutamate-oxaloacetate transaminase; 2 cytoplasmic malate dehydrogenase; 3 Mitochondrial glutamate-oxaloacetate transaminase; and 4 mitochondrial malate dehydrogenase. Aralar 1 or 2, the aspartate-glutamate exchanger; and OGC the 2-oxoglutarate (i.e. α-ketoglutarate) carrier.

The first clue was that in coupled mitochondria transfer of electrons in the respiratory chain results in a small pH gradient (inside alkaline) and a large electrical potential difference (inside negative) across the mitochondrial inner membrane (Safer 1975). The other related discovery was the observation that the exchange of the acidic amino acids aspartate and glutamate was mediated by a specific carrier in the inner membrane and was electrogenic in nature (LaNoue and Schoolwerth 1979). More specifically, a cotransported proton drove glutamate into the matrix, whilst an uncharged aspartate ion moved in the opposite direction (LaNoue 1974). The significance of this is that the other carrier involved in the shuttle and localised within the inner membrane, OGC (SLC20A4, Iacobazzi et al. 1992) mediates the electroneutral exchange of cytoplasmic malate for α-ketoglutarate in the matrix (Kaplan 2001). Hence the aspartate-glutamate exchanger is widely regarded as the rate-limiting step in the operation of the shuttle (Palmieri et al. 2001).

6.3.3. The Urea Cycle

Ammonia is produced as a by-product of protein and amino acid catabolism. It has been estimated that 505 mmoles of ammonia are generated and transferred to the liver each day (Jungas et al. 1992). Without further metabolism such levels would be toxic as exemplified by several disease states in which excess ammonia accumulates, resulting in for example brain edema, mental retardation and cerebral palsy (Butterworth 2001). Under normal conditions, this is prevented through a cyclical chain of reactions (Figure 6.3) wherein ammonia is converted by the liver into urea, which can then be excreted via the kidneys.

The urea cycle utilizes the amino acids, aspartate, arginine, citrulline and ornithine. The only one of these to be either consumed or produced is aspartate, which is required for the reaction catalyzed by argininosuccinate synthetase (ASS). Some of this aspartate may originate from transamination of oxaloacetate in the matrix followed by transport on the glutamate-aspartate exchanger, aralar 2 that is an integral part of the malate-aspartate shuttle (Palmieri et al. 2001, see above and Figures 6.1 and 6.2). Recently, it has been shown that in the fetal liver 90% and in adult liver 30% of ASS expression is located on the outer mitochondrial membrane although most texts designate this reaction as occurring in the cytoplasm (Husson et al. 2003).

The next amino acid, arginine is catabolised by arginase. In hepatocytes this reaction is mostly catalysed by a cytoplasmic enzyme, arginase I (Cheung and Raijman 1981). A second enzyme, arginase II, which is responsible for about 10% of the total hepatocyte arginase activity (Cheung and Raijman 1981), is present in the inner mitochondrial membrane. Arginase II has a wide tissue distribution (Colleluori et al. 2001). It has a minor role supplementing urea cycle activity in liver (Nissim et al. 2005). In addition it has been implicated in the biosynthesis of proline and polyamines (Wu and Morris 1998), and, through its influence on the intracellular arginine pool, as a regulator of NO_2 availability and smooth muscle relaxation in opossum anal internal sphincter (Baggio et al. 1999).

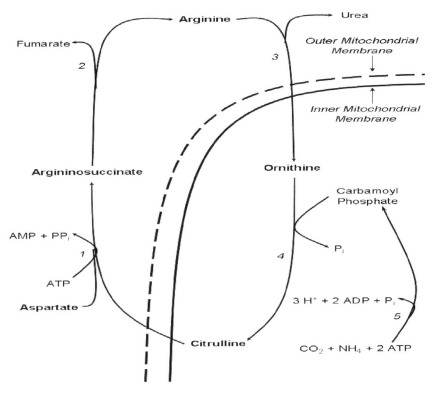

FIGURE 6.3. The urea cycle. Reactions involving the enzymes 1–3 occur in the cytoplasm, whilst reactions involving the enzymes 4–5 occur in the mitochondrial matrix. A heavy continuous line representing the mitochondrial inner membrane and a heavy dashed line representing the mitochondrial outer membrane indicate this segregation. Arrows indicate the direction of reactions. Enzymes are numbered: 1 Argininosuccinate synthetase; 2 Argininosuccinase; 3 Arginase; 4 Ornithine transcarbamoylase; and 5 Carbamoyl phosphate synthetase.

Finally, the only reactions that are definitively assigned to the mitochondrial matrix are those involving the enzymes: ornithine transcarbomoylase and carbomoyl phosphate synthetase (Jungas et al. 1992, Nissim et al. 2005), which in essence add free NH_3 onto ornithine to yield citrulline.

6.4. Amino Acids in Mitochondria Under Pathological Conditions

6.4.1. Introduction

In this section the role of amino acids in three different pathological states will be examined namely: ischemia-reperfusion injury, hypertrophy and cancer.

For the sake of simplicity the first two conditions will be discussed as they pertain to cardiac mitochondria, whilst cancer will be discussed in relation to glutamine.

6.4.2. Ischemia-Reperfusion Injury

When blood is diverted away from the heart (ischemia) this causes a decrease in the supply of nutrients and oxygen to the myocardium. This leads to a reduction in energy production in the mitochondria, a fall in intracellular pH, an increase in $[Ca^{2+}]_i$ and $[Na^+]_i$ and an accumulation of abnormal metabolites e.g. lactic acid (reviewed in Suleiman et al. 2001). If the ischemic period is short, restoration of the blood flow (reperfusion) enables cellular recovery; however after prolonged ischemia, reperfusion may result in significant cellular and tissue damage.

The foundation to the proposal that amino acids may alleviate ischemia-reperfusion injury was largely based on their potential to convert to citric acid cycle intermediates with subsequent substrate level phosphorylation and ATP production in the mitochondria (Pisarenko 1996; Taegtmeyer 1995; Sveldjeholm et al. 1995). Logical, though this may be, the concept does not enjoy universal support (Buckberg 1996). A likely cause of this controversy is the relatively poor understanding of the mechanisms of amino acid transport across the cardiac sarcolemma. Thus, in the majority of cases where the addition of exogenous amino acids leads to an increase in their myocardial concentration the outcome has been beneficial (Pisarenko et al. 1985; Pisarenko et al. 1995).

Branched chain amino acids are transported across the cardiac sarcolemma by the L system of mediation (King et al. 2001). The initial stages of their metabolism occur exclusively in the mitochondria where the branched chain aminotransferase is located (Hutson et al. 1988). Schwalb et al. (1989) exposed working rat hearts to cardioplegic arrest followed by 68 min of global normothermic ischemia and 40 min reperfusion. The control group received a modified St Thomas solution whilst the test groups received the same solution with 11.1 mmoles glucose plus or minus 3.5 mmoles branched chain amino acids (BCAA) (Schwalb et al. 1989). During reperfusion ATP levels and the recovery of cardiac output were significantly higher in the group receiving BCAA than in either of the other two groups (Schwalb et al. 1989). Similarly, addition of acidic amino acids to crystalloid or blood cardioplegia has also been shown to improve functional recovery and ATP levels in various animal hearts (Haas et al., 1984; Lazar et al. 1980; Bittl and Shine 1983; King et. al. 2004).

Oxidative stress accompanied by the production of reactive oxygen species (ROS) is one of the major pathogenesis of ischemia-reperfusion (Suleiman et al. 2001; Zweier et al. 1988). Figure 6.4 illustrates the results of experiments carried out using freshly isolated adult rat cardiomyocytes, which show that complex III of the respiratory chain in the mitochondria can be a major source of these ROS. This experiment was performed using the inhibitors, antimycin A, which prolongs the lifespan of ubisemiquinone thereby promoting ROS production, and myxothiazol, which prevents electron transfer to ubisemiquinone consequently

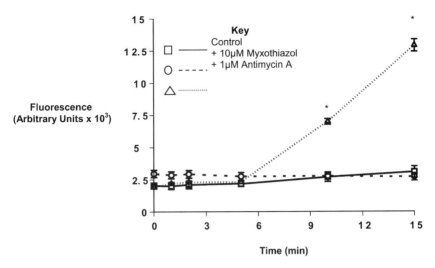

FIGURE 6.4. Mitochondrial production of reactive oxygen species in freshly isolated rat ventricular cardiomyocytes. The fluorescence of the probe, 5-(and-6)-chloromethyl-2', 7'-dichlorodihydrofluorescein diacetate (CM-DCFda) is proportional to the level of hydroxy radical generation. This graph shows a time course of the fluorescence measured in freshly isolated rat ventricular cardiomyocytes that have been loaded with 5 μmoles CM-DCFda then superfused with a control Tyrode solution or a Tyrode solution containing either 10 μmoles myxothiazol or 1 μmole antimycin A. The increase in fluorescence in the presence of antimycin A and the accompanying lack of effect of myxothiazol suggest that complex III of the mitochondrial respiratory chain can be a source of hydroxy radicals in cardiomyocytes. *p < 0.01 vs. control and + myxothiazol. Results shown are the means ±S.D (n = 4).

lowering ROS generation (King et al. 2003). When cardiomyocytes are isolated in the presence of glutamate this results in a higher $NAD^+/NADH$ ratio, higher ATP levels (Williams et al. 2001) and lower levels of ROS production during exposure to antimycin A or metabolic inhibition compared to non-glutamate loaded controls (King et al. 2003).

Although the exact mechanism behind the antioxidant effect of glutamate remains undetermined, it has been speculated that a pathway may exist linking the activity of glutamate dehydrogenase in the mitochondria through glutathione reductase to glutathione peroxidase and the detoxification of H_2O_2 to H_2O (Williams et al. 2001) (Figure 6.5).

6.4.3. Hypertrophy

Myocardial hypertrophy is an adaptive response to a chronic increase in the heart's workload. It occurs in several pathophysiological conditions including sustained hypertension, aortic constriction, valvular insufficiency and hyperthyroidism (Rossi and Lortet 1996). Its onset and development is associated with morphological and physiological restructuring of the heart that will affect all

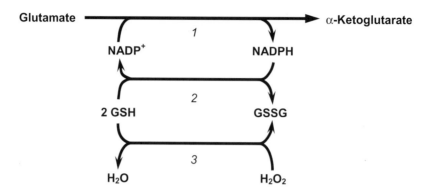

FIGURE 6.5. Hypothetical scheme linking glutamate metabolism in the mitochondria to the detoxification of hydrogen peroxide. Glutamate deamination in the mitochondria through 1: glutamate dehydrogenase using the $NADP^+$/NADPH couple is linked through the activities of 2: glutathione reductase and 3: glutathione peroxidase to the detoxification of hydrogen peroxide to water. Arrows indicate the directions of reactions. 2 GSH reduced glutathione, GSSG oxidized glutathione.

of the organelles and cellular functions (Rossi and Lortet 1996). The following discusses one of the ways in which these changes influence amino acids in the cardiac mitochondria.

In several models, the progression of cardiac hypertrophy is marked by increased glucose transport through GLUT1 (Montessuit and Thorburn 1999), higher activity levels of the enzymes involved in glycolysis (Seymour et al. 1990; Valadares et al. 1969) and a down-regulation of fatty acid metabolism (El Alaoui-Talibi and Moravec 1989). This shift towards a more anaerobic pattern of ATP production might also be expected to impact upon the activity of the malate-aspartate shuttle. This indeed appears to be the case in hypertrophy induced by chronic exposure to the thyroid hormone, triiodothyronine (T_3). In these hearts the capacity of the malate-aspartate shuttle is significantly increased by 33% compared to control (Scholz et al., 2000). This is accompanied in the T_3 treated hearts by a significantly increased expression of the aspartate-glutamate carrier and mitochondrial malate dehydrogenase in comparison to control (Scholz et al. 2000). A more recent study by the same group, which benefited from the molecular identification of the glutamate-aspartate carriers (Palmieri et al. 2001), suggested that the acidic amino acid carrier whose expression was upregulated in hypertrophy was in fact EAAT1 (Ralphe et al. 2005). This supports work carried out in a model of chronic hypertension, the Spontaneously Hypertensive Rat, where increased EAAT expression was shown in hypertrophic hearts compared to control (King et al., 2004). The last mentioned investigation also showed that the addition of 0.5 mmoles aspartate to isolated and perfused hypertrophic hearts led to a significantly improved functional recovery upon reperfusion and higher ATP levels at the end of ischemia compared to hypertrophic hearts not receiving aspartate (King et al. 2004).

6.4.4. Glutamine and Cancer

A curious feature of tumor cells, as opposed to their normal counterparts, is their appetite for the neutral amino acid, glutamine (reviewed in Medina 2001 and Souba 1993). Rapidly growing cancer cells exhibiting high rates of nucleotide and protein synthesis require a continuous source of nitrogen (Medina 2001). Glutamine, as the most abundant amino acid in the body, is the main vehicle for circulating ammonia in a non-toxic form and is therefore a handy conduit for enabling this supply (Medina et al. 1992). Thereafter, the nitrogen can be liberated through the actions of glutaminase. This enzyme is located exclusively in the mitochondria where it catalyzes the conversion of glutamine to glutamate and NH_3 (Queseda et al. 1988). There is a good correlation between the activity of glutaminase and the extent of malignant proliferation in various rat hepatomas (Linder-Horowitz et al. 1969) and breast cancer cells (Gómez-Fabre et al. 2000).

Glutamine may also be the major respiratory fuel in tumor cells. In order to maintain optimum growth rates in conditions where the oxygen tension is variable and often low (Griffiths et al. 2001), tumor cells rely more heavily on glycolysis than on oxidative phosphorylation for ATP synthesis (Rossignol et al., 2004). This adaptation is characterized by a shift away from the expression of pyruvate kinase type L to the M_2 type (Hacker et al. 1998, Ibsen, 1977). If glucose levels are very reduced a further metabolic revision occurs leading to the catabolism of glutamine for energy. Some studies go one step further and suggest that tumor cells can survive and grow in the absence of glucose as long as glutamine is present (Reitzer et al. 1979; Rossignol et al. 2004; Eagle et al. 1958).

The pathway of glutamine catabolism in tumor cells is referred to as glutaminolysis (Figure 6.6) by analogy to glycolysis (McKeehan 1982). The second step in this pathway is a transamination reaction catalyzed by mitochondrial glutamate-oxaloacetate transaminase. This is more favorable than metabolism through glutamate dehydrogenase, because the high levels of NADH and lactate found in tumor cells tend to force this enzyme towards glutamate synthesis (Eigenbrodt et al. 1998). For example, malignant transformation of the oval cell line OC/CDE 22 resulted in decreased cytoplasmic glutamate-oxaloacetate transaminase activity with concomitantly increased mitochondrial glutamate-oxaloacetate transaminase activity and no change in the activity of glutamate dehydrogenase compared to control cells (Mazurek et al. 1999). The final step in the glutaminolysis pathway involves oxidative decarboxylation catalyzed by the NAD^+ dependent mitochondrial malic enzyme to yield pyruvate and CO_2 (Loeber et al. 1991).

This importance of glutamine to tumor cell metabolism and proliferation led to the attempted development in the 1980s of glutamine related anticancer therapies (Medina 2001). These approaches largely involved the use of amino acid analogs to reduce glutamine availability. Clinical trials were not however successful due to toxic effects, lack of specificity and/or ineffectiveness (Medina et al. 1992; Souba et al. 1993). More recently it has been shown that glutamine starvation leads to apoptosis of murine hybridoma cells following activation of

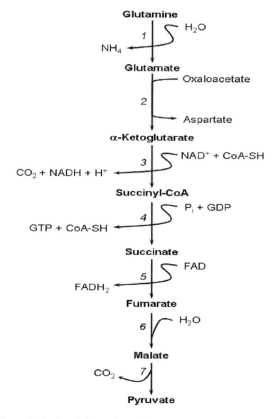

FIGURE 6.6. Glutaminolysis. Schematic pathway showing the possible production of pyruvate from glutamine in the mitochondria of cancer cells. Arrows indicate the direction of reactions. Enzymes are numbered: 1 glutaminase; 2 mitochondrial glutamate-oxaloacetate transaminase; 3 α-ketoglutarate dehydrogenase; 4 succinyl-CoA synthetase; 5 succinate dehydrogenase; 6 fumarase and 7 malic enzyme.

the caspases –9 and –3 (Paquette et al. 2005). Perhaps as our understanding of the complexities of cellular signaling and the pathways of apoptosis (reviewed elsewhere in this volume) continue to advance this may open up new avenues for a glutamine-based anticancer therapy.

References

Atlante A, Passarella S, Pierro P, Quagliariello E (1994) Proline transport in rat kidney mitochondria. Arch Biochem Biophys 309:139–148

Baggio R, Emig FA, Christianson DW, Ash DE, Chakdar S, Rattan S (1999) Biochemical and functional profile of a newly developed potent and isozyme-selective arginase inhibitor. J Pharmacol Exp Therapeut 290:1409–1416

Bittl JA, Shine KI (1983) Protection of ischemic rabbit myocardium by glutamic acid. Am J Physiol 245:H406–H412

Borst P (1963) Hydrogen transport and transport metabolisms. In Karlson, P. (ed.), Functionelle und Morphologische Organisation der Zelle, Springer-Verlag, Berlin, pp. 137–158

Bradford NM, McGivan JD (1973) Quantitative aspects of glutamate transport in rat liver mitochondria. Biochem J 134:1023–1029

Buckberg GD (1996) Invited editorial on "Effects of glutamate and aspartate on myocardial substrate oxidation during potassium arrest". J Thorac Cardiovasc Surg 112:1661–1663.

Butterworth RF (2001) Glutamate transporter and receptor function in disorders of ammonia metabolism. Ment Retard Dev Disabil Res Rev 7:276–279

Cheung CW, Raijman L (1981) Arginine, mitochondrial arginase, and the control of carbamyl phosphate synthesis. Arch Biochem Biophys 209:643–649

Colleluori DM, Morris SM Jr, Ash DE (2001) Expression purification and characterization of human type II arginase. Archiv Biochem Biophys 389:135–143

Dolinska M, Albrecht J (1998) L-Arginine uptake in rat cerebral mitochondria. Neurochem Int 33:233–236

Eagle H, Barban S, Levy M, Schulz HO (1958) Utilization of carbohydrates by human cell cultures. J Biol Chem 233:551–558

Eigenbrodt E, Kallinowski F, Ott M, Mazurek S, Vaupel P (1998) Pyruvate kinase and the interaction of amino acid and carbohydrate metabolism in solid tumors. Anticancer Res 18:3267–3274

El Alaoui-Talibi Z, Moravec J (1989) Limitation of long chain fatty acid oxidation in volume over-loaded rat hearts. Adv Exp Med Biol 248:491–497

Fiermonte G, Palmieri L, Todisco S, Agrimi G, Palmieri F, Walker JE (2002) Identification of the mitochondrial glutamate transporter. J Biol Chem 277: 19289–19294

Gómez-Fabre PM, Aledo JC, Del Castillo-Olivares A, Alonso FJ, Núñezde Castro I, Campos JA, Márquez J (2000) Molecular cloning, sequencing and expression studies of the human breast cancer cell glutaminase. Biochem J 345:365–375

Griffiths JR (2001) Causes and consequences of hypoxia and acidity in tumor microenvironments. Bioessays 23:295–296

Haas GS, DeBoer LWV, O'Keefe DD, Bodenhamer RM, Geffin MB, Drop LJ, Teplick RS, Daggett WM (1984) Reduction of postischemic myocardial dysfunction by substrate repletion during reperfusion. Circ 70(suppl I):I-65–I73

Hacker HJ, Steinberg P, Bannasch P (1998) Pyruvate kinase isoenzyme shift from L-type to M2-type is a late event in hepatocarcinogenesis induced in rats by a choline-deficient/DL-ethionine-supplemented diet. Carcinogenesis 19:99–107

Husson A, Brasse-Lagnel C, Fairand A, Renouf S, Lavoinne A (2003) Argininosuccinate synthetase from the urea cycle to the citrulline-NO cycle. Eur J Biochem 270: 1887–1899

Hutson SM, Fenstermacher D, Mahar C (1988) Role of mitochondrial transamination in branched chain amino acid metabolism. J Biol Chem 263:3618–3625

Hutson SM, Sweatt AJ, LaNoue KF (2005) Branched-chain amino acid metabolism: Implications for establishing safe intakes. J Nutr 135:1557S–1564S

Iacobazzi V, Palmieri F, Runswick MJ, Walker JE (1992) Sequences of the human and bovine genes for the mitochondrial 2-oxoglutarate gene. DNA Seq 3:79–88

Ibsen KH (1977) Interrelationships and functions of the pyruvate kinase isoenzymes and their variant forms: a review. Cancer Res 37:341–353

Indiveri C, Abruzzo G, Stipani I, Palmieri F (1998) Identification and purification of the reconstitutively active glutamine carrier from rat kidney mitochondria. Biochem J 333:285–290

Jungas RL, Halperin ML, Brosnan JT (1992) Quantitative analysis of amino acid oxidation and related gluconeogenesis in humans. Physiol Rev 72:419–448

Kanai Y, Hediger MA (2004) The glutamate/neutral amino acid transporter family SLC1: molecular, physiological and pharmacological aspects. Pflügers Archiv 447:469–479

Kaplan RS (2001) Structure and function of mitochondrial anion transport proteins. J Membrane Biol 179:165–183

King N, Lin H, McGivan JD, Suleiman MS (2004) Aspartate transporter activity in hypertrophic rat heart and ischaemia-reperfusion injury. J Physiol 556: 849–858

King N, McGivan JD, Griffiths EJ, Halestrap AP, Suleiman MS (2003) Glutamate loading protects freshly isolated and perfused adult rat cardiomyocytes from intracellular ROS generation. J Mol Cell Cardiol 35: 975–984

King N, Suleiman MS (2001) L-Leucine transport in rat heart under normal conditions and effect of simulated hypoxia. Mol Cell Biochem 221: 99–108

King N, Williams H, McGivan JD, Suleiman MS (2001) Characteristics of L-aspartate transport and expression of EAAC-1 in sarcolemmal vesicles and isolated cells from rat heart. Cardiovasc Res 52: 84–94

Kugler P (2004) Expression of glutamate transporters in rat cardiomyocytes and their localization in the T-tubular system. J Histochem. Cytochem 52:1385–1392

LaNoue KF, Tischler ME (1974) Electrogenic characteristics of the mitochondrial glutamate-aspartate antiporter. J Biol Chem 249:7522–7528

LaNoue KF, Schoolwerth AC (1979) Metabolite transport in mitochondria. Ann Rev Biochem 48:871–922

Lazar HL, Buckberg GD, Manganaro AM, Becker H (1980) Myocardial energy replenishment and reversal of ischemic damage by substrate enhancement of secondary blood cardioplegia with amino acids during reperfusion. J Thorac Cardiovasc Surg 80:350–359

Lehninger AL (1951) Phosphorylation coupled to oxidation of dihydrodi-phosphopyridine nucleotide. J Biol Chem 190:345–359

Linder-Horowitz M, Knox WE, Morris HP (1969) Glutaminase activities and growth rates of rat hepatomas. Cancer Res 29:1195–1199

Mathews CK, Van Holde KE (1990) Biochemistry. Benjamin Cummings Publishing Company, Redwood City, California

McKeehan WL (1982) Glycolysis, glutaminolysis and cell-proliferation. Cell Biol Int Rep 6: 635–650

Medina MA (2001) Glutamine and cancer. J Nutr 131:2539S–2542S

Medina MA, Sánchez-Jiménez F, Márquez J, Queseda AR, Núñez de Castro I (1992) Relevance of glutamine metabolism to tumor cell growth. Mol Cell Biochem 113:1–15

Montessuit C, Thorburn A (1999) Transcriptional activation of the glucose transporter GLUT1 in ventricular cardiomyocytes by hypertrophic agonists. J Biol Chem 274: 9006–9012

Nissim I, Luhovyy B, Horyn O, Daihin Y, Nissim I, Yudkoff M (2005) The role of mitochondrially bound arginase in the regulation of urea synthesis. J Biol Chem 280:17715–17724

Palmieri L, Pardo B, Lasorsa FM, Del Arco A, Kobayashi K, Lijima M, Runswick MJ, Walker JE, Saheki T, Satrústegui J, Palmieri F (2001) Citrin and aralar1 are Ca^{2+}-stimulated aspartate/glutamate transporters in mitochondria. EMBO J 20: 5060–5069

Paquette JC, Guérin PJ, Gauthier ER (2004) Rapid induction of the intrinsic apoptotic pathway by glutamine starvation. J Cell Physiol 202:912–921

Passarella S, Atlanate A, Valenti D, De Bari L (2003) The role of mitochondrial transport in energy metabolism. Mitochondrion 2:319–343

Pisarenko OI (1996) Mechanisms of myocardial protection by amino acids: facts and hypotheses. Clin Exptl Pharmacol 23:627–633

Pisarenko OI, Novikova EB, Serebryakova LI, Tskitishvili OV, Ivanov VE, Studneva IM (1985) Function and metabolism of dog heart in ischaemia and in subsequent reperfusion: effect of exogenous glutamic acid. Pflügers Archiv 405:377–383

Pisarenko OI, Studneva IM, Shulzhenko VS, Korchazhkina OV, Kapelko VI (1995) Substrate accessibility to cytosolic aspartate aminotransferase improves posthypoxic recovery of isolated rat heart. Biochemical Mol Med 55:138–148

Porter RK (2000) Mammalian mitochondrial inner membrane cationic and neutral amino acid carriers. Biochim Biophys Acta 1459:356–362

Queseda AR, Sánchez-Jiménez F, Perez-Rodriguez J, Márquez J, Medina MA, Núñez de Castro I (1988) Purification of phosphate-dependent glutaminase from isolated mitochondria of Ehrlich ascites tumor cells. Biochem J 255:1031–1036

Ralphe JC, Bedell K, Segar JL, Scholz TD (2005) Correlation between myocardial malate/aspartate shuttle activity and EAAT1 protein expression in hyper- and hypothyroidism. Am J Physiol. 288:H2521–H2526

Ralphe JC, Segar JL, Schutte BC, Scholz TD (2004) Localization and function of the brain excitatory amino acid transporter type 1 in cardiac mitochondria. J Mol Cell Cardiol 37:33–41

Reitzer LJ, Wice BM, Kennell D (1979) Evidence that glutamine, not glucose, is the main energy-source for Hela-cells. J Biol Chem 254:2669–2676

Rossi A, Lortet S (1996) Energy metabolism patterns in mammalian myocardium adapted to chronic pathophysiological conditions. Cardiovasc Res 31:163–171

Rossignol R, Gilkerson R, Aggeler R, Yamagata K, Remington SJ, Capaldi RA (2004) Energy substrate modulates mitochondrial structure and oxidative capacity in cancer cells. Cancer Res 64: 985–993

Safer B (1975) The metabolic significance of the malate-aspartate cycle in heart. Circ Res 37, 527–533

Scholz TD, TenEyck CJ, Schutte BC (2000) Thyroid hormone regulation of the NADH shuttles in liver and cardiac mitochondria. J Mol Cell Cardiol 32:1–10

Schwalb H, Yaroslavsky E, Borman JB, Uretzky G (1989) The effect of amino acids on the ischemic heart. J Thorac Cardiovasc Surg 98:551–556

Seymour AML, Eldar H, Radda GK (1990) Hyperthyroidism results in increased glycolytic capacity in the rat heart. A ^{31}P-NMR study. Biochim Biophys Acta 1055:107–116

Souba WW (1993) Glutamine and cancer. Ann. Surg 218:715–728

Suleiman MS, Halestrap AP, Griffiths EJ (2001) Mitochondria: a target for myocardial protection. Pharmacol Therapeut 89:29–46

Sveldjeholm R, Håkanson E, Vanhanen I (1995) Rationale for metabolic support with amino acids and glucose-insulin-potassium (GIK) in cardiac surgery. Ann Thorac Surg 59:S15–S22

Taegtmeyer H (1995) Metabolic support for the postischaemic heart. Lancet 345: 1552–1555

Townsend DM, Tew KD, Tapiero H (2004) Sulfur containing amino acids and human disease. Biomed Pharmacother 58:47–55

Valdares JRE, Singhai RL, Parulekar MR, Beznak M (1969) Influence of aortic coarctation on myocardial glucose-6-phosphate dehydrogenase. Can J Physiol Pharmacol47: 388–391

Williams H, King N, Griffiths EJ, Suleiman MS (2001) Glutamate-loading stimulates metabolic flux and improves cell recovery following chemical hypoxia in isolated cardiomyocytes. J Mol Cell Cardiol 33: 2109–2119

Williamson DH, Lund P, Krebs HA (1967) Redox state of free nicotinamide-adenine dinucleotide in cytoplasm and mitochondria of rat liver. Biochem J 103:514–527

Wu GY, Morris SM (1998) Arginine metabolism: nitric oxide and beyond. Biochem J 336:1–17

Yasuda T, Yamaguchi N, Kobayashi K, Nishi I, Horinouchi H, Jalil A, Li MX, Ushikai M, Lijima M, Kondo I, Saheki T (2000) Identification of two novel mutations in the SLC25A13 gene and detection of seven mutations in 102 patients with adult-onset type II citrullinaemia. Hum Genet 107:537–545

Zweier JL (1988) Measurement of superoxide-derived free radicals in the reperfused heart. J Biol Chem 263:1353–1357

Part 2
The Dynamic Nature of the Mitochondria

7
Mechanotransduction of Shear-Stress at the Mitochondria

Abu-Bakr Al-Mehdi

7.1. Mechanotransduction of Shear–Stress

7.1.1. Physical Forces due to Blood Flow

Endothelial cells in situ experience a variety of physical forces caused by hemodynamics. Tensile stress (a perpendicular force of pressure and stretch) affects mostly the smooth muscle cells, and shear-stress (a tangential, frictional force) affects specifically the endothelial cells.

Endothelial cells that are flow-adapted in vivo or in flow-chambers are different from their statically grown counterparts. However, the common practice of growing endothelial cells in Petri dishes initially overlooked this natural environmental physical factor until its role was convincingly recognized in endothelial cell physiology in the early eighties (Dewey et al. 1981). These physical factors and their role in endothelial cell function have drawn increased attention recently (Lansman 1988; Davies and Tripathi 1993). Hemodynamic forces generated by the flow of blood along the endothelial surface are now known to regulate both the morphology and function of endothelial cells in different physiological and patho-physiological settings. Endothelial cells transform the mechanical stimuli into electrical and biochemical signals in a process called mechano-transduction.

Although the precise mechanisms of cellular mechanotransduction have yet to be elucidated, tensile and shear-stress responsive ion channels and cytoskeletal elements may be involved. Once the role of physical forces in endothelial cell biology was recognized, researchers started studying the effects of shear-stress by applying flow to statically cultured endothelial cells in a variety of devices, like cone viscometers, parallel plate chambers, and artificial capillary tubes. These studies of the effect of increased laminar, turbulent and oscillatory flow on endothelial cell function have relevance for elucidating the pathogenetic mechanisms of hypertension, arteriosclerosis, and other intravascular disorders. Time-dependent responses of endothelial cells subjected to laminar shear-stress include: (1) rapid (within 1 min) hyperpolarization of the cell membrane and activation of G proteins, (2) a slower (1–60 min) activation of MAP kinases and NF-κB, cytoskeletal rearrangement, induction of connexin 43 and cell adhesion proteins, (3) realignment of focal adhesions by 1–6 h, and (4) a delayed (6–24 h) reorganization of cell surface and cellular alignment in the direction of flow (Davies and Tripathi, 1993). After 6 to 24 hours of shear-stress exposure, these cells can be considered 'flow adapted' analogous to their natural counterparts *in vivo*.

Strikingly, some responses of the endothelial cells were shown to be similar for both types of stimuli, i.e., increased shear-stress in non-adapted cells or loss of shear-stress in adapted cells. These responses include pinocytosis (Davies et al. 1984), intracellular calcium (James et al. 1995; Tozawa et al. 1999), reactive oxygen species generation (Hsieh et al. 1998; Wei et al. 1999), and nitric oxide generation (Mashour and Boock 1999; Al-Mehdi et al. 2000a). That is, the response of flow adapted endothelial cells to loss of shear-stress is similar in direction to the response following onset of flow for cells adapted to static conditions. These responses represent a paradigm of alteration from a state of adaptation (Davies 1995), that is, homeostasis at a new baseline level after the subsidence of acute responses to flow or no-flow. The acute responses themselves are the result of sensing increased shear-stress when flow is instituted over endothelial cells adapted to no-flow; or they represent its inverse, the result of sensing loss of shear-stress when flow is stopped for endothelial cells adapted to shear-stress, as would happen during ischemia in vessels in vivo. The high degree of adaptability of endothelial cells allows them to cope with the challenges of hemodynamic alterations that they are constantly exposed to even under physiological conditions and thereby contributing to their fitness for survival.

It is theorized that the sensing of shear-stress proceeds via specialized sensors, which may be specific plasma membrane molecules and microdomains, or via alterations in general biophysical forces such as membrane fluidity and cytoskeletal tension (Davies et al. 2003). The forces of shear-stress may act locally as the sensors detect them, or they may be transmitted to distant sites mechanically by means of the cytoskeleton, to elicit responses from organellar, nuclear, cell-cell junction, and focal adhesion mechanotransducers.

The cis-acting DNA sequences that confer shear-mediated transcriptional activation are not well characterized. Two potential candidates for such a shear-stress response element have been suggested. The first is the SSRE (shear-stress response element) identified as a 6-bp consensus sequence GAGACC, first described for the gene encoding platelet-derived growth factor B (Resnick and Gimbrone 1995). The second is a CT-rich Sp1 binding site within the vascular endothelial growth factor receptor (i.e., Flk-1/KDR) that showed increased occupancy after 3 hours of fluid shear-stress (Abumiya et al. 2002).

It is known that shear-stress affects nuclear gene expression; hundreds of genes are up- or down-regulated under a variety of shear-stress conditions (Chen et al. 2001; Garcia-Cardena et al. 2001; McCormick et al. 2001; Brooks et al.

2002; Kallmann et al. 2002; Peters et al. 2002). It is not known however, if shear-stress affects the expression of mitochondrial genes.

7.1.2. The Mechanotransduction Hypothesis of Endothelial Cell Response to Ischemia

Ischemia or termination of blood flow in organs with systemic circulation not only results in loss of oxygen, reduced substrate supply, and accumulation of metabolites, but also the loss of shear-stress for the endothelial cells. The models of hypoxia/reoxygenation performed on statically grown endothelial cells do not take into account the effects caused by the lack of shear-stress. In the lung, occlusion of the pulmonary artery in the presence of ventilation (as in the case of clinical pulmonary thromboembolism) would not lead to tissue anoxia, but would result in the loss of shear-stress. In fact, during ischemia, local lung tissue pO_2 will rise because oxygen is no longer removed by capillary blood. Therefore, an isolated, air-ventilated, artificial-medium perfused rat or mouse lung not only serves as an excellent model to study endothelial cells *in situ*, but also allows the effects of anoxia, non-hypoxic ischemia, and the loss of shear to be separated. Our experience with this isolated lung model and the development of an intact organ fluorescence imaging technique to study changes in endothelial cell function *in situ* led credence to the mechanotransduction hypothesis of ischemia (Al-Mehdi et al. 1997a; Al-Mehdi et al. 1998a; Al-Mehdi et al. 1998b). This hypothesis provides an explanation for the generation of reactive oxygen species (ROS) by the flow-deprived, but non-hypoxic endothelial cells, (Al-Mehdi et al. 1994; Al-Mehdi et al. 1996; Al-Mehdi et al. 1997a; Al-Mehdi et al. 1997b; Al-Mehdi et al. 1998a; Al-Mehdi et al. 1998b; Wei et al. 1999; Al-Mehdi et al. 2000a; Al-Mehdi et al. 2000b; Song et al. 2001; Wei et al. 2001; Fisher et al. 2002; Fisher et al. 2003; Wei et al. 2004). The basic premise of the hypothesis is that ischemia is sensed by the endothelial cells of the non-hypoxic lung as a loss of shear-stress, an inverse phenomenon of sensing the imposition of shear-stress by statically cultured endothelial cells upon resumption of flow. Therefore, the ischemic response of endothelial cells is a mechanotransduction event.

Recent data suggest a direct role of the mitochondria in flow-sensing and the responses to shear-stress. We postulate that the mitochondrial response is more important for the long-term effects of flow or no-flow, such as cellular and organellar adaptation. Apparently, the endothelial cells must acquire the ability to respond to the loss of shear-stress during flow-adaptation.

7.2. Mitochondrial Mechanotransduction

7.2.1. Why can Mitochondria be Mechanotransducers?

The frictional force that is produced by blood flow is localized at the plasma membrane. It is readily apparent that this force can act locally on membrane-associated mechanosensors. Membrane proteins such as integrins, G-proteins, ion channels and localized rigid domains (cholesterol rich caveolae and lipid rafts) can serve as mechanosensors. As opposed to the localized sensing via molecular or domain-specific mechanisms, global sensing mechanisms are thought to be mediated via alterations in plasma membrane viscosity (or fluidity) and cytoskeletal tension ('tensegrity' as in Stamenovic 2005). The mechanical force of shear-stress can be transmitted to non-apical parts of the cell via cytoskeletal elements. The cytoskeleton has been shown to transmit the mechanical force of shear-stress to intracellular sites. Indeed, the alignment of actin filaments of macrovascular endothelial cells in the direction of flow is responsible for overall cellular alignment and elongation (Malek and Izumo 1996).

Association of mitochondria with the cytoskeleton makes them potential sites of mechanotransduction. Mitochondrial movement uses the molecular motors kinesin and dynein that associate with microtubules and tether mitochondria to them. Microtubules are the thickest and most rigid of all cytoskeletal fibers; therefore, they are more suitable for transmission of mechanical force. The transmitted force of shear-stress, acting on points of contact between microtubules and organelles or basolateral focal adhesion sites, may be transduced locally to produce a signaling event. Mechanotransmission via microtubules, but not via intermediate or microfilaments, has been implicated in nuclear morphology changes with shear-stress (Stamatas and McIntire 2001).

The mechanotransduction of shear-stress can be modeled analogous to ligand-receptor signal transduction. For flow-mediated signaling, the 'ligand' or primary signal is the mechanical force and the 'receptor' of this force is a mechanotransducer. The downstream components of mechanotransduction signaling share the same second messengers and biochemical pathways as the conventional ligand-receptor systems.

Once the mechanical force has been transmitted to the mitochondrial membrane, a mechanotransducer on the mitochondrial membrane (inner or outer) would be required to convert the force into a biochemical signal. The identity of mitochondrial mechanotransducers is less clear than the mechanotransducers of the plasma membrane. The mtK_{ATP} channel, the Na^+/H^+, Na^+/Ca^{2+}, and K^+/H^+ exchangers, and the mitochondrial permeability transition pore can be considered as potential mechanotransducers. Stretch, a mechanical force that is experienced by vascular cells upon distension, has been shown to affect cellular responses with mitochondrial participation (Ali et al. 2004; Liao et al. 2004).

Actin filaments are the other cytoskeletal elements that need to be considered for mechanotransmission. If microtubules are tracks for movement, then actin filaments are stations for mitochondrial clustering or stopping (Boldogh and Pon 2006). Although actin filaments have lower elasticity than microtubules, their bundled and networked structure may be strong enough to transmit mechanical forces from the plasma membrane to the stationary mitochondria. The actin network has been shown to significantly contribute to cellular rigidity (Janmey et al. 1991).

7.2.2. Effect of Flow and Flow Cessation on Mitochondrial Function

Exposure of endothelial cells to flow or ischemia leads to changes in mitochondrial shape, size, position, number, and functional status. A shear-stress of 10 dyn/cm^2 increases mitochondrial membrane potential by 30%; but a higher shear-stress (60 dyn/cm^2) decreases mitochondrial membrane potential by 20% (Kudo et al. 2000).

Mitochondria of lung pulmonary microvascular endothelial cells subjected to 1–10 dynes/cm^2 laminar flow rapidly hyperpolarize (increased JC-1 fluorescence) upon cessation of flow, suggesting that the mitochondria sense and functionally respond to the loss of shear-stress (Figure 7.1). The response is seen in both pulmonary macro- and microvascular cell types after flow-adaptation. Acquisition of the hyperpolarization response to flow-cessation in flow-adapted macro- and macrovascular cell types is a paradigm of common functional adaptation. In contrast, only pulmonary macrovascular cells seem to exhibit signs of morphological adaptation to flow by elongation and alignment in the direction of the flow. This dissociation of morphological and functional adaptive responses to flow in the two cell types suggests the presence of separate mechanisms of flow-mediated signal transduction.

Microtubule disruption with 1 μM colchicine prevents flow cessation-induced mitochondrial hyperpolarization, suggesting that microtubules are involved in mechanotransmission (Figure 7.2). Moreover, the depolarization response to flow-cessation suggests that mitochondrial membrane potential is coupled to shear-stress via some other mechanisms. A role for the actin cytoskeleton may explain this 'paradoxical' depolarization response to the loss of shear-stress. A microtubular role for maintenance of mitochondrial membrane potential is

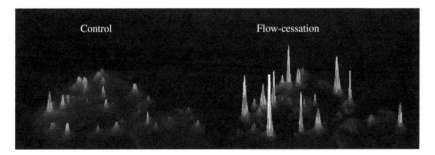

FIGURE 7.1. The left panel shows intensity plot of a rat pulmonary artery endothelial cell flow-adapted for 20 hours at 10 dynes/cm^2 and stained with JC-1 (Control). Right panel (Flow-cessation) shows the same cell after flow was stopped for 1 min while maintaining oxygenation. The increase in JC-1 red aggregate formation (indicated by increase in TRITC-channel fluorescence) indicates mitochondrial hyperpolarization. A purple-to white lookup table applied for pseudocolor rendering of intensity. Increases in green, yellow and red colors indicate that mitochondrial membrane hyperpolarization accompanies the loss of shear-stress.

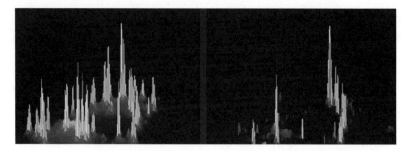

FIGURE 7.2. Flow-adapted endothelial cells were treated with 1 µM colchicine for 30 min before being subjected to flow-cessation. Panel on the left shows intensity plots of JC-1 aggregate fluorescence during flow showing polarized mitochondria. Flow-cessation in the presence of colchicine not only prevented mitochondrial hyperpolarization, but also led to a depolarization response.

TABLE 7.1. Changes in Ψ_m with microtubule stabilization and disruption

Condition	JC-1 Aggregate Fluorescence, Arbitrary Units	JC-1 Monomer Fluorescence, Arbitrary Units
Control	1376 ± 89 (n=4)	768 ± 56 (n=4)
Colchicine	756 ± 101 (n=3)*	1024 ± 65 (n=3)*
Paclitaxel	2215 ± 176 (n=3)#	460 ± 37 (n=3)#

Note: Mitochondria were imaged with a JC-1 filter set. The ratio of aggregate to monomer fluorescence, which increases at more negative Ψ_m, was 1.79 for control vs 0.73 and 4.8 upon colchicine and paclitaxel treatment, respectively. *p < 0.05 vs control; #p < 0.5 vs control and colchicine.

indicated by the contrasting effects of colchicine and paclitaxel (a microtubule stabilizer) (Table 7.1).

7.2.3. Mitochondrial Morphology and Function Under Shear-stress

It has been demonstrated using MitoTracker Green that endothelial cells at branch points of lung capillaries contain two-fold more mitochondria than their midsegment counterparts. The spatial pattern of *in situ* pulmonary endothelial cell intracellular Ca^{2+} levels and reactive oxygen generation in response to TNF-α stimulation positively correlate with the mitochondrial content at those sites (Parthasarathi et al. 2002). The major difference in the environment of midsegment and branch point cells is the magnitude and pattern of shear-stress. In the midsegment, flow is thought to be laminar and the shear-stress of consistent magnitude; while in the branch points, flow is thought to be turbulent and the shear-stress – highly variable. These observations suggest that shear-stress not only regulates the endothelial cell phenotype, but also affects the mitochondrial mass and their function in individual cells.

TABLE 7.2. Mitochondrial morphological parameters

Cell type and growth condition	Mitochondrial number/cell	Mitochondrial mass, μm^3	Mitochondrial length, μm
PAEC in static culture	257 ± 43	404 ± 34	8 ± 3
PAEC flow-adapted *in vitro*	145 ± 26*	199 ± 18*	7 ± 2
PAEC *in situ* in rat lung	36 ± 6	35 ± 6	5 ± 2
PMVEC in static culture	167 ± 32	557 ± 66	17 ± 6
PMVEC flow-adapted *in vitro*	95 ± 14*	279 ± 36*	15 ± 4
PMVEC *in situ* in rat lung	24 ± 7	19 ± 5	4 ± 2

Note: Mitochondrial mass was calculated from volume measurements, assuming a uniform cylindrical shape (diameter 0.5 μm). The number of cells analyzed was 5–15. *p < 0.05 vs static culture of same cell type.

The most striking finding was the relative paucity of mitochondria compared to their content in endothelial cells in static culture. Endothelial cells *in situ* are naturally flow-adapted. Indeed, Table 7.2 reveals that flow-adapted cells *in vitro* and *in vivo* have less mitochondria, compared to their statically cultured counterparts. The decreased mitochondrial content may be related to decreased energy requirements of flow-adapted endothelial cells that are more quiescent than cells under static culture. Figures 7.3 and 7.4 show mitochondrial disposition in endothelial cells in static culture and *in vivo* (naturally flow-adapted) respectively. Mitochondria in naturally flow-adapted pulmonary precapillary endothelial cells in intact rat lung exhibit a paucity of mitochondria, all with an apparent perinuclear distribution. Average mitochondrial diameter of the flow-adapted cells is approximately 0.5 μm. (Figure 7.3).

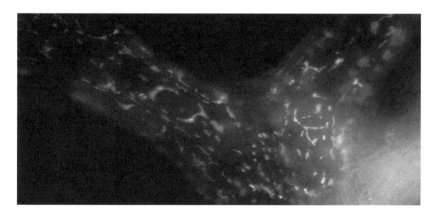

FIGURE 7.3. Intact organ fluorescence imaging of endothelial mitochondria in a subpleural pulmonary precapillary vessel in the rat lung. Isolated rat lung was perfused with 100 nM MitoTracker Green FM for 30 min and then the dye was washed out. Images of endothelial cell mitochondria in situ were acquired by transpleural imaging of the perfused lung with a 60x objective and Hamamatsu ORCA 100 ER digital camera using a Nikon 2000-TE microscope. Mitochondria appear as dots and thread and are located perinuclearly.

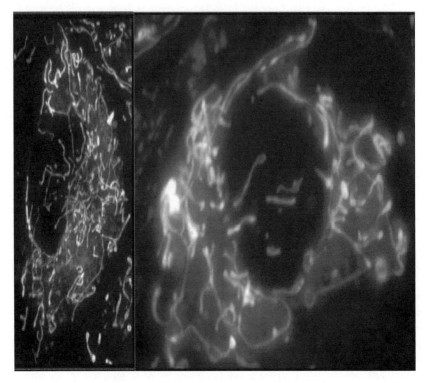

FIGURE 7.4. Mitochondria in pulmonary microvascular endothelial cells were labeled with 100 nM Mitotracker Red 633, and then stacks of z-axis images were acquired at 0.5 μm intervals. A single projection from a 3-D reconstructed image was obtained after PSF-based deconvolution using MetaMorph (Molecular Devices, Inc) software.

In contrast to the endothelial cells *in situ* which contain a few mitochondria, rat pulmonary microvascular endothelial cells in static culture contain numerous mitochondria per cell with different shapes and sizes, which are scattered throughout the cytoplasm (Figure 7.5). However, when pulmonary microvascular cells were cultured *in vitro* under 20 dyn/cm^2 shear-stress for 24 h, the cells assume the phenotype seen under in vivo flow. The mitochondria gradually acquire spherical shape, and after an initial increase, their velocity diminishes (Figure 7.6). These alterations indicate that the mitochondria may be an important site of mechanotransduction. This idea is also supported by the observation that disruption of microtubules, which interferes with mechanotransduction, prevents the maintenance of mitochondrial shape and structure (Figure 7.7). They remain straighter. This pattern of microtubule changes may alter the pattern of mitochondrial movement under flow, with the mitochondrial translocation being more restrictive when the microtubules assume the loop configuration (Figure 7.8).

Kudo et al. (2000) have shown that endogenous ATP content increases 70% with acute imposition of shear-stress on statically cultured cells (Kudo et al. 2000). When endothelial cells were adapted first to 12 h of flow at 10 dynes/cm^2

7. Mechanotransduction of Shear-stress at the Mitochondria   177

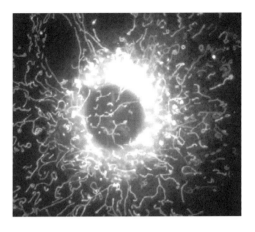

FIGURE 7.5. Mitochondria in a pulmonary microvascular endothelial cell were labeled with 100 nM MitoTracker Green FM for 30 min and then the dye was washed out. Images were acquired with a 100x objective and Hamamatsu ORCA 100 ER digital camera using a Nikon 2000-TE microscope. Numerous mitochondria appear as elongated threads. The picture was acquired with a monochrome camera, therefore, no pseudocolor rendition was obtained.

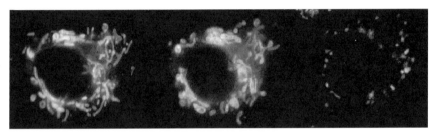

FIGURE 7.6. Reduction in mitochondrial number and other morphological changes in a pulmonary microvascular endothelial cell under flow. Cells were plated on coverslips and subjected to 20 dyn/cm^2 for up to 24 h in a parallel plate chamber. Pictures were taken after 2 h (left panel), 6 h (middle), and 24 h (right). Shape changes and gradual decrease in mitochondrial mass is observed with flow-adaptation. Labeling was with MitoTracker Green. The picture was acquired with a monochrome camera, therefore, no pseudocolor rendition was obtained.

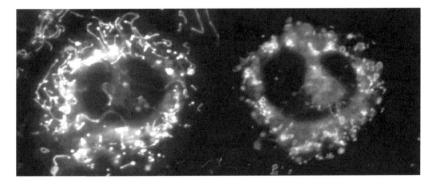

FIGURE 7.7. Colchicine treatment led to disruption of mitochondrial structure and hazy labeling with MitoTracker Red 633, indicating loss of mitochondrial membrane integrity (bottom right). The picture was acquired with a monochrome camera, therefore, no pseudocolor rendition was obtained.

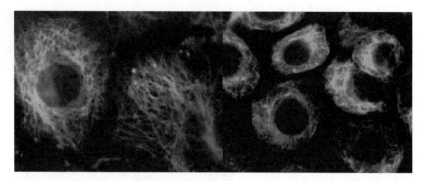

FIGURE 7.8. Endothelial cells from rat pulmonary microvasculature (left panel) and pulmonary artery (right panel) were plated on coverslips and subjected to 20 dyn/cm^2 for 3 h in a parallel plate chamber. Microtubules were stained with Oregon Green paclitaxel and mitochondria with MitoTracker Red 633.

in a dual-chamber parallel plate flow chamber, and then subjected to 6 h of non-hypoxic flow-cessation, ATP generation increased compared with statically cultured cells (Control, Figure 7.9). Hyperpolarization of mitochondria with flow-cessation (Figure 7.1) is compatible with the increased ATP generation. Twenty-four hour of flow-adaptation led to significant decrease in intracellular ATP levels compared to control (Figure 7.9).

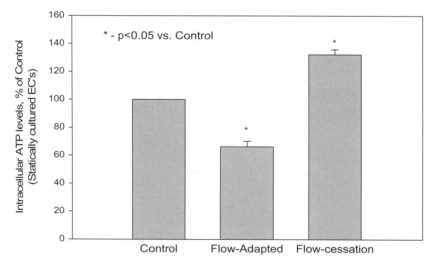

FIGURE 7.9. ATP generation with flow-adaptation and flow-cessation in pulmonary artery endothelial cells. Control: statically cultured cells; Flow-adapted: 24 h flow-adaptation at 10 dynes/cm^2; Flow-cessation: 6 h flow-cessation following a 12 h flow-adaptation period.

7.3. Conclusions

While the importance of mechanical forces, including shear-stress, in a variety of human diseases is now recognized, organellar mechanosensing has only begun to receive some attention. The mitochondrial responses to mechanical forces are distinct from the general cellular responses because of their specific role in ATP generation, Ca^{2+} homeostasis, apoptosis, aging, atherogenesis, and intermediary metabolism. Although the downstream pathways of shear-stress-induced cell signaling appear to mimic a ligand-receptor-second messenger model, the nature of the mechanosensor still remains unknown. Despite the recognized role of a shear-stress response element (the GAGACC SSRE) in flow regulated genes, the mechanism of flow regulation of transcription factors that bind to the SSRE is an area of speculation. Future studies that define the organellar mechanosensors, mechanotransduction pathways, and characterize the nature of pathogenetic mechanostimuli, might be helpful in the prevention of vascular disease.

References

Abumiya T, Sasaguri T, Taba Y, Miwa Y, Miyagi M (2002) Shear stress induces expression of vascular endothelial growth factor receptor Flk-1/KDR through the CT-rich Sp1 binding site. Arterioscler Thromb Vasc Biol 22: 907–13

Al-Mehdi A, Shuman H, Fisher AB (1994) Fluorescence microtopography of oxidative stress in lung ischemia-reperfusion. Lab Invest 70: 579-87

Al-Mehdi AB, Ischiropoulos H, Fisher AB (1996) Endothelial cell oxidant generation during K(+)-induced membrane depolarization. J Cell Physiol 166: 274–80

Al-Mehdi AB, Shuman H, Fisher AB (1997a) Intracellular generation of reactive oxygen species during nonhypoxic lung ischemia. Am J Physiol 272: L294–300

Al-Mehdi AB, Shuman H, Fisher AB (1997b) Oxidant generation with K(+)-induced depolarization in the isolated perfused lung. Free Radic Biol Med 23: 47–56

Al-Mehdi AB, Song C, Tozawa K, Fisher AB (2000a) Ca2+- and phosphatidylinositol 3-kinase-dependent nitric oxide generation in lung endothelial cells in situ with ischemia. J Biol Chem 275: 39807–10

Al-Mehdi AB, Zhao G, Dodia C, Tozawa K, Costa K, Muzykantov V, Ross C, Blecha F, Dinauer M, Fisher AB (1998a) Endothelial NADPH oxidase as the source of oxidants in lungs exposed to ischemia or high K+. Circ Res 83: 730–7

Al-Mehdi AB, Zhao G, Fisher AB (1998b) ATP-independent membrane depolarization with ischemia in the oxygen-ventilated isolated rat lung. Am J Respir Cell Mol Biol 18: 653–61

Al-Mehdi AB, Zhao G, Tozawa K, Fisher AB (2000b) Depolarization-associated iron release with abrupt reduction in pulmonary endothelial shear stress in situ. Antioxid Redox Signal 2: 335–45

Ali MH, Pearlstein DP, Mathieu CE, Schumacker PT (2004) Mitochondrial requirement for endothelial responses to cyclic strain: implications for mechanotransduction. Am J Physiol Lung Cell Mol Physiol 287: L486–96

Boldogh IR, Pon LA (2006) Interactions of mitochondria with the actin cytoskeleton. Biochim Biophys Acta

Brooks AR, Lelkes PI, Rubanyi GM (2002) Gene expression profiling of human aortic endothelial cells exposed to disturbed flow and steady laminar flow. Physiol Genomics 9: 27–41

Chen BP, Li YS, Zhao Y, Chen KD, Li S, Lao J, Yuan S, Shyy JY, Chien S (2001) DNA microarray analysis of gene expression in endothelial cells in response to 24-h shear stress. Physiol Genomics 7: 55-63

Davies PF (1995) Flow-mediated endothelial mechanotransduction. Physiol Rev 75: 519–60

Davies PF, Dewey CF, Jr., Bussolari SR, Gordon EJ, Gimbrone MA, Jr. (1984) Influence of hemodynamic forces on vascular endothelial function. In vitro studies of shear stress and pinocytosis in bovine aortic cells. J Clin Invest 73: 1121–9

Davies PF, Tripathi SC (1993) Mechanical stress mechanisms and the cell. An endothelial paradigm. Circ Res 72: 239–45

Davies PF, Zilberberg J, Helmke BP (2003) Spatial microstimuli in endothelial mechanosignaling. Circ Res 92: 359–70

Dewey CF, Jr., Bussolari SR, Gimbrone MA, Jr., Davies PF (1981) The dynamic response of vascular endothelial cells to fluid shear stress. J Biomech Eng 103: 177–85

Fisher AB, Al-Mehdi AB, Manevich Y (2002) Shear stress and endothelial cell activation. Crit Care Med 30: S192–7

Fisher AB, Al-Mehdi AB, Wei Z, Song C, Manevich Y (2003) Lung ischemia: endothelial cell signaling by reactive oxygen species. A progress report. Adv Exp Med Biol 510: 343–7

Garcia-Cardena G, Comander J, Anderson KR, Blackman BR, Gimbrone MA, Jr. (2001) Biomechanical activation of vascular endothelium as a determinant of its functional phenotype. Proc Natl Acad Sci U S A 98: 4478–85

Hsieh HJ, Cheng CC, Wu ST, Chiu JJ, Wung BS, Wang DL (1998) Increase of reactive oxygen species (ROS) in endothelial cells by shear flow and involvement of ROS in shear-induced c-fos expression. J Cell Physiol 175: 156–62

James NL, Harrison DG, Nerem RM (1995) Effects of shear on endothelial cell calcium in the presence and absence of ATP. Faseb J 9: 968–73

Janmey PA, Euteneuer U, Traub P, Schliwa M (1991) Viscoelastic properties of vimentin compared with other filamentous biopolymer networks. J Cell Biol 113: 155–60

Kallmann BA, Wagner S, Hummel V, Buttmann M, Bayas A, Tonn JC, Rieckmann P (2002) Characteristic gene expression profile of primary human cerebral endothelial cells. Faseb J 16: 589–91

Kudo S, Morigaki R, Saito J, Ikeda M, Oka K, Tanishita K (2000) Shear-stress effect on mitochondrial membrane potential and albumin uptake in cultured endothelial cells. Biochem Biophys Res Commun 270: 616–21

Lansman JB (1988) Endothelial mechanosensors. Going with the flow. Nature 331: 481–2

Liao XD, Wang XH, Jin HJ, Chen LY, Chen Q (2004) Mechanical stretch induces mitochondria-dependent apoptosis in neonatal rat cardiomyocytes and G2/M accumulation in cardiac fibroblasts. Cell Res 14: 16–26

Malek AM, Izumo S (1996) Mechanism of endothelial cell shape change and cytoskeletal remodeling in response to fluid shear stress. J Cell Sci 109 (Pt 4): 713–26

Mashour GA, Boock RJ (1999) Effects of shear stress on nitric oxide levels of human cerebral endothelial cells cultured in an artificial capillary system. Brain Res 842: 233–8

McCormick SM, Eskin SG, McIntire LV, Teng CL, Lu CM, Russell CG, Chittur KK (2001) DNA microarray reveals changes in gene expression of shear stressed human umbilical vein endothelial cells. Proc Natl Acad Sci USA 98: 8955–60

Parthasarathi K, Ichimura H, Quadri S, Issekutz A, Bhattacharya J (2002) Mitochondrial reactive oxygen species regulate spatial profile of proinflammatory responses in lung venular capillaries. J Immunol 169: 7078–86

Peters DG, Zhang XC, Benos PV, Heidrich-O'Hare E, Ferrell RE (2002) Genomic analysis of the immediate/early response to shear stress in human coronary artery endothelial cells. Physiol Genomics

Resnick N, Gimbrone MA, Jr. (1995) Hemodynamic forces are complex regulators of endothelial gene expression. FASEB J 9: 874–82

Song C, Al-Mehdi AB, Fisher AB (2001) An immediate endothelial cell signaling response to lung ischemia. Am J Physiol Lung Cell Mol Physiol 281: L993–1000

Stamatas GN, McIntire LV (2001) Rapid flow-induced responses in endothelial cells. Biotechnol Prog 17: 383–402

Stamenovic D (2005) Microtubules may harden or soften cells, depending of the extent of cell distension. J Biomech 38: 1728–32

Tozawa K, Al-Mehdi AB, Muzykantov V, Fisher AB (1999) In situ imaging of intracellular calcium with ischemia in lung subpleural microvascular endothelial cells. Antioxid Redox Signal 1: 145–54

Wei Z, Al-Mehdi AB, Fisher AB (2001) Signaling pathway for nitric oxide generation with simulated ischemia in flow-adapted endothelial cells. Am J Physiol Heart Circ Physiol 281: H2226–32

Wei Z, Costa K, Al-Mehdi AB, Dodia C, Muzykantov V, Fisher AB (1999) Simulated ischemia in flow-adapted endothelial cells leads to generation of reactive oxygen species and cell signaling. Circ Res 85: 682–9

Wei Z, Manevich Y, Al-Mehdi AB, Chatterjee S, Fisher AB (2004) Ca^{2+} flux through voltage-gated channels with flow cessation in pulmonary microvascular endothelial cells. Microcirculation 11: 517–26

Part 3
Mitochondria as Initiators of Cell Signaling

8
Formation of Reactive Oxygen Species in Mitochondria

Julio F. Turrens

8.1. Mitochondrial Sources of Reactive Oxygen Species

Living organisms obtain energy from the oxidation of various biomolecules, including carbohydrates, lipids and the carbon skeletons of amino acids. Under aerobic conditions, the reducing coenzymes produced during these reactions are re-oxidized in the electron transport chain, transferring electrons to molecular oxygen ($E°= +800$ mV) through a series of electron carriers in the respiratory chain. This electrochemical energy is converted into a proton gradient which, in turn, operates a rotor-type enzymatic complex (ATP synthase or Complex V), inducing conformational changes which cause ADP and inorganic phosphate to bind to the active site and ATP to be released (Noji and Yoshida 2001).

Although most biological oxidations involve removal of electron pairs, the electronic configuration of the outer layer of molecular oxygen prevents it from accepting more than one electron at a time. Thus, the complete reduction of oxygen to water at the end of the respiratory chain must occur as a series of four monovalent steps. Two of the partially reduced intermediates formed in this process, superoxide anion and hydrogen peroxide, are quite stable and eventually may become substrates in other redox reactions instead of continuing their reduction to water. Although the terminal electron carrier in the respiratory chain, cytochrome oxidase, retains all partially reduced intermediates until oxygen is completely reduced to water, other redox centers in the respiratory chain thermodynamically capable of reducing oxygen to superoxide anion ($E°= +310$ mV) may divert some electrons towards the formation of this radical. The resulting superoxide anion, in turn, may dismute (spontaneously or enzymatically) producing hydrogen peroxide. It has been estimated that the proportion of oxygen utilized in the production of these partially reduced species may account for up to 1–2% of the total oxygen used in the respiratory chain.

Two main factors control the rate of mitochondrial superoxide formation: oxygen concentration and the rate of electron flow. As electron carriers in the respiratory chain become more reduced (for example, in the absence of ADP to phosphorylate) their actual reduction potential becomes more negative favoring superoxide anion formation at sites different from the terminal oxidase. Similarly, in the presence of inhibitors of the electron transport chain, some (not all) electron carriers reduce oxygen to superoxide anion at a faster rate.

Various mitochondrial electron carriers reduce molecular oxygen to superoxide anion on both sides of the inner mitochondrial membrane (Lenaz 2001; Turrens 2003) Some of the sources include most of the sites of electron entry: Complex I (NADH dehydrogenase, Turrens and Boveris 1980; Turrens 2003), Complex II (succinate dehydrogenase, Lenaz 2001), glycerol-phosphate dehydrogenase (Lenaz 2001) and Complex III (ubiquinol-cytochrome c reductase, Cadenas et al. 1977; Turrens et al. 1985; Turrens 2003; Muller et al. 2004). Most of the superoxide produced by Complex III results from the reaction between molecular oxygen and ubisemiquinone produced during the Q-cycle, a reaction that occurs on both sides of the inner mitochondrial membrane (Muller et al. 2004). While most of the superoxide will eventually be converted into hydrogen peroxide by two different superoxide dismutases, part of the superoxide released into the inter-membrane space directly reacts with cytochrome c and it is reoxidized to molecular oxygen (see below). The resulting reduced cytochrome c is, in turn, reoxidized by Complex IV, thus contributing energy for ATP production.

Ubisemiquinone formation in Complex III requires one of the two electrons in ubiquinol to be transferred through cytochrome c to the terminal oxidase (Complex IV). Therefore, this reaction is inhibited by cyanide and other cytochrome oxidase inhibitors (Turrens et al. 1985; Turrens 2003). This fact is sometimes overlooked by investigators not familiar with the Q-cycle, who tend to think that because inhibition of Complex IV increases the steady state reduction level of all respiratory components, superoxide formation in Complex III should also increase. In summary, inhibitors blocking cytochrome b, like antimycin, stimulate superoxide formation in Complex III while inhibitors blocking ubiquinol oxidation (myxothiazol, cyanide, and carbon monoxide) inhibit this process. Yet, all these inhibitors stimulate superoxide formation in Complex I, because they all cause this complex to be fully reduced. The contribution of each mitochondrial complex to the intracellular steady state concentration of hydrogen peroxide varies from tissue to tissue, depending on the relative stoichiometry among the various respiratory components.

It is not clear which of the many redox centers found in Complex I is responsible for superoxide formation. The proposed sources of reactive oxygen species (ROS) include two iron-sulfur centers that are part of the Complex (either N2 (Genova et al. 2001) or N1a (Kushnareva et al. 2002) as well as the semiquinone form of FMN, the prosthetic group of NADH-dehydrogenase (Vinogradov and Grivennikova 2005). Although center N1a has a very low reduction potential (-370 mV) and does not participate in electron transport within Complex I, it has been proposed that it may act as a sink of electrons when Complex I becomes highly reduced, preventing the autoxidation of FMN (Hinshliffe and Sazanov 2005). This model would also explain the observation that Complex I becomes a significant source of superoxide anion in the presence of mitochondrial inhibitors (for example, rotenone, antimycin, Turrens and Boveris 1980, Turrens et al. 1982a).

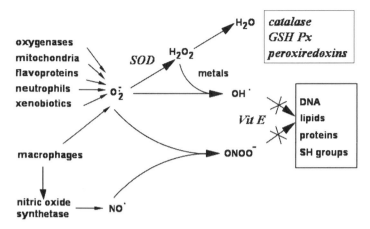

FIGURE 8.1. Intracellular sources of ROS and antioxidant defenses. The main ROS produced in biological systems is superoxide anion. This radical can dismute to hydrogen peroxide which, in turn, may be reduced to water by various peroxidases (catalase, glutathione (GSH) peroxidase (Px) and peroxiredoxins. Unless superoxide anion is eliminated, it may reduce transition metals and these transition metals in turn may catalyze the formation of hydroxyl radical ($\cdot$OH) a very powerful oxidant. Superoxide may also react with nitric oxide to generate peroxynitrite, another strong oxidant. These strong oxidants are not eliminated by enzymes but rather quenched by low molecular weight scavengers such as vitamin E and ubiquinone.

spontaneously or through the reaction catalyzed by superoxide dismutase, producing hydrogen peroxide and oxygen. Hydrogen peroxide may further react with reduced transition metals through a Fenton reaction, producing hydroxyl radicals, one of the most powerful oxidants in nature. Superoxide also plays a direct role in this reaction as a reductant for transition metals, in particular iron. Moreover, superoxide may also reduce ferric iron bound to ferritin, the protein responsible for intracellular iron storage, increasing the pool of available iron for these reactions. Finally, superoxide anion is also a precursor of peroxynitrite, a third ubiquitous oxidant produced as a result of the diffusion-controlled reaction between superoxide and nitric oxide. Nitric oxide is produced from arginine by a group of enzymes known as nitric oxide synthases or NOS. Two independent laboratories identified one of these isozymes in the mitochondrial matrix (Giulivi et al. 1998; Richter et al. 1999). This oxidant, sometimes considered a "reactive nitrogen species", is both an oxidant and a nitrating species which has been linked to a various pathological processes (Skinner et al. 1997).

8.3. Mitochondrial Antioxidant Defenses

The relative rates of ROS production and decomposition determine their steady state concentration and, ultimately, their potential to cause tissue injury. Whether enzymes or low molecular weight antioxidants are responsible for the elimination

The enzyme dihydroorotate dehydrogenase is another respiratory component capable of producing superoxide anion although its contribution is less important given the very specific role of this enzyme in pyridine metabolism (Forman and Kennedy 1976; Lenaz 2001).

In addition to the respiratory chain complexes mentioned above, several other mitochondrial oxidoreductases can also transfer electrons directly to oxygen, producing superoxide anion. One of these enzymes, monoamino oxidase, is located in the outer mitochondrial membrane of many tissues and catalyzes the oxidative deamination of various amines. In some tissues such as the nervous system, this reaction contributes substantially to the intracellular steady state concentration of hydrogen peroxide (Cohen et al. 1997).

Recent studies have shown that the α-ketoglutarate dehydrogenase complex also produce ROS, but in this case the rate of hydrogen peroxide production by this system depends on the NADH/NAD ratio (Tretter and Adam-Vizi 2004). Another study has shown that both α-ketoglutarate and pyruvate dehydrogenase complexes can produce ROS, and proposed that the flavoprotein dihydrolipoyl dehydrogenase, one of the enzymes common to both complexes, is responsible for this activity (Starkov et al. 2004).

The α-ketoglutarate dehydrogenase complex is not only a source of ROS but also a target. Tretter and Adam-Vizi (2000) also showed that this complex was inhibited by hydrogen peroxide and proposed that the physiological role of this reversible process could be to decrease the supply of NADH to the respiratory chain when the steady state concentration of hydrogen peroxide starts to increase, thus decreasing the source of electrons for ROS formation. It was later shown that the reversible inhibition of this complex resulted from the covalent binding of glutathionyl groups to critical sulfhydryl groups in the protein, and that this process could be reverted by the action of a recently identified mitochondrial glutaredoxin (Nulton-Persson and Szweda 2003).

In addition to all the mitochondrial sources of reactive oxygen species described above, some of the mitochondrial redox centers may transfer electrons to various xenobiotics, which in turn may react with oxygen generating superoxide anion. As an example, the antitumor agent doxorubicin may accept electrons from the mitochondrial Complex I, causing cardiotoxicity (Kotamraju et al. 2004).

8.2. Relative Reactivity of Various Reactive Oxygen Species

How reactive or toxic are these ROS? Many investigators new to the field tend to assume that the term ROS involves a series of highly reactive oxidants. Yet, not all these species are equally reactive. It is important to determine which are the species produced in every case before proposing any models to explain experimental observations (Figure 8.1).

Superoxide anion is the product resulting from the monovalent reduction of oxygen. Although superoxide is not a powerful oxidant, it plays a major role in oxidative stress through three different reactions. First, superoxide dismute of ROS depends on the reactivity of each specific ROS toward biomolecules. The elimination of strong oxidants (singlet oxygen and hydroxyl radical) is a non-enzymatic process since these oxidants react indiscriminately with any biological molecule, including proteins. Instead, strong oxidants are eliminated through direct reactions with small molecular weight antioxidants (for example, vitamin E, vitamin C and ubiquinone (Beyer 1994)). Less reactive species (superoxide anion, hydrogen peroxide and hydroperoxides) are eliminated by specific enzymes (Figure 8.1). The mitochondrion has its own set of enzymes and other antioxidants, many of which are different from those catalyzing the same reaction elsewhere in the cell.

Given the role of superoxide anion as a precursor of most oxidative reactions, it is important that this species be kept at the lowest possible steady state concentration. Superoxide is the substrate of various enzymes and proteins located in all mitochondrial compartments. The mitochondrial matrix contains Mn-superoxide dismutase (SOD-2), an enzyme of bacterial origin, which, like most mitochondrial proteins, is coded in the cell nucleus. This enzyme can be induced by a variety of cellular stresses including hypoxia (Sjostrom and Crapo 1981; Privale et al. 1989), hyperoxia (Clerck et al. 1998), radiation (Oberley et al. 1987), interleukins (Das et al. 1995) and many more forms of stress (MacMillan-Crow and Cruthirds et al. 2001).

The intermembrane space contains the cytosolic isoform of superoxide dismutase (Cu,Zn-SOD or SOD-1) (Okado-Matsumoto and Fridovich 2001). In addition to SOD, cytochrome c, one of the respiratory chain electron carriers located in the intermembrane space, provides a second line of protection against superoxide anion produced in this compartment. In the oxidized form, this respiratory chain electron carrier can oxidize superoxide anion, generating molecular oxygen. As a result, in this unique scenario, superoxide anion returns the electron to the respiratory chain, contributing to oxidative phosphorylation (Xu 2005).

Mitochondrial hydrogen peroxide and other peroxides are eliminated by a group of enzymes called peroxidases. Glutathione peroxidases are a group of selenium-containing proteins which reduce hydrogen peroxide and lipoperoxides at the expense of reduced glutathione, thereby acting as a hydrogen donor (Ursini et al. 1995). The resulting oxidized glutathione is re-reduced by glutathione reductase, using NADPH as a co-substrate. Mitochondria contain various isoforms of this peroxidase with different specificity. The L-form of phospholipid hydroperoxide glutathione peroxidase (PHGPx, an enzyme that specifically reduces oxidized phospholipids in the membrane) contains the N-terminus mitochondrial signal and is only found in this organelle (Imai and Nakagawa 2003). It has been proposed that this enzyme plays an important role in preventing apoptosis by blocking the release of cytochrome c in response to the oxidation of cardiolipin (Imai and Nakagawa 2003).

Mitochondria also contain a second type of thiol-dependent small molecular weight peroxidase, known as peroxiredoxin (Rhee et al. 2001). This enzyme decomposes hydrogen peroxide using a different type of thiol-containing hydrogen donor, known as thioredoxin (Rhee et al. 2001). The predominant peroxiredoxins found in mitochondria are peroxiredoxin-III and V, the latter also found in other organelles (Chang et al., 2004; Banmayer et al. 2005). Catalase, a peroxidase normally found in peroxisomes of most cells and in the cytoplasm of erythrocytes is also found in the matrix of heart mitochondria (Radi et al. 1991) but it is absent in mitochondria from all other tissues tested, including skeletal muscle (Phung et al. 1994).

Some mitochondrial coenzymes involved in various metabolic processes also serve as low molecular weight antioxidants. For example, ubiquinone (coenzyme Q), the quinone responsible for carrying electrons between various dehydrogenases and Complex III, is both a source and a scavenger of ROS, depending on other metabolic conditions (Beyer 1994). Lipoic acid, a coenzyme in the dihydrolipoyl transacylase component of both pyruvate and α-ketoglutarate dehydrogenase complexes, is not only responsible for transferring acetyl or succinyl groups to Coenzyme A, but also is a source of free sulfhydryl groups that may act as free radical scavengers. Similarly, reduced glutathione acts as a direct scavenger of hydroxyl radicals in addition to its role as the hydrogen donor for glutathione peroxidase. Finally, mitochondria contain vitamin E whose exclusive role is to terminate the chain of free radical-dependent reactions within membranes (Cooper and Shapira 2003; Fariss et al. 2005).

8.4. Physiological and Pathological Scenarios Associated with Mitochondrial Reactive Oxygen Species Metabolism

8.4.1. Metal-Mediated Pathology

Oxidative stress is a result of either an increased formation of ROS or limited antioxidant defenses, since both processes are responsible for the steady state concentration of ROS. The two most common processes leading to an increased ROS formation in mitochondria are: a) inhibition of electron flow in the respiratory chain, thus increasing the reduction level of electron carriers that transfer one electron to oxygen and b) increased redox cycling by xenobiotics or metals that may catalyze or propagate free radical formation. In many cases, the increased mitochondrial oxidative cascade triggers an opening of the permeability transition pore, allowing cytochrome c to be released into the cytoplasm, starting the process of apoptosis. This topic is discussed elsewhere in this book.

An increased mitochondrial steady concentration of reactive oxygen species plays an important role in various genetic diseases. For example, Friedreich's ataxia is associated with a genetic deficiency in a nuclear encoded protein (frataxin), whose function is not clear but appears to be involved in mitochondrial iron metabolism (Shapira 2002; Cooper and Shapira 2003). This mutation alters the assembly of iron sulfur proteins (important prosthetic groups in Complex I, II and III as well as in the enzyme fumarase in the Krebs' cycle) inhibiting respiration. Moreover, the pool of available free iron is also increased, providing a catalyst for the formation of hydroxyl radicals through a Fenton-type reaction with hydrogen peroxide.

Similarly, Wilson disease affects copper homeostasis and increases the availability of free copper that may catalyze lipid peroxidation while inhibiting cytochrome oxidase, the terminal component in the respiratory chain which contains two copper atoms (Shapira 2002).

8.4.2. Diabetes and Parkinson's Disease

In recent years several research groups have proposed a role of mitochondrial ROS formation in relatively more common diseases such as diabetes or Parkinson's disease. In the case of diabetes, hyperglycemia has been associated with oxidative stress through a complex process that includes protein kinase C activation, cytochrome c release and apoptosis (Ye et al. 2004; Fariss et al. 2005). There are also rare cases of diabetes that appear to be triggered by a mitochondrial DNA mutation of the leucine tRNA, but in this case the correlation between mutation and disease is not clear since mutations in other mitochondrial tRNAs are not linked to diabetes (Maasen et al. 2005).

In the case of Parkinson's disease, various studies have shown an increase in mitochondrial ROS formation by Complex I as one of the initial steps leading to this disease. One of the most common animal models of Parkinson's disease involves exposing rodents to MPTP (1-methyl-4-phenyl-1,2,3,6-tetrahydropyridine), a toxic compound that is first metabolized to MPP^+ (1-Methyl-4-phenylpyridinium) which in turn inhibits the respiratory chain at the level of Complex I. This hypothesis is further supported by the fact that chronic exposure to small concentrations of rotenone, a Complex I inhibitor, also causes destruction of the substantia nigra, and has recently been developed as another animal model for Parkinson's disease (Shapira et al. 1993; Trojanowski 2003; Sherer et al. 2003). In addition to Complex I, ROS produced by monoamine oxidase have also been implicated in the pathogenesis of Parkinson's disease (Cohen et al. 1997).

8.4.3. Hyperoxia and Hypoxia

As mentioned above, oxygen concentration plays a central role in the rate ROS formation as most of these reactions are non-enzymatic and are controlled by mass action. In studies carried out almost 25 years ago, we showed that the formation of superoxide by the respiratory chain was linearly dependent on oxygen concentration (Turrens et al. 1982a). Similarly, the release of hydrogen peroxide from intact mitochondria increased with oxygen concentration, yet exhibited a biphasic behavior. Between 0 and 60% oxygen, the rate of hydrogen peroxide production increased linearly with pO_2 but between 60% and 100% oxygen, the rate of release, although linear, showed a steeper slope suggesting that the intramitochondrial antioxidant defenses were overwhelmed (Turrens et al. 1982b). This result was very interesting because it correlated with the histological observation that oxygen causes structural damage to the lungs at pO_2 higher than 60% (Crapo et al. 1983).

All tissues are damaged by exposure to hyperoxic conditions. Under normobaric hyperoxic conditions, only tissues directly exposed to atmospheric oxygen such as the lungs (Pagano and Barazzone-Argiroffo 2003) and eyes (Lucey and Dangman 1984) are injured, because the limited solubility of oxygen in the plasma prevents other tissues from being exposed to high oxygen concentrations. Hyperbaric oxygen, however, causes more oxygen to be dissolved in the plasma, resulting in toxic oxygen concentrations in peripheral tissues. Prolonged exposure to hyperbaric oxygen causes convulsions resulting from brain damage (Gershman et al. 1954).

It has been proposed that hypoxia is also associated with increased formation of ROS (Schumacker 2002). It is difficult to propose a simple mechanism to connect hypoxia with oxidative stress, since the main substrate for the process, oxygen, is also the limiting species. Different scenarios may contribute to ROS formation under hypoxic conditions. For example, nitric oxide formation triggered by hypoxia may inhibit electron flow in the electron transport chain, increasing the reduction state of some respiratory components (Poderoso et al. 1999). Part of the process may involve the induction of critical proteins triggered by the activation of hypoxia inducible factors (HIF), which control the expression of various glycolytic enzymes and other intracellular signaling factors (Semenza 2001; Semenza 2002).

8.5. Mitochondrial Oxidative Stress and Aging

Aging is a complex phenomenon likely to be caused by a variety of independent processes. Several studies have implicated the accumulation of damage resulting from mitochondrial ROS as one of the main factors responsible for aging. One of the earliest pieces of evidence came from studies carried out in Bruce Ames' laboratory showing that the age-dependent accumulation of base modifications in mitochondrial DNA vastly exceeds the rate of base modification in nuclear DNA (Richter et al. 1988). In a recent study using single molecule PCR of mitochondrial DNA the authors showed that aged human substantia nigra neurons present a substantial amount of deletions. These deletions were more abundant in cytochrome oxidase-deficient neurons, suggesting that they may be responsible for inhibition of respiration (Kraytsberg et al. 2006).

Sohal and collaborators in the late 80s and early 90s, also showed that the rate of hydrogen peroxide formation in mitochondria increases with age (Sohal and Sohal 1991) providing a mechanism to explain, at least in part, the correlation between oxidized DNA bases and aging described by Ames.

In another interesting study using MnSOD knock-out transgenic animals, the investigators showed that heterozygous mutants showed the same mitochondrial defects (including opening of the permeability transition pore and apoptosis) as the normal animals, but at a much earlier age (Kokoszca et al. 2001). In other words, the increased mitochondrial steady state concentration of superoxide resulting from MnSOD deficiency is associated with premature aging.

In most mammalian species females live longer than males. Comparisons in mitochondrial metabolism between female and male rats provided another indirect piece of evidence connecting ROS formation in mitochondria with aging. Mitochondria isolated from female rats released hydrogen peroxide at half the rate as the same organelles isolated from males, due in part to increased activities of antioxidant enzymes in the mitochondrial matrix. The lower rate of mitochondrial ROS formation in females was estrogen dependent, and increased to the same level as in males in ovariectomized animals (Sastre et al. 2002). By using pharmacological inhibitors of signaling pathways the authors showed that these differences were estrogen-dependent, caused by an upregulation of antioxidant enzymes (Mn-SOD and glutathione peroxidase) via the activation of MAP kinases and NF-κB (Viña et al. 2005).

A recent study provided one of the most striking pieces of evidence in support of this hypothesis. The investigators created transgenic mice overexpressing catalase in the mitochondria by transfecting animals with a vector containing the peptide signal for translocation into the mitochondrion. Increased expression of catalase in the mitochondrial matrix (an enzyme that is not normally in this organelle) correlated with an average of 5-month increase in the lifespan of mice, which also correlated with decreased incidence of cataracts and other markers of aging (Schriner et al. 2005).

8.6. Conclusions

The formation of superoxide anion and other ROS is an unavoidable consequence of aerobic metabolism. Although the proportion of oxygen partially reduced to ROS is relatively small, continuous production under normal and pathological conditions leads to a variety of processes, ranging from aging to apoptosis and disease.

References

Banmeyer I, Marchand C, Clippe A, Knoops B (2005) Human mitochondrial peroxiredoxin 5 protects from mitochondrial DNA damages induced by hydrogen peroxide. FEBS Lett 579:2327–2333

Beyer RE (1994) The role of ascorbate in antioxidant protection of biomembranes: Interaction with vitamin E and coenzyme Q. J Bioenerg Biomembr 26:349–358

Cadenas E, Boveris A, Ragan CI, Stoppani AOM (1977) Production of superoxide radicals and hydrogen peroxide by NADH-ubiquinone reductase and ubiquinol-cytochrome c reductase from beef heart mitochondria. Arch Biochem. Biophys 180:248–257

Chang TS, Cho CS, Park S, Yu S, Kang SW, Rhee SG (2004) Peroxiredoxin III, a mitochondrion-specific peroxidase, regulates apoptotic signaling by mitochondria. J Biol Chem 279:41975–41984

Clerch LB, Massaro D, Berkovich A (1998) Molecular mechanisms of antioxidant enzyme expression in lung during exposure to and recovery from hyperoxia. Am J Physiol Lung Cell Mol Physiol 274:L313–L319

Cohen,G., Farooqui,R.,, Kesler,N. (1997) Parkinson disease: A new link between monoamine oxidase and mitochondrial electron flow. Proc Natl Acad Sci USA 94:4890–4894

Cooper JM, Schapira AH (2003) Friedreich's Ataxia: disease mechanisms, antioxidant and Coenzyme Q10 therapy. Biofactors 18:163–171

Crapo JD, Freeman BA, Barry BE, Turrens JF, Young SL (1983) Mechanisms of hyperoxic injury to the pulmonary microcirculation. Physiologist, 26:170–176

Das KC, Lewis-Molock Y, White CW (1995) Thiol modulation of TNFa and IL-1 induced MnSOD gene expression and activation of NF-kappaB. Mol Cell Biochem 148:45–57

Fariss MW, Chan CB, Patel M, Van Houten B, Orrenius S (2005) Role of mitochondria in toxic oxidative stress. Mol Interv 5:94–111

Forman HJ, Kennedy J (1976) Dihydroorotate-dependent superoxide production in rat brain and liver. A function of the primary dehydrogenase. Arch Biochem Biophys 173:219–224

Genova ML, Ventura B, Giuliano G, Bovina C, Formiggini G, Parenti CG, Lenaz, G (2001) The site of production of superoxide radical in mitochondrial Complex I is not a bound ubisemiquinone but presumably iron-sulfur cluster N2. FEBS Lett 505: 364–368

Gerschman R, Gilbert DL, Nye SW, Dwyer P, Fenn WO (1954) Oxygen poisioning and X-irradiation: a mechanism in common. Science 119:623–626

Giulivi C, Poderoso JJ, Boveris A (1998) Production of nitric oxide by mitochondria. J Biol Chem 273:11038–11043

Granger DN, Hollwarth ME, Parks DA (1986) Ischemia-reperfusion injury: role of oxygen-derived free radicals. Acta Physiol Scand 126(Suppl. 548):47–63

Hinchliffe P, Sazanov LA (2005) Organization of iron-sulfur clusters in respiratory Complex I. Science 309:771–774

Imai H, Nakagawa Y (2003) Biological significance of phospholipid hydroperoxide glutathione peroxidase (PHGPx, GPx4) in mammalian cells. Free Rad Biol Med 34:145–169

Kokoszca JE, Coskun P, Esposito LA, Wallace DC (2001) Increased mitochondrial oxidative stress in the Sod2 (+/–) mouse results in the age-related decline of mitochondrial function culminating in increased apoptosis. Proc Natl Acad Sci USA 98: 2278–2283

Kotamraju S, Kalivendi SV, Konorev E, Chitambar CR, Joseph J, Kalyanaraman B (2004) Oxidant-induced iron signaling in Doxorubicin-mediated apoptosis. Methods Enzymol 378:362–382

Kraytsberg Y, Kudryavtseva E, McKee AC, Geula C, Kowall NW, Khrapko K (2006) Mitochondrial DNA deletions are abundant and cause functional impairment in aged human sustantia nigra neurons. Nature, in press

Kushnareva Y, Murphy A.N, Andreyev A. (2002) Complex I-mediated reactive oxygen species generation: modulation by cytochrome c and NAD(P)+ oxidation-reduction state. Biochem J 368, 545–553

Lenaz G (2001) The mitochondrial production of reactive oxygen species: mechanisms and implications in human pathology. IUBMB Life 52:159–164

Lucey JF, Dangman B (1984) A reexamination of the role of oxygen in retrolental fibroplasia. Pediatrics 73:82–96

Maassen JA, Janssen GM, Hart LM (2005) Molecular mechanisms of mitochondrial diabetes (MIDD). Ann Med 37:213–221

MacMillan-Crow LA, Cruthirds DL (2001) Invited review – Manganese superoxide dismutase in disease. Free Radic Res 34:325–336

Muller FL, Liu Y, Van Remmen H (2004) Complex III releases superoxide to both sides of the inner mitochondrial membrane. J Biol Chem 279:49064–49073

Noji H, Yoshida M (2001) The rotary machine in the cell, ATP synthase. J Biol Chem 276:1665–1668

Nulton-Persson,A.C., Szweda LI (2003) Reversible inactivation of alpha-ketoglutarate dehydrogenase in response to alterations in the mitochondrial glutathione status. Biochemistry 42:4235–4242

Oberley LW, St.Clair DK, AutorAP, Oberley TD (1987) Increase in manganese superoxide dismutase activity in the mouse heart after X-irradiation. Arch Biochem Biophys 254:69–80

Okado-Matsumoto A Fridovich,I (2001) Subcellular distribution of superoxide dismutases (SOD) in rat liver: Cu,Zn-SOD in mitochondria. J Biol Chem 276:38388–38393

Pagano A, Barazzone-Argiroffo C (2003) Alveolar cell death in hyperoxia-induced lung injury. Ann NY Acad Sci 1010:405–416

Phung CD, Ezieme JA, Turrens JF (1994) Hydrogen peroxide metabolism in skeletal muscle mitochondria. Arch Biochem Biophys 315:479–482

Poderoso JJ, Lisdero C, Schöpfer F, Riobó N, Carreras MC, Cadenas E, Boveris A (1999) The regulation of mitochondrial oxygen uptake by redox reactions involving nitric oxide and ubiquinol. J Biol Chem 274:37709–37716

Privalle CT, Beyer WF Jr, Fridovich I (1989) Anaerobic induction of proMn-superoxide dismutase in Escherichia coli. J Biol Chem 264:2758–2763

Radi R, Turrens JF, Chang LY, Bush KM, Crapo JD, Freeman BA (1991) Detection of catalase in rat heart mitochondria. J Biol Chem 266:22028–22034

Rhee SG, Kang SW, Chang TS, Jeong W, Kim K (2001) Peroxiredoxin, a novel family of peroxidases. IUBMB Life 52:35–41

Richter C, Park JW, Ames BN (1988) Normal oxidative damage to mitochondrial and nuclear DNA is extensive. Proc Natl Acad Sci USA 85:6465–6467

Richter C, Schweizer ,M, Ghafourifar P (1999) Mitochondria, nitric oxide, and peroxynitrite. Methods Enzymol 301:381–393

Sastre J, Borras C, Gracia-Sala B, Lloret A, Pallardo FV, Viña J (2002) Mitochondrial damage in aging and apoptosis. Ann NY Acad Sci 959:448–451

Schapira AH (2002) Primary and secondary defects of the mitochondrial respiratory chain. J Inherit Metab Dis 25:207–214

Schapira AHV, Hartley A, Cleeter MWJ, Cooper JM (1993) Free radicals and mitochondrial dysfunction in Parkinson's disease. Biochem Soc Trans 21:367–370

Schriner SF, Linford NJ, Martin GM, Treuting P, Ogburn CF, Emond M, Coskun PE, Ladiges W, Wolf N, Van Remmen H, Wallace DC, Rabinovitch PS (2005) Extension of murine life span by overexpression of catalase targeted to mitochondria. Science 308:1909–1911

Schumacker PT (2002) Hypoxia, anoxia, and O2 sensing: the search. Am J Physiol Cell Mol Physiol 283:L918–L921

Semenza G (2001) HIF-1, O2, and the 3 PHDs: how animal cells signal hypoxia to the nucleus. Cell 107:1–3

Sherer TB; Kim JH; Betarbet R; Greenamyre JT (2003) Subcutaneous rotenone exposure causes highly selective dopaminergic degeneration and alpha-synuclein aggregation. Exp Neurol 179:9–16

Skinner Skinner KA, Crow JP, Skinner HB, Chandler RT, Thompson JA, Parks DA (1997) Free and protein-associated nitrotyrosine formation following rat liver preservation and transplantation. Arch Biochem Biophys 342:282–288

Semenza G (2002) Signal transduction to hypoxia-inducible factor 1. Biochemical Pharmacology 5-6:993–998

Sjostrom K, Crapo JD (1981) Adaptation to oxygen by preexposure to hypoxia: enhanced activity of mangani superoxide dismutase. Bull Europ Physiopath Resp 17(suppl.): 111–116

Sohal RS, Sohal BH (1991) Hydrogen peroxide release by mitochondria increases during aging. Mech Ageing Dev 57:187–202

Starkov AA, Fiskum G, Chinopoulos C, Lorenzo BJ, Browne SE, Patel MS, Beal MF (2004) Mitochondrial alpha-ketoglutarate dehydrogenase complex generates reactive oxygen species. J Neurosci 24:7788

Tretter L, Adam-Vizi V (2000) Inhibition of Krebs cycle enzymes by hydrogen peroxide: A key role of a-ketoglutarate dehydrogenase in limiting NADH production under oxidative stress. J Neurosci 20:8972–8979

Tretter L, Adam-Vizi V (2004) Generation of reactive oxygen species in the reaction catalyzed by alpha-ketoglutarate dehydrogenase. J Neurosci 24:7771–7778

Trojanowski JQ (2003) Rotenone neurotoxicity: a new window on environmental causes of Parkinson's disease and related brain amyloidoses. Exp Neurol 179:6–8

Turrens JF (2003) Mitochondrial formation of reactive oxygen species. J Physiol (London) 552:335–344.

Turrens JF, Alexandre A, Lehninger AL (1985) Ubisemiquinone is the electron donor for superoxide formation by complex III of heart mitochondria. Arch Biochem Biophys 237:408–414

Turrens JF, Boveris A (1980) Generation of superoxide anion by the NADH dehydrogenase of bovine heart mitochondria. Biochem J 191:421–427

Turrens JF, Freeman BA, Crapo JD (1982a) Hyperoxia increases hydrogen peroxide formation by lung mitochondria and microsomes. Arch Biochem Biophys 217:411–419

Turrens JF, Freeman BA, Levitt JG, Crapo JD (1982b) The effect of hyperoxia on superoxide production by lung submitochondrial particles. Arch Biochem Biophys 217:401–410.

Ursini F, Maiorino M, Brigelius-Flohé R, Aumann KD, Roveri A, Schomburg D, Flohé,L (1995) Diversity of glutathione peroxidases. Methods Enzymol 252:38–53

Viña J, Borras C, Gambini J, Sastre J, Pallardo FV (2005) Why females live longer than males? Importance of the upregulation of longevity-associated genes by oestrogenic compounds. FEBS Lett 579: 2741–2745

Vinogradov AD, Grivennikova VG (2005) Generation of superoxide-radical by the NADH:ubiquinone oxidoreductase of heart mitochondria. Biochemistry (Moscow) 70:120–7

Xu JX (2004) Radical metabolism is partner to energy metabolism in mitochondria. Ann NY Acad Sci 1011:57–60

Ye G, Metreveli MS, Donthi RV, Xia S, Xu M, Carlson EM, Epstein PN (2004) Catalase protects cardiomyocyte function in models of type 1 and type 2 diabetes. Diabetes 53:1336–1343.

9
Mitochondrial Calcium: Role in the Normal and Ischaemic/Reperfused Myocardium

Elinor J. Griffiths, Christopher J. Bell, Dirki Balaska and Guy A. Rutter

9.1. Introduction

The role of intramitochondrial free $[Ca^{2+}]$ ($[Ca^{2+}]_m$) in regulating energy production in the heart is now well-accepted: An increase in $[Ca^{2+}]_m$, such as occurs during increased workload or adrenaline release, activates the mitochondrial dehydrogenases to increase NADH and hence ATP production (reviewed in (Hansford 1994; McCormack et al. 1990)). But $[Ca^{2+}]_m$ could also potentially regulate whole-cell cell Ca^{2+} signalling, by reducing the free $[Ca^{2+}]$ available for contraction, or by ion channel regulation.

In addition to this physiological role, an increase in $[Ca^{2+}]_m$ has long been associated with the transition from reversible to irreversible injury during ischaemia/reperfusion (Shen and Jennings 1972a; Shen and Jennings 1972b). It has also been known for some time that Ca^{2+} uptake by mitochondria can, under certain conditions, promote a non-specific increase in the permeability of the inner mitochondrial membrane (Hunter et al. 1976) – later found to be inhibited by cyclosporine A (Crompton et al. 1988), and referred to as the mitochondrial permeability transition pore (MPTP). We now know the MPTP opens upon reperfusion, but not ischaemia, of the heart (Griffiths and Halestrap 1995), and plays a pivotal role in cell death by both necrosis and apoptosis (see the chapter by Halestrap in this volume).

Thus understanding how $[Ca^{2+}]_m$ is regulated is important in both normal and pathological heart function. In this chapter, we will focus on recent developments in the role of $[Ca^{2+}]_m$ in energy production and Ca^{2+} signalling, the pathways of mitochondrial Ca^{2+} transport during ischaemia/reperfusion (IR), and finally the Ca^{2+} transporters as possible targets of therapeutic intervention.

9.2. Physiological Role of Mitochondrial [Ca^{2+}]

9.2.1. Mitochondrial Ca^{2+} Transport Pathways

The classical properties of these channels were derived mainly from studies on isolated mitochondria (reviewed in (Gunter and Pfeiffer 1990; Nicholls and Crompton 1980)): In summary, mitochondrial uptake occurs via a Ca^{2+}-uniporter (CaUP), and efflux via a Na^+/Ca^{2+} exchanger (mNCX). An Na^+-independent pathway exists in non-excitable cells, such as hepatocytes, but there appears to be no equivalent pathway in heart cells (Nicholls and Crompton 1980). The Ca^{2+} transporters of mitochondria are depicted schematically in Figure 9.1, which also summarises the known inhibitors of the pathways: the CaUP is inhibited by ruthenium red (RUR) (Moore 1971) and Ru360 (Ying et al. 1991), and the mNCX by diltiazem, clonazepam, and CGP37157 (Cox et al. 1993; Cox and Matlib 1993).

The kinetics of these channels in isolated mitochondria indicated that they were too slow to play any role in intracellular Ca^{2+} signalling during excitation-contraction (EC) coupling of the heart. The CaUP exhibits low affinity, high capacity transport of Ca^{2+}, whereas the mNCX has a much lower Vmax, saturating at $[Ca^{2+}]_m$ below 1µM. In fact, it was predicted that net Ca^{2+} influx would occur only when external [Ca^{2+}] rose above about 500 nM (Gunter and Pfeiffer 1990; Nicholls and Crompton 1980), much higher than the resting, diastolic cytosolic free [Ca^{2+}] ($[Ca^{2+}]_c$) of 100-200 nM. This, together with the

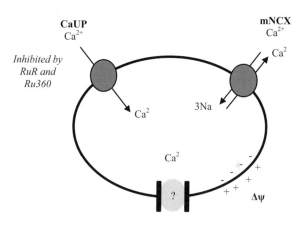

FIGURE 9.1. Pathways of mitochondria Ca^{2+} transport in the heart. The model depicts the classical pathways of mitochondrial Ca^{2+} transport elucidated from studies on isolated mitochondria (see text for details). The mitochondrial permeability transition pore (MPTP) is not thought to play a role under normal physiological conditions. CaUP – the mitochondrial Ca^{2+}-uniporter; mNCX – the mitochondrial Na^+/Ca^{2+}-exchanger.

sigmoidal nature of mitochondrial Ca^{2+} uptake (in presence of physiological $[Mg^{2+}]$) suggested that the CaUP responded to a "weighted average" of $[Ca^{2+}]_c$; and so systolic increases in $[Ca^{2+}]_c$ (typically about 1 µM) would contribute much more to mitochondrial Ca^{2+} uptake than diastolic $[Ca^{2+}]_c$ (Miyata et al. 1991). However, later studies in intact (non-cardiac) cells indicated that in vivo, the situation may be very different: mitochondria located in close proximity to Ca^{2+}-release channels on the endoplasmic reticulum (ER) are exposed to a much higher level of $[Ca^{2+}]$ than those elsewhere in the cytosol (Rizzuto et al. 1992; Visch et al. 2004). These localised regions of high $[Ca^{2+}]_c$ would therefore overcome the problem of the low affinity of the CaUP for Ca^{2+}, and possibly give rise to more rapid uptake rates at physiological $[Ca^{2+}]_c$ than those predicted from the early studies on isolated mitochondria.

The driving force for Ca^{2+}-uptake is provided by the mitochondrial membrane potential ($\Delta\Psi_m$), and so Ca^{2+} competes with ATP synthesis in using the proton motive force generated by the respiratory chain. Although under physiological conditions, any beat-to-beat changes in $[Ca^{2+}]_m$ would not be expected to produce a decrease in $\Delta\Psi_m$ sufficient to inhibit ATP synthesis (Nicholls and Crompton 1980), under pathological conditions, accumulation of large amounts of Ca^{2+} by mitochondria would firstly reduce ATP synthesis, and secondly induce the MPTP. This is discussed further in section 9.3.

As well as taking into account their subcellular proximity to other organelles, mitochondria from different areas of the cell may have intrinsic differences in Ca^{2+} transport activity: for example, McMillin-Wood et al. found that mitochondria isolated from interfibrillar regions of the myocyte had a greater capacity for Ca^{2+} uptake than mitochondria isolated from subsarcolemmal regions (McMillin-Wood et al. 1980). A separate study, using electron probe microanalysis (EPMA) in adult myocytes, revealed more rapid increases in $[Ca^{2+}]_m$ in a population of sub-sarcolemmal mitochondria than in more central mitochondria during cell contraction (Gallitelli et al. 1999). However, EPMA only detects changes in total $[Ca^{2+}]_m$, and so does not reflect dynamic changes in free $[Ca^{2+}]_m$ that are relevant to the activation of intramitochondrial enzymes, or that would play a role in EC coupling. Rates of mitochondrial Ca^{2+} uptake may also differ between species (this is considered in more detail in the next section), and between sexes; mitochondria isolated from male rat hearts were found to have a higher rate of mitochondrial Ca^{2+} uptake than those from female rats (Arieli et al. 2004).

But investigating the role of mitochondrial Ca^{2+} transport in the heart *in situ* has been problematic due to the difficulties of measuring $[Ca^{2+}]_m$ in living cells, and due to the lack of specific inhibitors of the transporters. These areas are discussed below.

9.2.2. $[Ca^{2+}]_m$ and ATP Production

Early studies on isolated mitochondria showed that the main regulator of ATP production was likely to be ADP (Chance and Williams 1956), so it seemed probable that this parameter had to increase, and ATP levels fall, before any stimulation of respiration occurred (Brown 1992). However, studies using ^{31}P-NMR in whole hearts showed that ATP levels were maintained during increased workload (Katz et al. 1989), implicating alternative regulatory mechanisms in the control of mitochondrial respiration. One such mechanism is provided by Ca^{2+}: in many mammalian cell types, changes in the free Ca^{2+} concentration in the mitochondrial matrix ($[Ca^{2+}]_m$), by agonists that increase cytosolic free $[Ca^{2+}]$ ($[Ca^{2+}]_c$), stimulate ATP production by activating three mitochondrial dehydrogenases (McCormack et al. 1990), and also possibly by activating the F_0F_1ATPase (Lim et al. 2002; Territo et al. 2000).

The ability of $[Ca^{2+}]_m$ to increase ATP on a rapid timescale, as would have to occur in the heart to meet sudden changes in energy demand, has not been investigated, and exactly how ATP supply and demand are so well-matched in the heart remains controversial. This is at least partly due to the difficulties in measuring free $[Ca^{2+}]_m$ and ATP on a rapid timescale in living cells.

NAD(P)H levels have been measured in whole hearts (Heineman and Balaban 1993; Katz et al. 1987), myocytes (White and Wittenberg 1993) or trabeculae (Brandes and Bers 1996) as an indirect indicator of increases in the rate of ATP synthesis. Several studies have concluded that when cells or whole hearts are excited with ultraviolet light, 90% of cellular autofluorecence originates from NADH in mitochondria (Carry et al. 1989; Duchen and Biscoe 1992). One problem with this view is that any change in oxygen tension will tend to increase NAD(P)H due to inhibition of respiration (Balaban 2002). However, in well-oxygenated hearts, no change in NADH was observed in response to physiological increases in workload (Heineman and Balaban 1993). Additionally, altering the substrate from glucose to pyruvate did increase NADH, and this was associated with an increase in the ATP/ADP + P_i, and increases in O_2 consumption (Scholz et al. 1995). However, although lactate gave the same increase in NADH, there was no rise in ATP/ADP + P_i. So similar levels of NADH were associated with different ATP/ADP + P_i, and so an increase in NADH cannot be taken as an unambiguous indicator of increased ATP production (Scholz et al. 1995).

In myocytes, conflicting results have been obtained as to whether NADH changes upon rapid stimulation of cells – and increases, no change and decreases have all been reported (Griffiths et al. 1997b; White and Wittenberg 1993; White and Wittenberg 1995). In isolated trabeculae, NADH was more carefully measured by using an internal reference that negated motion artifacts. When the muscle strips were stimulated at 3Hz from rest, an initial drop in NADH occurred within 5 sec, followed by an increase to the initial or higher levels over 60 sec (Brandes and Bers 1997). However, under physiological conditions, the heart is never likely to be subjected to such a sudden and steep increase in energy demand (i.e. from rest to beating rapidly).

We have preliminary data suggesting that ATP levels in myocytes that are already beating remain constant upon addition of an adrenergic agonist (to produce an increase in contraction and hence ATP demand). This was done by expressing luciferase in either the cytosolic or mitochondrial compartment of adult rat myocytes. However, similar to the results using NADH measurements in trabeculae, if the myocytes were suddenly stimulated to beat rapidly from rest, then an initial fall in [ATP] was observed followed by recover, over a time-course that paralleled Ca^{2+} uptake by the mitochondria.

Unfortunately Ru360 and RuR did not appear to be cell-permeant in our hands, and so we could not determine whether uptake of Ca^{2+} by mitochondria was essential in maintaining a constant ATP supply.

9.2.3. Measurement of $[Ca^{2+}]_m$ in Living Cells

The most common methods for continuously monitoring Ca^{2+} in living cells or tissues have involved using fluorescent indicators (Tsien, 1980), such as fura-2 (Grynkiewicz et al. 1985), or luminescent proteins, such as aequorin (Ridgway and Ashley 1967). But in order to measure $[Ca^{2+}]$ specifically in mitochondria, several problems must be overcome: the dye or protein must be able to cross the cell membrane but be retained within the cell, it must act as a reporter for Ca^{2+} without significantly perturbing cell function, and it must be specifically located to mitochondria. In single heart cells these problems have been particularly difficult to surmount: In order to localise fluorescent indicators to the mitochondrial compartment, three approaches have been employed: a) loading the indicator, e.g. indo-1, as the acetoxymethyl ester, which enters both cytosol and mitochondria, and then quenching the cytosolic indo-1 fluorescence by superfusion with Mn^{2+} (Miyata et al. 1991), b) manipulating loading conditions of the indicator so that it predominantly loads into mitochondria (Griffiths et al. 1997a) – for this a cold-warm loading protocol is used where cells are loaded with the indicator at room temperature and then incubated for an extended period at 37^oC, which promotes loss of cytosolic indo-1 through sarcolemmal anion channels, and c) use of indicators such as rhod-2 that should preferentially accumulate into mitochondria of respiring cells due to the negative mitochondrial membrane potential ($\Delta\Psi_m$) (Trollinger et al. 1997).

The validity of all these techniques has been questioned to some extent: The Mn^{2+}-quench technique was subject to the main criticism that it was impossible to be certain how much of the cytosolic indo-1 fluorescence had been quenched, and also that Mn^{2+} might interfere with cell Ca^{2+} transport systems. Therefore the second method, where cytosolic indo-1 is allowed to leave the cytosol via an anion exchanger would seem more reliable, but again it is difficult to be certain that all the cytosolic dye had been removed. However, both these techniques gave identical results in adult rat cardiomyocytes, in that $[Ca^{2+}]_m$ did not change during a single contraction of the cell, but did increase significantly over tens of seconds when cells were stimulated to beat rapidly in the presence of noradrenaline (Griffiths et al. 1997a). The final method where the indicator should accumulate in mitochondria due to their negative $\Delta\Psi_m$ is not that straightforward, since incubation conditions again affect the distribution of the dye. Lemasters and colleagues used rhod-2 and cold-warm loading conditions to localise the indicator to the mitochondria of rabbit myocytes (Trollinger et al. 1997). The difference that loading conditions make is highlighted in Figure 9.2, where we found that

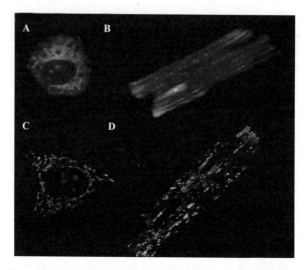

FIGURE 9.2. Neonatal and adult rat cardiomyocytes loaded with rhod-2. Myocytes were loaded with rhod-2 (a Ca^{2+}-sensitive fluorescent indicator) using different loading conditions. Panels A and B – cells were loaded for 15 min at 37°C. Panels C and D – cells were loaded at 4°C for 30 min followed by incubation at 37°C for 2 hours. The latter procedure results in localisation of dye to the mitochondrial compartment only, whereas in the former rhod-2 is distributed throughout the cell.

rhod-2 is localised to all regions of the cell under standard loading conditions at 37°C, but if the cold-warm procedure is followed, a characteristic mitochondrial pattern results (Balaska and Griffiths, unpublished data).

A technique that has been successful in measuring $[Ca^{2+}]_m$ in non-cardiac cells is that of using the photoprotein aequorin targeted specifically to the mitochondria (Rizzuto et al. 1992). This is achieved by fusing the cDNA for aequorin to the leader sequence of cytochrome c oxidase subunit VIII. The construct also expresses green fluorescent protein (GFP) for detecting cells that have been successfully transfected. Robert et al (Robert et al. 2001) used this technique in neonatal cardiomyocytes, where the "fugene" reagent was used to transfect cells with the cDNA of the aequorin construct. They observed beat-to-beat changes in $[Ca^{2+}]_m$ using this method, and also used untargeted aequorin to measure $[Ca^{2+}]_c$. The kinetics of mitochondrial Ca^{2+} transients were broadly similar to those of the cytosol, although of a reduced amplitude. The aequorin appeared to be correctly targeted to mitochondria, as evidence by immunolocalisation studies, and co-release of aequorin and the mitochondrial marker enzyme, citrate synthase upon cell permeabilisation with digitonin. Neither Ru360 nor ruthenium red could be used in intact cells, but both were capable of inhibiting mitochondrial Ca^{2+} uptake in permeabilised cells. One limitation of this study is that only about 10% of cells were transfected with aequorin, and single cell measurements could not be performed.

However adult cardiomyocytes cannot be transfected by reagents such as fugene or lipofectin, and neither are they amenable to microinjection (Griffiths, *unpublished observations*). So we made an adenovirus containing the aequorin construct, targeted to either mitochondria or cytosol (mtAq and cAq, respectively). This appears to be the only method currently for transfecting adult cardiomyocytes. We found that in adult rat myocytes, $[Ca^{2+}]_m$ did change during a single cell contraction, but only when cells were exposed to high external $[Ca^{2+}]$ (4mM), and isoproterenol (a β-adrenergic agonist) (Bell et al. 2004). But the level of aequorin expression achieved was insufficient to be detected in a single cell, at least for the mtAq construct.

The most recent group of compounds to be used for measuring mitochondrial Ca^{2+} are those based on Ca^{2+}-sensitive derivatives of green fluorescent protein (Pozzan et al. 2003). These have the advantage that they can be engineered as ratiometric probes, and can, in theory, be targeted to any subcellular compartment. However, these are not without problems, especially in terms of correct targeting to mitochondria. Pozzan's group recently compared a variety of fluorescent proteins, and found that the amount successfully targeted to mitochondria varied considerably, with between 10 and 35% of the protein mistargeted (Filippin et al. 2005). The most efficient targeting was achieved, somewhat bizarrely, using a cDNA of a GFP-based protein fused with a double leader-sequence of cytochrome c oxidase. This yellow cameleon (YC) is a FRET-based Ca^{2+} indicator (it is excited by one wavelength and Ca^{2+} binding causes opposite effects at two emission wavelengths) and is an ideal ratiometric indicator for confocal microscopy and in particular for two-photon confocal microscopy techniques. However, even in this YC 10% of the aequorin was mistargeted. Also, if the targeting sequence of any of the probes was not removed from inside the mitochondria, the fluorescent properties were "grossly" affected (Filippin et al. 2005).

So none of the techniques currently available is without problems, and although the GFP-based proteins would seem to give the best results in terms of mitochondrial targeting and being fluorescent and ratiometric, at present at least 10% of signal is mistargeted and non-mitochondrial. Also, the GFP-based proteins still have to be expressed in adult myocytes using a virus; so they are not straightforward to use, and time consuming to make. At present, the fluorescent indicators such as rhod-2, fura-2 or indo-1 still seem the best way forward given that they can be used in single cells, can be localised to mitochondria providing correct loading protocols are followed, and can be detected using relatively simple photometry (as well as imaging if desired).

9.2.4. *Role of $[Ca^{2+}]_m$ in Whole-Cell Ca^{2+}-Signalling*

9.2.4.1. Beat-To-Beat Changes in $[Ca^{2+}]_m$?

In addition to their role in transmitting changes in $[Ca^{2+}]_c$ to the mitochondrial matrix to bring about increases in ATP supply, mitochondria have the potential to act as local buffering systems, modulating local $[Ca^{2+}]_c$, and so regulating

Ca^{2+}-dependent processes (Jacobson and Duchen 2004). But in order to play any role in EC coupling in the heart, the mitochondrial Ca^{2+} transporters must be able to respond on a millisecond timescale to changes in $[Ca^{2+}]_c$. Initially, predictions from isolated mitochondria were against this (as discussed above). And studies of ours and others on isolated rat, cat and ferret myocytes agreed with this prediction, i.e. that $[Ca^{2+}]_m$ did not change during a single contraction of the cell, but rather increased over tens of seconds (Griffiths et al. 1997a; Miyata et al. 1991; Zhou et al. 1998). Since this was still rapid enough to allow increased synthesis of ATP upon an increased demand, it fitted nicely with the original predictions based on isolated mitochondrial studies; that the role of the mitochondrial Ca^{2+} transporters is to convey changes in $[Ca^{2+}]_c$ brought about by increased workload adrenaline-stimulation to the mitochondrial matrix, and hence increase ATP synthesis by activating the dehydrogenases (McCormack et al., 1990).

However, results using different indicators, and different species, have varied; subsequent to the studies described above, we found that mitochondrial Ca^{2+} transients did occur on a beat-to-beat basis in guinea-pig myocytes (using indo-1 and the cold-warm mitochondrial loading procedure) (Griffiths 1999b), in contrast to our previous observations in rat myocytes (Griffiths et al. 1997a). Similar observations had previously been made on rabbit myocytes, using either fluo-3 or rhod-2 and confocal microscopy (Chacon et al. 1996; Trollinger et al. 1997).

We have very recently found, using targeted aequorin described above, that $[Ca^{2+}]_m$ can change on a beat-to-beat basis in adult rat cardiomyocytes, but only when cells are stimulated under conditions of high extracelluar $[Ca^{2+}]$ (4mM, compared with 1mM in previous studies) and in the presence of isoproterenol (Bell et al. 2004). In neonatal cardiomyocytes, mitochondrial Ca^{2+} transients have also been observed using targeted aequorin and the kinetics were broadly similar to those observed in the cytosol (Robert et al. 2001). Thus, there appears to be an emerging consensus that increases in $[Ca^{2+}]_m$ can occur over timescales relevant to energy production – and for these purposes it does not matter whether the increase in $[Ca^{2+}]_m$ occurs over a timescale of milliseconds or seconds. Also, there are differences in Ca^{2+} transport systems, both between species and during development. It is not yet known whether these differences are due to variations in the kinetics of the transporters themselves, or in their subcellular location and interaction with other Ca^{2+} transport systems. However, differences have been shown for other Ca^{2+} transporters, for example, the contribution of the SR and sarcolemmal Ca^{2+} fluxes during EC coupling varies both during development (Tibbits et al. 2002), and across species (Bers 1991).

9.2.4.2. Use of Inhibitors of the Mitochondrial Ca^{2+} Transporters

To elucidate a role of the mitochondrial Ca^{2+} transporters in cell signalling, in theory inhibitors of the transporters could be used. Two inhibitors of the CaUP have been commonly used, firstly ruthenium red (RuR) and secondly

Ru360 (a derivative of RuR). RuR is useful in isolated mitochondria, but in intact cells and tissues, it also inhibits the SR Ca^{2+} release channel (ryanodine receptor) (Gupta et al. 1989). Thus at low concentrations in both whole hearts and isolated myocytes, RuR lowers $[Ca^{2+}]_c$ (Griffiths 2000; Gupta et al. 1988); an effect that indirectly affects $[Ca^{2+}]_m$. However, the effect of RuR on $[Ca^{2+}]_c$ is biphasic in that at higher concentrations, RuR causes spontaneous contractions and eventually contracture, indicative of cytosolic Ca^{2+} overload (Griffiths 2000; Gupta et al. 1988). The reason for this is not known, but may include inhibition of respiration and ATP synthesis – this has been shown in liver mitochondria (Vasington et al. 1972) – and would indirectly increase $[Ca^{2+}]_c$ by disrupting ion-controlling mechanisms. Also, RuR is a polycation, and so could conceivably displace bound intracellular $[Ca^{2+}]$ (Griffiths 2000).

Ru360, appears to be of more practical use since at concentrations up to 10 μM, it appears to inhibit mitochondrial Ca^{2+} uptake (induced by cell depolarisation) in intact myocytes without affecting SR Ca^{2+} transport, L-type Ca^{2+}-channel, or sarcolemmal NCX (Matlib et al. 1998). Unfortunately, in our hands Ru360 did not appear to enter either adult or neonatal rat cardiomyocytes, since it had no effect on mitochondrial Ca^{2+} transients (Bell, Rutter, Griffiths, *unpublished data*). A lack of effect on $[Ca^{2+}]_m$ in neonatal rat myocytes was also reported by Robert et al. (2001) But several other studies have now reported effects of Ru360 on cellular Ca^{2+}-handling. For example, Ru360 was found to reduce the speed of Ca^{2+} waves in isolated rat cardiomyocytes in a similar manner to thapsigargin (an inhibitor of SERCA) (Landgraf et al. 2004), suggesting that blockade of mitochondrial Ca^{2+} uptake prolongs the cytosolic Ca^{2+} signal. Ru360 also enhanced the duration of the cytosolic Ca^{2+} transient, though did not affect its amplitude (Landgraf et al. 2004). Mitochondrial Ca^{2+} uptake may also play a role in controlling Ca^{2+} fluxes across the sarcolemma by L-type Ca^{2+}-channels (Sanchez et al. 2001). Inhibiting CaUP by Ru360 reduced the amplitude of cytosolic Ca^{2+} transients when cells were stimulated at a frequency of 3 Hz, but not at 0.1 Hz. The authors suggested that mitochondria played a role in sequestering Ca^{2+}, and so removed Ca^{2+}-dependent inactivation of L-type Ca^{2+} channels. A schematic diagram of how the mitochondrial Ca^{2+} transporters may interact with those of the SR and sarcolemma is shown in Figure 9.3. In contrast to studies using rat myocytes, toxic effects of Ru360 have been reported in one study in isolated mouse ventricular myocytes, where Ru360 induced a dose-dependent rise in resting $[Ca^{2+}]_c$ (significant at 1.0 μM), with contracture of myocytes occurring at = 10μM (Seguchi et al. 2005).

As with RuR, the concentrations of Ru360 that have been used in whole hearts appear to differ from those used in isolated myocytes. In isolated perfused rat hearts, Ru360 had no effect on contractile function at < 5μM; above this concentration, contractile force generation was depressed and resting tension elevated (de Jesus Garcia-Rivas et al. 2005) – whether this was attributed to mitochondrial or non-mitochondrial effects cannot be determined from this study. Thus it is difficult to know whether inhibiting mitochondrial Ca^{2+} uptake in

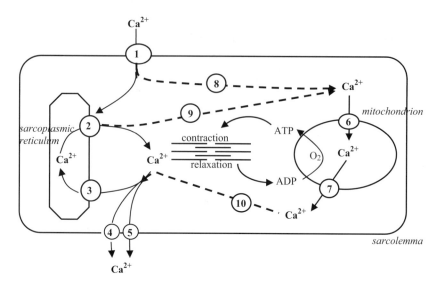

FIGURE 9.3. Schematic diagram of Ca^{2+} handling during normal excitation-contraction coupling in the heart, showing putative role of the mitochondrial Ca^{2+} transporters.
1. Voltage-dependent L-type Ca^{2+}-channels.
2. SR Ca^{2+} release pathway (ryanodine receptor).
3. SR Ca^{2+} uptake pathway (SERCA ATPase)
4. Sarcolemmal Na^+/Ca^{2+} exchanger (NCX)
5. Sarcolemmal Ca^{2+}-ATPase (Ca^{2+} efflux)
6. Mitochondria Ca^{2+}-uniporter (CaUP)
7. Mitochondrial Na^+/Ca^{2+} exchanger (mNCX)
8-10. Possible routes of Ca^{2+} movement between L-type Ca^{2+}-channels, SR efflux and uptake pathways, and sarcolemmal Ca^{2+}-efflux pathways.

a normal beating heart has any detrimental effect, although it is tempting to speculate that partial inhibition of mitochondrial Ca^{2+} uptake achieved by low concentrations of Ru360 has no adverse effect, whereas total inhibition, which would occur when using concentrations > 5µM, depresses cardiac function. Even a partial inhibition of CaUP may have a deleterious effect if the heart were stimulated with adrenaline and/or subjected to increased workload, assuming that an increase in $[Ca^{2+}]_m$ under these conditions was essential for increasing ATP supply. However, to the best of our knowledge, this type of experiment has yet to be attempted.

Similarly, use of inhibitors of the mNCX in intact cells and hearts is not without problems: of the 3 inhibitors that are effective in isolated mitochondria, namely diltiazem, clonazepam and CGP37157, we found only clonazepam to be effective in rat myocytes (Griffiths et al. 1997b). Diltiazem cannot be used because of its effects on L-type Ca^{2+}-channels, and we found that CGP37157 had variable effects on myocyte contractility and $[Ca^{2+}]_m$, possibly suggesting problems with cell permeability, or a non-mitochondrial effect, as in neuronal cells where CGP37157 was reported to inhibit voltage-gated Ca^{2+} channels (Baron and Thayer 1997). Clonazepam could be used as a relatively specific inhibitor of mNCX in adult rat myocytes, and at concentrations up to 100 µM had no significant effect on normal myocytes contraction or the cytosolic Ca^{2+} transient (Griffiths et al., 1997b). However, in adult guinea-pig myocytes, and in neonatal rat myocytes, clonazepam totally inhibited contraction (Bell & Griffiths, *unpublished observations*). A likely explanation for this is that clonazepam at these concentrations inhibits the sarcolemmal NCX, which is known to contribute to a much greater extent to Ca^{2+} removal from the cytosol during normal EC coupling in these cell types: in adult rat cells the SR accounts for over 95% of cytosolic Ca^{2+} removal (Bers 1991), so any effect of clonazepam on the NCX in these cells would go unnoticed.

9.2.5. Is There More than One Pathway for Ca^{2+} Influx?

Whereas for many years there was thought to only be one Ca^{2+}-uptake pathway in mitochondria – the classical CaUP – more recently the existence of a rapid uptake mode (RAM) has been proposed (Buntinas et al. 2001; Sparagna et al. 1995). However, whether this represented a separate pathway or different state of the CaUP is unclear. Neither is it obvious how this pathway is reconciled with the original studies on isolated mitochondria, indicating that, although the CaUP can take up large amounts of Ca^{2+}, uptake is relatively slow. Another recent finding proposed the existence of a ryanodine receptor in the mitochondrial membrane (mRyR) (Beutner et al. 2001; Beutner et al. 2005); this appears to allow rapid uptake of Ca^{2+} at relatively low concentrations, and so could account for physiological beat-to-beat Ca^{2+} uptake. Since this pathway is inhibited as external $[Ca^{2+}]$ increases, it could have easily been missed in earlier studies on isolated mitochondria, the majority of which used relatively high external $[Ca^{2+}]$, and indicators that were not sensitive to sub-micromolar $[Ca^{2+}]$. The authors suggest that as the ryanodine-sensitive pathway becomes inhibited, the classical CaUP takes over and accounts for mitochondrial Ca^{2+} uptake at higher external $[Ca^{2+}]$, and so would be the more pathologically relevant one (Beutner et al. 2005). These authors also found that ryanodine at concentrations of 2–20 µM could inhibit large-amplitude swelling of mitochondria, a measure of MPTP opening (Beutner et al. 2005). < 2 µM ryanodine enhanced swelling, which was in agreement with its effects on the RyR-1 channel of skeletal muscle i.e. that at lower concentrations ryanodine activates the channel. It is still unclear as to whether the mRyR is the same as the CaUP – they are both inhibited by RuR and Mg^{2+}, suggesting they may be the same channel. But having two separate pathways is an attractive hypothesis, and would explain many of the controversies or discrepancies in results to date. For example, Ca^{2+} transients occur in adult guinea-pig and neonatal rat myocytes, but not in adult rat myocytes. Although highly speculative at present, a different expression of the Ca^{2+} transport pathways between species, and/or stage of development, could explain these results.

9.3. Role of Mitochondrial Ca^{2+} in Ischaemia/Reperfusion Injury

9.3.1. Alterations in $[Ca^{2+}]_m$ During Ischaemia/Reperfusion and Hypoxia/Reoxygenation Injury.

Mitochondria have the capacity to take up huge amounts of Ca^{2+} (Gunter and Pfeiffer 1990; Nicholls and Crompton 1980) and thus could potentially remove toxic levels of Ca^{2+} from the cytosol. Unfortunately, such accumulation of Ca^{2+} can eventually damage mitochondria both by competing for ATP production and more importantly by inducing the mitochondrial permeability transition pore (MPTP). It has been known for many years that Ca^{2+}-induced mitochondrial dysfunction during ischaemia is associated with the transition from reversible to irreversible cell damage (Bush et al. 1980; Fleckenstein et al. 1983; Sordahl and Stewart 1980): Mitochondria accumulate deposits of calcium phosphate, become swollen, and eventually rupture. This sequence of events, whereby mitochondrial Ca^{2+} uptake leads to mitochondrial dysfunction (and hence prevents oxidative phosphorylation) was later found to be due to Ca^{2+}-induced opening of the MPTP. The MPTP is discussed in detail in the chapter by Halestrap et al. in this volume. Similarly, there are several reviews detailing evidence for mitochondrial dysfunction in reperfusion injury (Di Lisa et al. 2003; Halestrap et al. 2004; Honda et al. 2005; Piper et al. 1994; Suleiman et al. 2001). In this chapter we will concentrate on the pathways of mitochondrial Ca^{2+} transport under pathological conditions, whether excessive mitochondrial Ca^{2+} uptake is beneficial or harmful under conditions of ischaemia/reperfusion, and whether the mitochondrial Ca^{2+} transport pathways could be targets for protective interventions.

Mitochondrial Ca^{2+} uptake upon reperfusion of the heart has the potential to be protective, since it would effectively remove Ca^{2+} from the cytosol, and prevent cellular Ca^{2+} overload. Unfortunately, excessive Ca^{2+} uptake by mitochondria is damaging, firstly by inhibiting ATP synthesis (by lowering $\Delta\Psi_m$), and secondly by inducing the MPTP.

With regard to the first point, $\Delta\Psi_m$ appears to remain constant during the contractile cycle in isolated myocytes; implying that any beat-to-beat changes in $[Ca^{2+}]_m$ are too small to significantly affect $\Delta\Psi_m$ under physiological conditions (Di Lisa et al. 1995; Griffiths 1999a), and so Ca^{2+} uptake would not be expected to affect ATP synthesis. In isolated mitochondria, respiration-driven Ca^{2+} uptake causes a dissipation of $\Delta\Psi_m$ (Akerman 1978), which begins with external $[Ca^{2+}]$ as low as 1-2μM. However, $\Delta\Psi_m$ recovers once Ca^{2+} uptake stops. With regard to induction of the MPTP by Ca^{2+}, heart mitochondria are more resistant than other tissues such as liver and kidney (Griffiths and Halestrap 1993), and can accumulate large amounts of Ca^{2+}, (20 nmol/mg protein, equivalent to approximately 10-20 μM) without triggering of the MPTP (Crompton et al. 1978). But if mitochondria are first depleted of adenine nucleotides, or subjected to oxidative stress, conditions that prevail upon reperfusion, then the MPTP is induced at a far lower $[Ca^{2+}]_m$ (Griffiths and Halestrap 1995). The decrease in $\Delta\Psi_m$ due to MPTP opening causes release of accumulated Ca^{2+} from mitochondria (Crompton and Heid 1978), a process inhibited by cyclosporine A (Crompton et al. 1988).

$[Ca^{2+}]_m$ increases upon reperfusion of whole hearts, and hypoxia/reoxygenation (or models of metabolic inhibition) of myocytes (Chacon et al. 1994; Delcamp et al. 1998), and Miyata et al. (1992) found that whether or not rat cardiomyocytes recovered from hypoxia depended on the level of $[Ca^{2+}]_m$ achieved at the end of the hypoxic period: cells having $[Ca^{2+}]_m$ greater than 250-300 nM invariably hypercontracted upon reperfusion (Miyata et al. 1992). This value of 300 nM $[Ca^{2+}]_m$ is well within the normal physiological range of $[Ca^{2+}]_m$, and lower than that required for maximum activation of the dehydrogenase enzymes (0.5–1 µM). However other factors upon reperfusion can lower the threshold of $[Ca^{2+}]_m$ needed for the MPTP to open; for example low adenine nucleotides and free radical generation. The values for increases in $[Ca^{2+}]_m$ during hypoxia of isolated myocytes are in good agreement with those found in whole hearts: Allen et al (Allen et al. 1993) measured $[Ca^{2+}]$ in mitochondria isolated following hypoxia or reoxygenation of rat hearts perfused with fura-2. They found that 80 min of hypoxia caused no change in total tissue calcium, whereas free $[Ca^{2+}]_m$ rose from pre-hypoxic values of 160 nM, to 360 nM and 570 nM after 50 and 80 min of hypoxia, respectively. Reperfusion after 50 min hypoxia resulted in no further change in $[Ca^{2+}]_m$, but after 80 min hypoxia, reperfusion led to a 10-fold increase in $[Ca^{2+}]_m$ (Allen et al. 1993). This also highlights the fact that measurements of total tissue calcium cannot be extrapolated to measurements of free $[Ca^{2+}]$.

Mitochondria from different species also have different Ca^{2+} uptake capacities, which may influence their recovery from hypoxia/ischaemia; for example, the rate of Ca^{2+} uptake was greater in muskrat (a diving mammal) compared with guinea-pig mitochondria, and the former were also more resistant to Ca^{2+}-induced MPTP opening (McKean 1991). The authors of this study suggested that this may aid the animal during recovery from hypoxia. Another factor to consider is the effect of gender: mitochondria from female rats had lower rates of mitochondria Ca^{2+} uptake than those from male rat heart, and the "female" mitochondria also had improved ability to repolarise following Ca^{2+} overload. It is well known that females are protected from some types of heart disease by oestrogen, and this may be one possible mechanism to explain the protective effects of oestrogen on ischaemia-reperfusion injury, although a mechanism for this has yet to be proposed (Arieli et al. 2004).

But it should be remembered that limited (rather than excessive) Ca^{2+} uptake by mitochondria on reperfusion has the capacity to be protective: for example, by activation of pyruvate dehydrogenase (PDH). Several studies have shown that increased glucose oxidation on reperfusion improves recovery of the heart, and that drugs like ranolazine, which increase PDH activity, can improve post-ischaemic recovery (Clarke et al. 1993). So in designing protective strategies, it is a matter of getting the balance right between beneficial versus harmful effects of mitochondrial Ca^{2+} uptake. The situation is further complicated by the fact that many other factors on reperfusion affect this balance.

9.3.2. Studies using Inhibitors of Mitochondrial Ca^{2+} Transport

9.3.2.1. RUR and Ru360

RuR has been reported to protect hearts or myocytes from reperfusion/reoxygenation damage at concentrations ranging 0.1 – 6 µM (Benzi and Lerch 1992; Carry et al. 1989; Figueredo et al. 1991; Grover et al. 1990; Leperre et al. 1995; Miyamae et al. 1996; Park et al. 1990; Peng et al. 1980; Stone et al. 1989). Most of these studies have suggested that protection is due to effects of RuR on $[Ca^{2+}]_m$. However, few of these studies also reported effects on $[Ca^{2+}]_c$, and it seems more likely that the protective effects were due to an energy sparing effect as a result of reducing $[Ca^{2+}]_c$; a conclusion reached by Benzi and Lerch (1992). We have found that much higher concentrations of RuR (about 20µM) are required to actually inhibit $[Ca^{2+}]_m$ uptake in cardiomyocytes, and this concentration has a deleterious, not a protective, effect (Griffiths et al. 1998). So although low concentrations of RuR do appear to be cardioprotective, this cannot be attributed to direct effects on mitochondrial Ca^{2+} uptake.

Ru360 has also been reported to protect whole hearts against ischaemia/ reperfusion injury: Pre-treatment of isolated rat hearts with 10 µM Ru360 provided protection against reperfusion injury, as determined from infarct size and enzyme release (Zhang et al. 2006). However another study reported that lower concentrations of Ru360 were required: functional recovery of rat hearts from ischaemia was optimal between 0.25–1 µM Ru360 (de Jesus Garcia-Rivas et al. 2005). Recovery declined at higher concentrations, and was not seen at 10 µM Ru360. This protective effect of Ru360 was maintained upon separation of mitochondria in terms of their ability to synthesise ATP and resist MPTP opening upon addition of external Ca^{2+}. $[Ca^{2+}]_m$ (as measured following isolation of mitochondria at the end of the perfusion protocol) was also decreased in Ru360 treated hearts (de Jesus Garcia-Rivas et al. 2005).

9.3.2.2. Inhibitors of the Na^+/Ca^{2+} Exchanger

Diltiazem can protect hearts from ischaemia/reperfusion damage, but this has mainly been attributed to its effects on sarcolemmal L-type Ca^{2+}-channels (Winniford et al. 1985); although a recent study suggested that diltiazem inhibits cytosolic Na^+ overload through Na^+ channels in the ischaemic heart, which would indirectly preserve mitochondrial integrity (Takeo et al. 2004).

Clonazepam cannot be used in whole hearts as it appears to inhibit contractility by a non-myocyte effect (Griffiths, *unpublished observation*), and there have been no reports of the effects of CGP37157 in whole hearts. But clonazepam has been used in isolated myocyte models of hypoxia/reoxygenation injury: In suspensions of rat myocytes, 100 µM clonazepam reduced total cell $[Ca^{2+}]$, Ca^{2+} uptake (mitochondrial and cytosolic $[Ca^{2+}]$ were not measured separately) and LDH release, indicating preservation of membrane integrity (Sharikabad et al. 2001). Zhu et al. (2000) subjected cultured adult rat myocytes to 5mM extracellular Ca^{2+}, a condition that produced an increase in $[Ca^{2+}]_m$, measured using rhod-2, and in necrotic cell death (Zhu et al. 2000). Pre-treatment with clonazepam for 30 min reduced both the increase in $[Ca^{2+}]_m$, and the number of necrotic cells (to about 50% of control). Interestingly, cyclosporine A (CsA, an inhibitor of the MPTP) and clonazepam together reduced $[Ca^{2+}]_m$ to similar levels as clonazepam alone, but the occurrence of necrotic cells was only 9% of control. This suggests that although clonazepam may have reduced $[Ca^{2+}]_m$ in a sub-population of cells, the MPTP still opened in other cells, and presence of CsA then protected these cells from MPTP opening, and necrosis. Another study in isolated myocytes found that clonazepam (100μM) inhibited total cellular $[Ca^{2+}]$ increase in response to oxidative stress induced by H_2O_2 (Sharikabad et al. 2004).

As well as the Ca^{2+} transport inhibitors themselves, intracellular acidification, such as occurs during ischaemia, may also prevent excessive mitochondrial Ca^{2+} uptake: in isolated rat mitochondria, lowering external pH to 6.8 resulted in mitochondrial acidification – this was measured in permeabilised myocytes that had been previously loaded with the pH-sensitive fluorescent indicator carboxy-SNARF, under conditions where it localised into mitochondria (Gursahani and Schaefer 2004). The acid pH decreased the rate of Ca^{2+} uptake, probably as a consequence of reduced $\Delta\Psi_m$ (Gursahani and Schaefer 2004) . Maintaining an acid pH upon reperfusion is known to delay intracellular Ca^{2+} accumulation and protect against reperfusion injury (Panagiotopoulos et al. 1990), and additionally inhibits opening of the MPTP (Halestrap 1991).

9.3.2.3. Do the Transporters Reverse Direction During Hypoxia?

Although several studies had reported an increase in $[Ca^{2+}]_m$ during hypoxia (see above), the mechanism of entry of the Ca^{2+} was either not discussed, or assumed to be the CaUP. But $\Delta\Psi_m$ depolarises during hypoxia (Di Lisa et al. 1995), and this would be expected to inhibit Ca^{2+} uptake through the CaUP. In order to investigate the mechanisms and time-course of $[Ca^{2+}]_m$ changes during cell injury, we used a model of single cell hypoxia/reoxygenation injury; this allowed continuous monitoring of $[Ca^{2+}]_m$, and correlation with cell functional recovery upon reoxygenation. We found that the increase in myocyte $[Ca^{2+}]_m$ during hypoxia could *not* be prevented by RuR but instead could be prevented by clonazepam. In contrast, upon reoxygenation, RuR once again inhibited Ca^{2+} uptake, whilst clonazepam inhibited efflux (Griffiths et al. 1998). This allowed us to propose the following model: during hypoxia Ca^{2+} entry into mitochondria occurs via mNCX, and the CaUP is largely inactive. Upon reoxygenation, however, the transporters regain their normal directionality. This model of hypoxia/reoxygenation injury is represented in Figure 9.4. Clonazepam (but not RuR), also protected the cells from reoxygenation induced damage, as indexed by the myocytes' ability to restore synchronised cell contractions (Griffiths et al. 1998). We have provided further evidence that entry of Ca^{2+} during hypoxia occurs via mNCX by showing, using a model of simulated hypoxia, that the

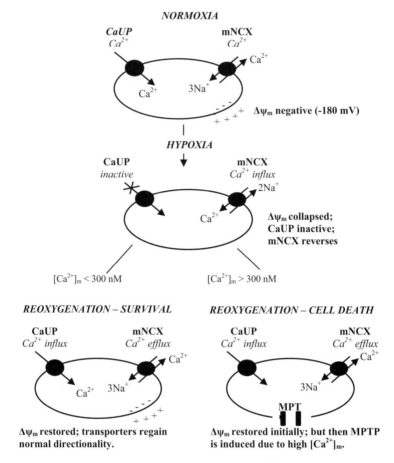

FIGURE 9.4. Changes in cardiomyocyte mitochondrial Ca^{2+} transport pathways during hypoxia/reoxygenation injury – implications for cell survival. Studies using a single-cell model of hypoxia-reoxygenation injury found that the level of $[Ca^{2+}]_m$ achieved at the end of hypoxia determined whether cells survived (recovered ability to respond synchronously to electrical stimulation), or hypercontracted (cell death). Cells only survived if $[Ca^{2+}]_m$ remained below about 250-300 nM. The accumulation of $[Ca^{2+}]_m$ during hypoxia also appeared to be via reversal of the mNCX, and not the CaUP (see text for further explanation). We proposed that upon reoxygenation, $\Delta\Psi_m$ initially repolarises. Then if $[Ca^{2+}]_m$ remains below about 300 nM, ATP synthesis is restored, and the Ca^{2+} transport pathways of the cell regain their normal function. However if $[Ca^{2+}]_m$ rises above about 300nM, this prevents ATP synthesis, and the MPTP is induced resulting in cell death.

process is Na^+-dependent (Griffiths 1999a). In support of this idea, a reversal of the mNCX under conditions of metabolic inhibition was also found in a study on cultured renal epithelial cells: CGP37157 inhibited the rise in $[Ca^{2+}]_m$ (rhod-2 fluorescence) induced by metabolic inhibition (2-deoxyglucose plus cyanide) (Smets et al. 2004).

9.3.2.4. Ischaemic Preconditioning

A discussion of the mechanism of ischaemic preconditioning (IPC) is beyond the scope of this chapter, but is dealt with elsewhere in this volume. However, one of the proposed protective mechanisms is that IPC leads to opening of a mK_{ATP} channel, and opening of this channel leads to membrane depolarisation (Murata et al. 2001). This, in turn, reduces Ca^{2+} uptake by mitochondria. A reduced $[Ca^{2+}]_m$ has been reported in a model of IPC (Murata et al. 2001), but it is far from clear that this was the result of mK_{ATP} opening. Furthermore, the effect of, for example, diazoxide, on $\Delta\Psi_m$ is controversial – some studies finding that it does produce a decrease (Murata et al. 2001), and others that it does not (Hanley et al. 2002; Katoh et al. 2002; Lawrence et al. 2001). The role of this mK_{ATP} channel has been almost entirely inferred indirectly based on studies using pharmacological modulation of the channel; by activators or inhibitors such as diazoxide and 5-hydroxdecanoate. An increasing number of studies recognise that these agents are not specific, and that their actions on the heart can be explained by actions other than on mK_{ATP} channels (Hanley et al. 2002; Katoh et al. 2002; Lawrence et al. 2001; Lim et al. 2002).

9.4. Therapeutic Implications

As well as possible benefits of modulating $[Ca^{2+}]_m$ during ischaemia/reperfusion injury, there is evidence that both RuR and Ru360 can reduce ventricular fibrillation (VF): In a pacing-induced model of VF in the isolated rat heart, perfusion with RuR (5µM), or Ru360 (10µM), resulted in conversion of VF to ventricular tachycardia (VT) (Kawahara et al., 2003). This is potentially beneficial since VT commonly occurs before VF, and sudden cardiac death (Myerburg et al. 1997).

However, the benzodiazepines like clonazepam and diltiazem cannot be used in the whole heart as specific antagonists of mitochondrial Ca^{2+} efflux, because of their effects on coronary vessels (diltiazem, for example, is used to reduce high blood pressure). However, design of more specific inhibitors of the mNCX may be of benefit in states where ATP synthesis is impaired such as in cardiomyopathies, since maintaining $[Ca^{2+}]_m$ at higher levels may in turn increase ATP production. There is some precedent for the idea that inhibiting the mNCX can enhance [ATP]: in pancreatic islets CGP37157 increased oxidative phosphorylation, and potentiated glucose-stimulated insulin release (Lee et al. 2003), prompting the authors to suggest it as a novel insulin secretagogue.

An intriguing study looked at human mitochondrial complex I deficiency – this is associated with numerous clinical symptoms, ranging from lethal encephalopathy's, of which Leigh disease is the most common, to neurodegenerative disorders, including Parkinson's disease (Triepels et al. 2001). This study used skin fibroblasts from children with Complex I deficiency, and measured cytosolic Ca^{2+} transients in response to bradykinin (Visch et al. 2004). They found impaired increases in $[Ca^{2+}]_c$, $[Ca^{2+}]_m$, and ATP production in response to bradykinin in the complex I deficient fibroblasts. Treatment with 1µM CGP37157 restored all these parameters. They suggested that (i) decreased ATP production in this disease is a result of reduced mitochondrial Ca^{2+} uptake, possibly due to a reduced $\Delta\Psi_m$ in these patients, and (ii) CGP37157 can indirectly promote enhanced release of Ca^{2+} from the ER in these cells by enhancing mitochondrial Ca^{2+} uptake so that there is less inhibitory Ca^{2+} available at the mouth of the IP_3-operated ER Ca^{2+} release channels.

Given the limitations on using the current mitochondrial Ca^{2+} transport inhibitors, it would be of great benefit if the transporters themselves could be cloned, for use in transgenic or knockout studies. Unfortunately there have only been a few sporadic attempts to purify the CaUP (Mironova et al. 1994; Saris et al. 1993; Zazueta et al. 1998) and mNCX (Li et al. 1992; Paucek and Jaburek 2004), and none has been more than partially successful. One of the main problems is that mitochondrial membrane proteins are notoriously difficult to purify, and the benefits of attempting their characterisation may require a lot of effort for little apparent gain. Nevertheless, a concerted effort to accomplish this may provide more benefit than can be foreseen at present.

References

Akerman KE (1978) Changes in membrane potential during calcium ion influx and efflux across the mitochondrial membrane. Biochim Biophys Acta 502(2): 359–66

Allen SP, Darley-Usmar VM, McCormack JG, Stone D (1993) Changes in mitochondrial matrix free calcium in perfused rat hearts subjected to hypoxia-reoxygenation. J Mol Cell Cardiol 25(8): 949–58

Arieli Y, Gursahani H, Eaton MM, Hernandez LA, Schaefer S (2004) Gender modulation of Ca(2+) uptake in cardiac mitochondria. J Mol Cell Cardiol 37(2): 507–13

Balaban RS (2002) Cardiac energy metabolism homeostasis: role of cytosolic calcium. J Mol Cell Cardiol 34(10): 1259–71

Baron KT, Thayer SA (1997) CGP37157 modulates mitochondrial Ca2+ homeostasis in cultured rat dorsal root ganglion neurons. Eur J Pharmacol 340(2-3):295–300

Bell CJ, Rutter GA, Griffiths EJ (2004) Calcium oscillations in mitochondria and cytosol of neonatal and adult rat cardiomyocytes detected using targeted aequorin. J Physiol 577P:PC72

Benzi RH, Lerch R (1992) Dissociation between contractile function and oxidative metabolism in postischemic myocardium. Attenuation by ruthenium red administered during reperfusion. Circ Res 71(3):567–76

Bers DM (1991) Species differences and the role of sodium-calcium exchange in cardiac muscle relaxation. Ann N Y Acad Sci 639:375–85

BeutnerG, Sharma VK, Giovannucci DR, Yule DI, Sheu SS (2001) Identification of a ryanodine receptor in rat heart mitochondria. J Biol Chem 276(24):21482–8

Beutner G, Sharma VK, Lin L, Ryu SY, Dirksen RT, Sheu SS (2005) Type 1 ryanodine receptor in cardiac mitochondria: Transducer of excitation-metabolism coupling. Biochim Biophys Acta 1717(1):1–10

Brandes R, Bers DM (1996) Increased work in cardiac trabeculae causes decreased mitochondrial NADH fluorescence followed by slow recovery. Biophys J 71(2): 1024–35

Brandes R, Bers DM (1997) Intracellular Ca2+ increases the mitochondrial NADH concentration during elevated work in intact cardiac muscle. Circ Res 80(1):82–7

Brown GC (1992) Control of respiration and ATP synthesis in mammalian mitochondria and cells. Biochem J 284 (Pt 1):1–13

Buntinas L, Gunter KK, Sparagna GC, Gunter TE (2001) The rapid mode of calcium uptake into heart mitochondria (RaM): comparison to RaM in liver mitochondria. Biochim Biophys Acta 1504(2–3):248-61

Bush LR, Shlafer M, Haack DW, Lucchesi BR (1980) Time-dependent changes in canine cardiac mitochondrial function and ultrastructure resulting from coronary occlusion and reperfusion. Basic Res Cardiol 75(4):555–71

Carry MM, Mrak RE, Murphy ML, Peng CF, Straub KD, Fody EP (1989) Reperfusion injury in ischemic myocardium: protective effects of ruthenium red and of nitroprusside. Am J Cardiovasc Pathol 2(4):335–44

Chacon E, Ohata H, Harper IS, Trollinger DR, Herman B, Lemasters JJ (1996) Mitochondrial free calcium transients during excitation-contraction coupling in rabbit cardiac myocytes. FEBS Lett 382(1-2):31–6

Chacon E, Reece JM, Nieminen AL, Zahrebelski G, Herman B, Lemasters JJ (1994) Distribution of electrical potential, pH, free Ca2+, and volume inside cultured adult rabbit cardiac myocytes during chemical hypoxia: a multiparameter digitized confocal microscopic study. Biophys J 66(4):942–52

Chance B, Williams GR (1956) Respiratory enzymes in oxidative phosphorylation. VI. The effects of adenosine diphosphate on azide-treated mitochondria. J Biol Chem 221(1):477–89

Clarke B, Spedding M, Patmore L, McCormack JG (1993) Protective effects of ranolazine in guinea-pig hearts during low-flow ischaemia and their association with increases in active pyruvate dehydrogenase. Br J Pharmacol 109(3):748-50

Cox DA, Conforti L, Sperelakis N, Matlib MA (1993) Selectivity of inhibition of Na(+)-Ca2+ exchange of heart mitochondria by benzothiazepine CGP-37157. J Cardiovasc Pharmacol 21(4):595–9

Cox DA, Matlib MA (1993) A role for the mitochondrial Na(+)-Ca2+ exchanger in the regulation of oxidative phosphorylation in isolated heart mitochondria. J Biol Chem 268(2):938–47

Crompton M, Ellinger H, Costi A (1988) Inhibition by cyclosporin A of a Ca2+-dependent pore in heart mitochondria activated by inorganic phosphate and oxidative stress. Biochem J 255(1):357–60

Crompton M, Heid I (1978) The cycling of calcium, sodium, and protons across the inner membrane of cardiac mitochondria. Eur J Biochem 91(2):599–608

Crompton M, Moser R, Ludi H, Carafoli E (1978) The interrelations between the transport of sodium and calcium in mitochondria of various mammalian tissues. Eur J Biochem 82(1):25–31

de Jesus Garcia-Rivas G, Guerrero-Hernandez A, Guerrero-Serna G, Rodriguez-Zavala JS, Zazueta C (2005) Inhibition of the mitochondrial calcium uniporter by the oxo-bridged dinuclear ruthenium amine complex (Ru360) prevents from irreversible injury in postischemic rat heart. FEBS J 272(13):3477–88

Delcamp TJ, Dales C, Ralenkotter L, Cole PS, Hadley RW (1998) Intramitochondrial [Ca2+] and membrane potential in ventricular myocytes exposed to anoxia-reoxygenation. Am J Physiol 275(2 Pt 2):H484–94

Di Lisa F, Blank PS, Colonna R, Gambassi G, Silverman HS, Stern MD, Hansford RG (1995) Mitochondrial membrane potential in single living adult rat cardiac myocytes exposed to anoxia or metabolic inhibition. J Physiol 486 (Pt 1):1–13

Di Lisa F, Canton M, Menabo R, Dodoni G, Bernardi P (2003) Mitochondria and reperfusion injury. The role of permeability transition. Basic Res Cardiol 98(4):235–41

Duchen MR, Biscoe TJ (1992) Mitochondrial function in type I cells isolated from rabbit arterial chemoreceptors. J Physiol 450:13–31

Figueredo VM, Dresdner KP, Jr., Wolney AC, Keller AM (1991) Postischaemic reperfusion injury in the isolated rat heart: effect of ruthenium red. Cardiovasc Res 25(4):337-42

Filippin L, Abad MC, Gastaldello S, Magalhaes PJ, Sandona D, Pozzan T (2005) Improved strategies for the delivery of GFP-based Ca2+ sensors into the mitochondrial matrix. Cell Calcium 37(2):129–36

Fleckenstein A, Frey M, Fleckenstein-Grun G (1983) Consequences of uncontrolled calcium entry and its prevention with calcium antagonists. Eur Heart J 4 Suppl H:43–50

Gallitelli MF, Schultz M, Isenberg G, Rudolf F (1999) Twitch-potentiation increases calcium in peripheral more than in central mitochondria of guinea-pig ventricular myocytes. J Physiol 518 (Pt 2):433–47

Griffiths EJ (1999a) Reversal of mitochondrial Na/Ca exchange during metabolic inhibition in rat cardiomyocytes. FEBS Lett 453(3):400–4

Griffiths EJ (1999b) Species dependence of mitochondrial calcium transients during excitation-contraction coupling in isolated cardiomyocytes. Biochem Biophys Res Commun 263(2):554–9

Griffiths EJ (2000) Use of ruthenium red as an inhibitor of mitochondrial Ca(2+) uptake in single rat cardiomyocytes. FEBS Lett 486(3):257–60

Griffiths EJ, Halestrap AP (1993) Pyrophosphate metabolism in the perfused heart and isolated heart mitochondria and its role in regulation of mitochondrial function by calcium. Biochem J 290 (Pt 2):489–95

Griffiths EJ, Halestrap AP (1995) Mitochondrial non-specific pores remain closed during cardiac ischaemia, but open upon reperfusion. Biochem J 307 (Pt 1):93–8

Griffiths EJ, Ocampo CJ, Savage JS, Rutter GA, Hansford RG, Stern MD, Silverman HS (1998) Mitochondrial calcium transporting pathways during hypoxia and reoxygenation in single rat cardiomyocytes. Cardiovasc Res 39(2):423–33

Griffiths EJ, Stern MD, Silverman HS (1997a) Measurement of mitochondrial calcium in single living cardiomyocytes by selective removal of cytosolic indo 1. Am J Physiol 273(1 Pt 1):C37–44

Griffiths EJ, Wei SK, Haigney MC, Ocampo CJ, Stern MD, Silverman HS (1997b) Inhibition of mitochondrial calcium efflux by clonazepam in intact single rat cardiomyocytes and effects on NADH production. Cell Calcium 21(4):321–9

Grover GJ, Dzwonczyk S, Sleph PG (1990) Ruthenium red improves postischemic contractile function in isolated rat hearts. J Cardiovasc Pharmacol 16(5):783–9

Grynkiewicz G, Poenie M, Tsien RY (1985) A new generation of Ca2+ indicators with greatly improved fluorescence properties. J. Biol. Chem. 260(6):3440–3450

Gunter TE, Pfeiffer DR (1990) Mechanisms by which mitochondria transport calcium. Am J Physiol 258(5 Pt 1):C755–86

Gupta MP, Dixon IM, Zhao D, Dhalla NS (1989) Influence of ruthenium red on rat heart subcellular calcium transport. Can J Cardiol 5(1):55–63

Gupta MP, Innes IR, Dhalla NS (1988) Responses of contractile function to ruthenium red in rat heart. Am J Physiol 255(6 Pt 2):H1413–20

Gursahani HI, Schaefer S (2004) Acidification reduces mitochondrial calcium uptake in rat cardiac mitochondria. Am J Physiol Heart Circ Physiol 287(6):H2659–65

Halestrap AP (1991) Calcium-dependent opening of a non-specific pore in the mitochondrial inner membrane is inhibited at pH values below 7. Implications for the protective effect of low pH against chemical and hypoxic cell damage. Biochem J 278 (Pt 3):715–9

Halestrap AP, Clarke SJ, Javadov SA (2004) Mitochondrial permeability transition pore opening during myocardial reperfusion–a target for cardioprotection. Cardiovasc Res 61(3):372–85

Hanley PJ, Mickel M, Loffler M, Brandt U, Daut J (2002) KATP channel-independent targets of diazoxide and 5-hydroxydecanoate in the heart. J Physiol (Lond) 542(3): 735–741

Hansford RG (1994) Physiological role of mitochondrial Ca2+ transport. J Bioenerg Biomembr 26(5):495–508

Heineman FW, Balaban RS (1993) Effects of afterload and heart rate on NAD(P)H redox state in the isolated rabbit heart. Am J Physiol 264(2 Pt 2):H433–40

Honda HM, Korge P, Weiss JN (2005) Mitochondria and Ischemia/Reperfusion Injury. Ann NY Acad Sci 1047(1):248–258

Hunter DR, Haworth RA, Southard JH (1976) Relationship between configuration, function, and permeability in calcium-treated mitochondria. J Biol Chem 251(16): 5069–77

Jacobson J, Duchen MR (2004) Interplay between mitochondria and cellular calcium signalling. Mol Cell Biochem 256–257(1-2):209-18

Katoh H, Nishigaki N, Hayashi H (2002) Diazoxide Opens the Mitochondrial Permeability Transition Pore and Alters Ca2+ Transients in Rat Ventricular Myocytes. Circulation 105(22):2666–2671

Katz LA, Koretsky AP, Balaban RS (1987) Respiratory control in the glucose perfused heart. A 31P NMR and NADH fluorescence study. FEBS Lett 221(2):270–6

Katz LA, Swain JA, Portman MA, Balaban RS (1989) Relation between phosphate metabolites and oxygen consumption of heart in vivo. Am J Physiol 256(1 Pt 2): H265–74

Kawahara K, Takase M, Yamauchi Y (2003) Ruthenium red-induced transition from ventricular fibrillation to tachycardia in isolated rat hearts: possible involvement of changes in mitochondrial calcium uptake. Cardiovasc Pathol 12(6):311–21

Landgraf G, Gellerich FN, Wussling MH (2004) Inhibitors of SERCA and mitochondrial Ca-uniporter decrease velocity of calcium waves in rat cardiomyocytes. Mol Cell Biochem 256-257(1–2):379-86

Lawrence CL, Billups B, Rodrigo GC, Standen NB (2001) The KATP channel opener diazoxide protects cardiac myocytes during metabolic inhibition without causing mitochondrial depolarization or flavoprotein oxidation. Br J Pharmacol 134(3):535–542

Lee B, Miles PD, Vargas L, Luan P, Glasco S, Kushnareva Y, Kornbrust ES, Grako KA, Wollheim CB, Maechler P, Olefsky JM, Anderson CM (2003) Inhibition of mitochondrial Na+-Ca2+ exchanger increases mitochondrial metabolism and potentiates glucose-stimulated insulin secretion in rat pancreatic islets. Diabetes 52(4):965–73

Leperre A, Millart H, Prevost A, Trenque T, Kantelip JP, Keppler BK (1995) Compared effects of ruthenium red and cis [Ru(NH3)4Cl2]Cl on the isolated ischaemic-reperfused rat heart. Fundam Clin Pharmacol 9(6):545–53

Li W, Shariat-Madar Z, Powers M, Sun X, Lane RD, Garlid KD (1992) Reconstitution, identification, purification, and immunological characterization of the 110-kDa Na+/Ca2+ antiporter from beef heart mitochondria. J. Biol. Chem. 267(25): 17983–17989

Lim KHH, Javadov SA, Das M, Clarke SJ, Suleiman MS, Halestrap AP (2002) The effects of ischaemic preconditioning, diazoxide and 5-hydroxydecanoate on rat heart mitochondrial volume and respiration. J Physiol (Lond) 545(3):961–974

Matlib MA, Zhou Z, Knight S, Ahmed S, Choi KM, Krause-Bauer J, Phillips R, Altschuld R, Katsube Y, Sperelakis N, Bers DM (1998) Oxygen-bridged dinuclear ruthenium amine complex specifically inhibits Ca2+ uptake into mitochondria in vitro and in situ in single cardiac myocytes. J Biol Chem 273(17):10223–31

McCormack JG, Halestrap A P, Denton RM (1990) Role of calcium ions in regulation of mammlian intramitochondrial metabolism. Physiol Rev 70(2):391–425

McKean TA (1991) Calcium uptake by mitochondria isolated from muskrat and guinea pig hearts. J Exp Biol 157:133–42

McMillin-Wood J, Wolkowicz PE, Chu A, Tate CA, Goldstein MA, Entman ML (1980) Calcium uptake by two preparations of mitochondria from heart. Biochim Biophys Acta 591(2):251–65

Mironova GD, Baumann M, Kolomytkin O, Krasichkova Z, Berdimuratov A, Sirota T, Virtanen I, Saris NE (1994) Purification of the channel component of the mitochondrial calcium uniporter and its reconstitution into planar lipid bilayers. J Bioenerg Biomembr 26(2):231–8

Miyamae M, Camacho SA, Weiner MW, Figueredo VM (1996) Attenuation of postischemic reperfusion injury is related to prevention of $[Ca2+]_m$ overload in rat hearts. Am J Physiol Heart Circ Physiol 271(5):H2145–2153

Miyata H, Lakatta EG, Stern MD, Silverman HS (1992) Relation of mitochondrial and cytosolic free calcium to cardiac myocyte recovery after exposure to anoxia. Circ Res 71(3):605–13

Miyata H, Silverman HS, Sollott SJ, Lakatta EG, Stern MD, Hansford RG (1991) Measurement of mitochondrial free Ca2+ concentration in living single rat cardiac myocytes. Am J Physiol 261(4 Pt 2):H1123–34

Moore CL (1971) Specific inhibition of mitochondrial Ca++ transport by ruthenium red. Biochem Biophys Res Commun 42(2):298–305

Murata M, Akao M, O'Rourke B, Marban E (2001) Mitochondrial ATP-sensitive potassium channels attenuate matrix Ca(2+) overload during simulated ischemia and reperfusion: possible mechanism of cardioprotection. Circ Res 89(10):891–8

Myerburg RJ, Interian A, Jr., Mitrani RM, Kessler KM, Castellanos A (1997) Frequency of sudden cardiac death and profiles of risk. Am J Cardiol 80(5B):10F–19F

Nicholls DG, Crompton M (1980) Mitochondrial calcium transport. FEBS Lett 111(2):261–8

Panagiotopoulos S, Daly MJ, Nayler WG (1990) Effect of acidosis and alkalosis on postischemic Ca gain in isolated rat heart. Am J Physiol Heart Circ Physiol 258(3):H821–828

Park Y, Bowles DK, Kehrer JP (1990) Protection against hypoxic injury in isolated-perfused rat heart by ruthenium red. J Pharmacol Exp Ther 253(2):628–35

Paucek P, Jaburek M (2004) Kinetics and ion specificity of Na(+)/Ca(2+) exchange mediated by the reconstituted beef heart mitochondrial Na(+)/Ca(2+) antiporter. Biochim Biophys Acta 1659(1):83–91

Peng CF, Kane JJ, Straub KD, Murphy ML (1980) Improvement of mitochondrial energy production in ischemic myocardium by in vivo infusion of ruthenium red. J Cardiovasc Pharmacol 2(1):45–54

Piper HM, Noll T, Siegmund B (1994) Mitochondrial function in the oxygen depleted and reoxygenated myocardial cell. Cardiovasc Res 28(1):1–15

Pozzan T, Mongillo M, Rudolf R (2003) The Theodore Bucher lecture. Investigating signal transduction with genetically encoded fluorescent probes. Eur J Biochem 270(11):2343–52

Ridgway EB, Ashley CC (1967) Calcium transients in single muscle fibers. Biochem Biophys Res Commun 29(2):229–34

Rizzuto R, Simpson AW, Brini M, Pozzan T (1992) Rapid changes of mitochondrial Ca2+ revealed by specifically targeted recombinant aequorin. Nature 358(6384):325–7

Robert V, Gurlini P, Tosello V, Nagai T, Miyawaki A, Di Lisa F, Pozzan T (2001) Beat-to-beat oscillations of mitochondrial [Ca2+] in cardiac cells. Embo J 20(17): 4998–5007

Sanchez JA, Garcia MC, Sharma VK, Young KC, Matlib MA, Sheu S-S (2001) Mitochondria regulate inactivation of L-type Ca2+ channels in rat heart. J Physiol (Lond) 536(2):387–396

Saris NE, Sirota TV, Virtanen I, Niva K, Penttila T, Dolgachova LP, Mironova GD (1993) Inhibition of the mitochondrial calcium uniporter by antibodies against a 40–kDa glycoproteinT. J Bioenerg Biomembr 25(3):307-12

Scholz TD, Laughlin MR, Balaban RS, Kupriyanov VV, Heineman FW (1995) Effect of substrate on mitochondrial NADH, cytosolic redox state, and phosphorylated compounds in isolated hearts. Am J Physiol 268(1 Pt 2):H82–91

Seguchi H, Ritter M, Shizukuishi M, Ishida H, Chokoh G, Nakazawa H, Spitzer KW, Barry WH (2005) Propagation of Ca2+ release in cardiac myocytes: role of mitochondria. Cell Calcium 38(1):1–9

Sharikabad MN, Ostbye KM, Brors O (2001) Increased [Mg2+]o reduces Ca2+ influx and disruption of mitochondrial membrane potential during reoxygenation. Am J Physiol Heart Circ Physiol 281(5):H2113–23

Sharikabad MN, Ostbye KM, Brors O (2004) Effect of hydrogen peroxide on reoxygenation-induced Ca2+ accumulation in rat cardiomyocytes. Free Rad Biol Med 37(4):531–538

Shen AC, Jennings RB (1972a) Kinetics of calcium accumulation in acute myocardial ischemic injury. Am J Pathol 67(3):441–52

Shen AC, Jennings RB (1972b) Myocardial calcium and magnesium in acute ischemic injury. Am J Pathol 67(3):417–40

Smets I, Caplanusi A, Despa S, Molnar Z, Radu M, vandeVen M, Ameloot M, Steels P (2004) Ca2+ uptake in mitochondria occurs via the reverse action of the Na+/Ca2+ exchanger in metabolically inhibited MDCK cells. Am J Physiol Renal Physiol 286(4):F784–794

Sordahl LA, Stewart ML (1980) Mechanism(s) of altered mitochondrial calcium transport in acutely ischemic canine hearts. Circ Res 47(6):814–20.

Sparagna GC, Gunter KK, Sheu SS, Gunter TE (1995) Mitochondrial calcium uptake from physiological-type pulses of calcium. A description of the rapid uptake mode. J Biol Chem 270(46):27510–5.

Stone D, Darley-Usmar V, Smith DR, O'Leary V (1989) Hypoxia-reoxygenation induced increase in cellular Ca2+ in myocytes and perfused hearts: the role of mitochondria. J Mol Cell Cardiol 21(10):963–73.

Suleiman MS, Halestrap AP, Griffiths EJ (2001) Mitochondria: a target for myocardial protection. Pharmacol Ther 89(1):29–46.

Takeo S, Tanonaka K, Iwai T, Motegi K, Hirota Y (2004) Preservation of mitochondrial function during ischemia as a possible mechanism for cardioprotection of diltiazem against ischemia/reperfusion injury. Biochem Pharmacol 67(3):565–574.

Territo PR, Mootha VK, French SA, Balaban RS (2000) Ca(2+) activation of heart mitochondrial oxidative phosphorylation: role of the F(0)/F(1)–ATPase. Am J Physiol Cell Physiol 278(2):C423-35.

Tibbits GF, Xu L, Sedarat F (2002) Ontogeny of excitation-contraction coupling in the mammalian heart. Comp Biochem Physiol A Mol Integr Physiol 132(4):691–8.

Triepels RH, Van Den Heuvel LP, Trijbels JM, Smeitink JA (2001) Respiratory chain complex I deficiency. Am J Med Genet 106(1):37–45.

Trollinger DR, Cascio WE, Lemasters JJ (1997) Selective loading of Rhod 2 into mitochondria shows mitochondrial Ca2+ transients during the contractile cycle in adult rabbit cardiac myocytes. Biochem Biophys Res Commun 236(3):738–42.

Tsien RY (1980) New calcium indicators and buffers with high selectivity against magnesium and protons: design, synthesis, and properties of prototype structures. Biochemistry 19(11):2396–404.

Vasington FD, Gazzotti P, Tiozzo R, Carafoli E (1972) The effect of ruthenium red on Ca 2+ transport and respiration in rat liver mitochondria. Biochim Biophys Acta 256(1):43–54.

Visch HJ, Rutter GA, Koopman WJ, Koenderink JB, Verkaart S, de Groot T, Varadi A, Mitchell KJ, van den Heuvel LP, Smeitink JA, Willems PH (2004) Inhibition of mitochondrial Na+-Ca2+ exchange restores agonist-induced ATP production and Ca2+ handling in human complex I deficiency. J Biol Chem 279(39):40328–36.

White RL, Wittenberg BA (1993) NADH fluorescence of isolated ventricular myocytes: effects of pacing, myoglobin, and oxygen supply. Biophys J 65(1):196–204.

White RL, Wittenberg BA (1995) Effects of calcium on mitochondrial NAD(P)H in paced rat ventricular myocytes. Biophys J 69(6):2790–9.

Winniford MD, Willerson JT, Hillis LD (1985) Calcium antagonists for acute ischemic heart disease. Am J Cardiol 55(3):116B–124B.

Ying WL, Emerson J, Clarke MJ, Sanadi DR (1991) Inhibition of mitochondrial calcium ion transport by an oxo-bridged dinuclear ruthenium ammine complex. Biochemistry 30(20):4949–52.

Zazueta C, Zafra G, Vera G, Sanchez C, Chavez E (1998) Advances in the purification of the mitochondrial Ca2+ uniporter using the labeled inhibitor 103Ru360. J Bioenerg Biomembr 30(5):489–98.

Zhang SZ, Gao Q, Cao CM, Bruce IC, Xia Q (2006) Involvement of the mitochondrial calcium uniporter in cardioprotection by ischemic preconditioning. Life Sci 78(7): 738–45.

Zhou Z, Matlib MA, Bers DM (1998) Cytosolic and mitochondrial Ca2+ signals in patch clamped mammalian ventricular myocytes. J Physiol 507 (Pt 2):379–403.

Zhu LP, Yu XD, Ling S, Brown RA, Kuo TH (2000) Mitochondrial Ca(2+)homeostasis in the regulation of apoptotic and necrotic cell deaths. Cell Calcium 28(2):107–17.

10
Mitochondrial Ion Channels

Brian O'Rourke

10.1. Introduction

The maintenance of a large electrochemical driving force for protons across the mitochondrial inner membrane is essential for the production of ATP through oxidative phosphorylation. At face value, the opening of energy dissipating ion channels in the mitochondria would be unfavorable for energy transduction, but a wealth of evidence now indicates that selective (and some non-selective) ion channels may become active under various physiological or pathophysiological conditions. This review summarizes recent investigations into the functional roles for mitochondrial ion channels and efforts to identify molecular correlates related to specific ion fluxes across the inner membrane. While it is clear that mitochondrial ion channels play important roles in cellular life and death, our understanding of their structure is limited, and seminal discoveries are eagerly awaited.

Interest in mitochondria has grown in recent years as the list of important cellular functions for this organelle has expanded from its fundamental role as a provider of energy to generator of reactive oxygen species (ROS), modulator of intracellular Ca^{2+} fluxes, initiator of cell death, mediator of cell protection, and participant in the processes of aging and disease. With regard to oxidative phosphorylation, the majority of the participating proteins have been identified at the molecular level, and in many cases, high resolution crystal structures are available. Importantly, a robust model of the energy transduction process, refined over the decades since the chemiosmotic principle was proposed, is available (for review, see Nicholls and Ferguson 2002).

In contrast, many of the important proteins involved in mediating ion transport across the mitochondrial inner membrane have not been identified, although they have been well characterized at the functional level. Extensive evidence is available demonstrating that selective electrophoretic K^+ (Brierley et al. 1971; Hansford and Lehninger 1972; Jung et al. 1977; Jung et al. 1984) and Ca^{2+} (reviewed by Gunter and Pfeiffer 1990) uptake by mitochondria occurs and that Na^+ is exchanged for H^+ (Douglas and Cockrell 1974) and Ca^{2+} (Jung et al. 1995). Similarly, K^+/H^+ exchange balances the K^+ uniport activity and modulates mitochondrial volume (Garlid 1996), yet no proteins have been definitively associated with any of these cation transport processes. Numerous transporters of anionic metabolites across the inner membrane, usually of the antiporter type, have also been identified at the molecular level (Jezek and Jezek 2003). However, apart from the inorganic phosphate carrier (Gerreira and Pedersen 1993) and the mitochondrial uncoupling proteins (UCP1, 2, and 3) (Rousset et al. 2004), little is known about the molecular structure of putative inner membrane anion channels that can mediate rapid anion flux.

Since ion channels are capable of transporting millions of ions per second, and the electrochemical driving forces for ion movement across the inner membrane are enormous, the knowledge gap regarding mitochondrial ion channel structure is perhaps understandable – these proteins must necessarily be present in extremely low abundance, or have a very low open probability, in order to maintain the low permeability to ions required to exploit the protonmotive force for the generation of ATP via the mitochondrial ATP synthase. Nevertheless, although the opening of mitochondrial ion channels may be brief and highly controlled, the significance of the effects of their activation cannot be overstated. Mitochondrial ion channels are crucial to the mechanism of energy supply and demand matching and are the decisive factor in determining whether a cell lives and dies.

Emerging evidence also suggests that the mitochondrial network can act as an intracellular signaling network. For example, the activation of redox sensitive transcription factors (e.g. HIF-1α or NF-κB) and signalling pathways (e.g. protein kinase C) is likely to involve the generation of reactive oxygen species by the mitochondria (Guzy et al. 2005). Ion channels may contribute to this function in direct (serving as free radical transport pathways) or indirect (by altering the rate of respiration and the leak of electrons to ROS) ways. These novel and important functions of mitochondria underscore the need to identify and study the structure and organization of the proteins in the mitochondrial outer and inner membranes. While much progress has been made in assaying the function of mitochondrial ion channels and exchangers, the challenge ahead will be to assign structures to the various ion transport pathways and to study how increasing or decreasing their expression affects the integrated cell function.

10.2. Fast Ion Movements Across the Inner Membrane

A basic tenet of the chemiosmotic hypothesis is that, apart from the movement of protons through the mitochondrial ATP synthase and through non-specific low conductance "proton leak" pathways, the mitochondrial inner membrane must be quite impermeable to ions in order to maintain the efficiency of oxidative phosphorylation. However, the large electrical driving force for ion movement ($\sim$180mV) would strongly favor ion flow through any open ion channels, even in the absence of a concentration gradient across the inner membrane. A classic and unquestionable example is Ca^{2+} influx through the Ca^{2+} uniporter. Under conditions that prevent (e.g. in the presence of nucleotides) the activation of the mitochondrial permeability transition pore (PTP), mitochondria can take up enormous amounts of Ca^{2+} (up to almost 1M total Ca^{2+}) although the matrix free

Ca^{2+} concentration remains in the μM range due to the reversible precipitation of Ca^{2+} with Pi. This Ca^{2+} sink property of the mitochondria is circumvented if PTP opening is triggered by the Ca^{2+} overload, allowing both the release of the accumulated Ca^{2+} and other matrix constituents with mass $<\sim 1.5$kDa. The complete breakdown of the ion permeability barrier upon opening of the PTP short circuits proton-coupled energy transduction and dissipates mitochondrial membrane potential ($\Delta\Psi_m$).

K^+ conductance can also be substantial in energized mitochondria (Brierley et al. 1971; Hansford and Lehninger 1972; Jung et al. 1977; Diwan et al. 1985) and recent efforts have focused on characterizing the regulation and structure of proteins involved in K^+ uniport activity. The importance of this endeavor is underscored by an accumulating body of evidence that mitochondrial K^+ uptake is tied to protection against ischemic- or oxidative stress-mediated cell injury (discussed below). The interplay between K^+ uniport activity and K^+/H^+ exchange is an important physiological mechanism for regulating mitochondrial volume (Garlid et al. 1996).

Although the ion transport rates for Na^+ movement across the mitochondrial membrane are slow compared to ion channel (uniporter)-mediated processes, both the Na^+/H^+ exchange and Na^+/Ca^{2+} exchangers of the inner membrane play key roles in counterbalancing the activity of the Ca^{2+} uniporter. Thus, their identification and characterization will be essential for developing a complete picture of ion homeostasis in mitochondria.

Mitochondrial swelling induced by cation and anion movements into the matrix compartment has been extensively employed to define ion permeabilities across the mitochondrial inner membrane (Nicholls and Ferguson 2002). A series of papers published in the 1980's postulated that an inner membrane anion channel (IMAC) was present under special conditions (e.g., divalent depletion) (Garlid and Beavis 1986; Beavis 1992; Beavis and Powers 2004). Subsequently, direct single channel patch-clamp methods revealed a number of partially anion selective conductances in the inner membrane (Kinnally et al. 1993; Borecky et al. 1997), with the most prominent being the so-called "centum pS" channel, a strong voltage-dependent outwardly rectifying current (Sorgato et al. 1987). Over the past few years, we have accumulated evidence suggesting that IMAC might underlie oscillatory mitochondrial depolarizations induced by substrate deprivation (O'Rourke 2000) or oxidative stress (Aon et al. 2003). We propose that these channels play a prominent role in post-ischemic electrical and contractile dysfunction in the heart (Akar et al. 2005). The mitochondrial benzodiazepine receptor (mBzR) appears to modulate this conductance, and is currently the only link to a defined structure that may be involved in this important process.

10.3. Physiological Roles of Mitochondrial Ion Channels

Clearly, the mitochondrial ion transport process of primary importance to the cell is the controlled transit of protons down their electrochemical gradient by the mitochondrial F_1F_o ATP synthase, coupled to ATP production. Less clear is the role of proton leak pathways across the inner membrane. It has been postulated that the leak pathways might optimize the thermodynamic efficiency of energy transduction (Stucki 1980). Alternatively, energy dissipation through proton leaks has been suggested to modulate the rate of ROS production by the electron transport chain (Demin et al. 1998). This is the rationale given to explain some of the effects of the mitochondrial uncoupling proteins, but this conjecture remains to be unequivocally proven. The actual structures responsible for proton leaks across the inner membrane have not been determined.

Another clear physiological role for mitochondrial ion channels is to mediate the uptake of Ca^{2+} into the matrix. The Vmax for influx through the mitochondrial Ca^{2+} uniporter is large (Bragadin et al. 1979) and Ca^{2+} movements are electrophoretically driven (Gunter and Gunter 1994). Ca^{2+} uptake by mitochondria plays an important role in stimulating oxidative phosphorylation through the activation of TCA cycle dehydrogenases (Denton and McCormack 1990) and perhaps other sites in the electron transport chain (Bender and Kadenbach 2000; Territo et al. 2000). Mitochondrial Ca^{2+} uptake during ischemia and reperfusion also is a determinant of cell injury. Recently, the properties of a highly selective Ca^{2+} channel in the mitochondrial inner membrane have been shown to match that of the Ca^{2+} uniporter, providing additional supportive evidence that it is truly an ion channel (Kirichok et al. 2004).

Since the mitochondria represent a restricted compartment bounded by the inner and outer membranes, any net ion movements, accompanied by water, have profound effects on the volume of the organelle. Thus, mitochondria have been suggested to resemble "perfect osmometers" (Halestrap 1987). Modulation of the mitochondrial matrix volume is likely to serve an important physiological role in optimizing the oxidation of substrates like fatty acids. K^+ uniport activity, counterbalanced by the actions of the K^+/H^+ exchanger, provides a mechanism by which mitochondrial volume may be regulated (Garlid 1996). It should be noted that anion movements must accompany cation transport in order to maintain electroneutrality, so the regulation of anion channels in the inner membrane might also be a physiological mechanism for mitochondrial volume regulation, as previously suggested for the IMAC (Garlid and Beavis 1986).

10.4. Protective K^+ Channels

More than 20 years ago, it was demonstrated that pharmacological agents capable of opening K^+ channels protect hearts against ischemia-reperfusion injury (Lamping and Gross 1985; Grover et al. 1989). This finding is now firmly established and K^+ channel openers have come to be viewed as "chemical preconditioners", i.e., compounds that can mimic the protective effects of brief cycles of ischemia and reperfusion. Moreover, the finding that K^+ channel inhibitors such as glibenclamide and 5-hydroxydecanoate could block the protective effects of either ischemic preconditioning (IPC) or K^+ channel openers (Grover and Garlid, 2000) suggested that K^+ channels were a native effector of preconditioning. While earlier studies naturally presumed that the target of the K^+ channel openers was the sarcolemmal ATP-sensitive K^+ (K_{ATP}) channel, the focus has shifted recently to the mitochondria as the primary cardioprotective target of these compounds.

10.5. Sarcolemmal K_{ATP}

Plasma membrane K_{ATP} channels, initially characterized in cardiac cells in 1983 (Noma 1983), have been extensively studied at the molecular level (Bryan et al. 2004). Although their physiological role is clear in the regulation of insulin release from the pancreas and in the modulation of vascular tone in smooth muscle, our understanding of why cardiac cells express a high density of sarcolemmal K_{ATP} channels (sarcK_{ATP}) remains cloudy. The recent availability of transgenic mouse models in which components of the sarcK_{ATP} channel have been knocked out provides a new opportunity to answer this question (Chutkow et al. 2001; Suzuki et al. 2002; Gumina et al. 2003; Suzuki et al. 2003). In the mouse, the primary physiological role of sarcK_{ATP} is apparently to help the animal cope with metabolic stress. Mice that lack the pore-forming subunit of the cardiac sarcK_{ATP} (Kir6.2) have a severely compromised ability to tolerate ischemia - even short periods of ischemia lead to rapid ischemic contracture of the heart (Suzuki et al. 2002; Suzuki et al. 2003). Similarly, when the K_{ATP} channel is pharmacologically inhibited in the mouse with HMR1098, a selective sarcK_{ATP} blocker, ischemic dysfunction is accentuated. The injury induced by sarcK_{ATP} inhibition in the mouse is so severe that the innate ability to protect hearts with preconditioning stimuli is lost. Function is also compromised with exercise, a more physiological form of metabolic stress (Zingman et al. 2002). From these data we can conclude that the mouse is highly dependent on sarcK_{ATP} channels for survival under conditions of high energy demand.

In contrast, in larger animal species (e.g., rabbits) (Sato et al. 2000), and in humans (Ghosh et al. 2000), sarcK_{ATP} appears to play a minor role in protecting the heart during ischemia. Selective pharmacological inhibition of sarcK_{ATP} has little or no effect on infarct size after ischemia and reperfusion or on the cardiac preconditioning response (Tanno et al. 2001). Rather, sarcK_{ATP} contributes to post-ischemic electrical dysfunction by increasing the dispersion of repolarization and the heterogeneity of electrical excitability (Billman et al. 1998; Akar et al. 2005). Moreover, as established in previous studies (Grover and Garlid 2000), the action potential shortening effects of sarcK_{ATP} activation during ischemia are not correlated with the extent of protection afforded by K^+ channel openers. Thus, other targets of these compounds, including the mitochondria, have been investigated in the context of protection against metabolic stress, as discussed below.

10.6. Mitochondrial K_{ATP}

In 1991, ATP-sensitive K^+ channels were reported to be present in the liver mitochondrial inner membrane, using the direct mitoplast patch-clamp method (Inoue et al. 1991). The mitochondrial K_{ATP} channel (mitoK_{ATP}) had properties similar to those observed in the sarcolemma of cardiac cells. Thus a link was established between the effects of K^+ channel opener compounds on mitochondrial function (Paucek et al. 1992; Szewczyk et al. 1993; Garlid 1996; Garlid et al. 1997) and a specific ion channel target that could be associated with protection against ischemic injury in intact hearts (Garlid et al. 1997) or isolated myocytes (Liu et al. 1998). The general hypothesis that an increase in mitochondrial inner membrane permeability to K^+ improves cellular tolerance to ischemia-reperfusion injury has found widespread support in the setting of a variety of tissues including liver, gut, brain, kidney and the heart.

Even before the description of mitoK$_{ATP}$, studies of the mitochondrial K^+ uniporter provided evidence that K^+ selective channels were present on the mitochondrial inner membrane. Since then, specific functional evidence supporting mitoK$_{ATP}$ has fallen into several categories, including: electrophysiological recordings of channels reconstituted into proteoliposomes with mitochondrial membrane proteins, measurements of K^+ uptake into mitochondria or reconstituted liposomes, mitochondrial swelling assays, changes in mitochondrial redox potential or alterations in mitochondrial energetic parameters such as respiration and $\Delta\Psi_m$, as described in recent reviews of the subject (Garlid et al. 2003; O'Rourke 2004).

As for the other mitochondrial channels mentioned above, the lack of a specific molecular entity associated with mitoK$_{ATP}$ has largely restricted the supportive arguments to the study of available pharmacological agents. A strong caveat to acknowledge is the fact that some of these agents have substantial non-specific effects on mitochondria that may or may not contribute to the effects of the compound on the response to ischemia-reperfusion injury. It therefore behooves the careful investigator to test several active channel openers or inhibitors when possible, to find the common effect from structurally dissimilar agents that are likely to have different non-specific actions. A variety of K^+ channel openers have been shown to activate mitoK$_{ATP}$, including diazoxide, nicorandil, BMS191095 (Grover et al. 2001), cromakalim, levcromakalim, EMD60480, EMD57970 (Garlid et al. 1996), pinacidil, RP66471, minoxidil sulfate, KRN2391 (Szewczyk et al. 1993) - only the first three show significant selectivity towards the mitochondrial versus the sarcolemmal isoform of the K$_{ATP}$ channel in cardiac myocytes. Another drawback is that there is only one widely available K^+ inhibitor, 5-hydroxydecanoate, that selectively inhibits mitoK$_{ATP}$ without blocking sarcK$_{ATP}$ (Sato et al. 1998). The classical K$_{ATP}$ channel inhibitor, glibenclamide, is a sulfonylurea that blocks both the sarcK$_{ATP}$ and mitoK$_{ATP}$ isoforms, while HMR1098 is usually found to be selective for the sarcolemmal channel.

Recent reports suggest that a number of other compounds may modulate mitoK$_{ATP}$, including sildenafil (Ockaili et al. 2002), levosimedan (Kopustinskiene et al. 2001), YM934 (Tanonaka et al. 1999), and MCC-134 (Sasaki et al. 2003). MCC-134 was capable of inhibiting mitoK$_{ATP}$ while activating sarcK$_{ATP}$. Importantly, diazoxide-mediated protection against simulated ischemia was prevented by MCC-134, supporting the argument that mitoK$_{ATP}$, rather than sarcK$_{ATP}$ channels, were the mediators of protection. The opposite effects have been reported for the antiarrhythmic drug Bepridil, which was shown in a recent study to activate mitoK_{ATP} while inhibiting sarcK_{ATP} [(Sato et al. 2006)]. While this combination of actions would be expected to be cardioprotective and prevent arrhythmias triggered by the opening of sarcK_{ATP}, Bepridil also blocks other well known targets, including the L-type Ca^{2+} channel, making interpretation of the overall results difficult to attribute specifically to mitoK_{ATP}.

Another emerging area of interest is how signal transduction pathways either activate, or are activated by, the mitoK_{ATP} channel. Signalling pathways linked to phosphoinositide hydrolysis, protein kinase C (PKC) activation or tyrosine kinases have been shown to be mediators and/or effectors of cellular protection (Cohen et al. 2000). In many cases, the downstream effects of receptor activation can be blocked not only by inhibitors of the kinases, but by inhibition of the mitoK_{ATP} channel (Uchiyama et al. 2003). This begs the question of whether the channel lies upstream or downstream of the post-translational modifications mediated by either PKC or other kinases. One plausible link between the activation of mitoK_{ATP} and signalling would be a change in redox-sensitive pathways as a result of an increase in mitochondrially-derived reactive oxygen species (Pain et al. 2000; Oldenburg et al. 2003). This could occur in response to the increase in respiratory rate (and consequent leak of electrons to superoxide) induced by the opening of the K^+ channel. A common effector, glycogen synthase kinase 3β (GSK-3β) has recently been proposed as the integrator of various preconditioning stimuli including the actions of K^+ channel openers (Juhaszova et al. 2004). Activation of GSK-3β blunts the effects of laser-induced oxidative stress on the activation of the PTP in isolated heart cells. The mitoK_{ATP} channel is likely to be both a target and effector in these pathways. For example, nitric oxide donors (Sasaki et al., 2000) and PKC activators (Sato et al. 1998) can enhance the activation of mitoK_{ATP} in isolated cardiac cells, and in some studies, the mitoK_{ATP} inhibitor 5-hydroxydecanoate could not only prevent IPC when applied during the preconditioning phase, but could also block the protection against infarction when given before the long index ischemia, for both early and delayed preconditioning protocols (Ockaili et al. 1999). More recently, the NO-cyclic GMP-G kinase signal cascade has been implicated in the activation of mitoK_{ATP} (Xu et al. 2004; Costa et al. 2005; Dang et al. 2005).

10.7. Mitochondrial K_{Ca}

In 2002, we identified another mitochondrial K^+ channel, the mitoK_{Ca} channel, in the cardiac mitochondrial inner membrane and linked it to protection against ischemia-reperfusion injury (Xu et al. 2002). In mitoplast patch-clamp experiments, this channel displayed properties resembling the Ca^{2+}-activated K^+ channel found in the plasma membrane of various cells, including smooth muscle myocytes. MitoK_{Ca} was inhibited by charybdotoxin and iberiotoxin and these toxins were shown to blunt K^+ uptake into mitochondria in partially permeabilized adult cardiac cells. Moreover, a K_{Ca} opener, NS-1619, accelerated mitochondrial K^+ uptake and decreased infarct size in rabbit hearts subjected to

30 minutes of global ischemia and 2 hours of reperfusion. We hypothesized that mitoK$_{Ca}$ might be activated under pathophysiological conditions when mitochondrial Ca^{2+} uptake is enhanced. The partial depolarization of $\Delta\Psi_m$ by mitoK$_{Ca}$ could act as a safeguard against excessive mitochondrial Ca^{2+} accumulation by decreasing the electrochemical driving force for Ca^{2+} entry. MitoK$_{Ca}$ might also play a physiological role to fine-tune mitochondrial volume and/or Ca^{2+} accumulation under conditions of increased cardiac workload. Ca^{2+} activation of this channel would be expected to cause a partial depolarization of $\Delta\Psi_m$, which would decrease the driving force for Ca^{2+} entry under conditions of positive inotropic stimulation or ischemia.

Subsequent reports have confirmed that K$_{Ca}$ channel openers protect hearts against ischemic injury (Shintani et al. 2004; Stowe et al. 2006). Similar to the effect of mitoK$_{ATP}$ activation, mitoK$_{Ca}$ opening has been implicated in early- and delayed-preconditioning (Wang et al, 2004) and may participate in the cardioprotection triggered by ischemia or receptor activation (Cao et al. 2005a; Cao et al. 2005b; Gao et al. 2005). NS-1619-mediated preconditioning can be prevented by blocking K$_{Ca}$ channels with paxilline or by scavenging ROS during the exposure to the opener (Stowe et al. 2006). The latter effect, as well as the finding that PKC activation may be upstream of mitoK$_{Ca}$ opening (Cao et al. 2005a), is reminiscent of the role of mitoK$_{ATP}$ in preconditioning and implies that mitoK$_{Ca}$ too may be both a trigger and an effector of protection. With regard to PKC activation of mitoK$_{Ca}$, it should be noted that one recent report (Sato et al. 2005) showed modulation of mitoK$_{Ca}$ by the cyclic AMP-activated protein kinase (PKA) pathway, but not by the PKC pathway, in contradistinction to the regulation of mitoK$_{ATP}$. Although both mitochondrial K$^+$ channels appear to have similar effects on mitochondrial function, each has a distinct and non-overlapping pharmacology (Wu et al. 2002; Sato et al. 2005), providing strong support for the idea that enhanced mitochondrial K$^+$ uptake is the common factor associated with resistance to cell injury.

Antibodies against the "Big K$^+$" (BK) type K$_{Ca}$ channel cross-react with purified mitochondrial membranes (Wu et al. 2002) and recently, the β subunit of the BK channel has also been reported in the mitochondria (Ohya et al. 2005). However, it is still not clear whether the mitochondrial protein is identical to, or just homologous with, the surface membrane channel. If it is the same, then the question arises as to how the channel might be targeted to the mitochondrial inner membrane.

10.8. Channels Activated by Metabolic Stress

The events leading up to necrotic or apoptotic cell death have been the subject of extensive investigation in the search for novel targets for treating disorders such as acute coronary syndrome, stroke, heart failure, diabetes, and Alzheimer's disease. Mitochondria play a central role in the mechanism of cell death, and channels on both the inner and outer membranes represent untapped targets for therapeutic intervention. To date, few compounds have been developed with the specific intent of modifying mitochondrial function, although it is now clear that many drugs strongly interact with mitochondrial targets. While our knowledge of the molecular structure of mitochondrial channels is rudimentary, their importance cannot be denied, thus providing a powerful incentive to pursue studies in this area.

With regard to the mitochondrial outer membrane, it is now well accepted that a variety of stimuli can alter the permeability of the outer membrane to allow cytochrome c and apoptosis-inducing factors to enter the cytoplasm, triggering a well-defined cell death program. Recent evidence has implicated a number of Bcl-2 family proteins in the formation of mitochondrial apoptosis-induced channels (MAC) in the outer membrane (Pavlov et al. 2001; Martinez-Caballero et al. 2005). The formation of these cytochrome c permeable channels is correlated with the translocation of pro-apoptotic proteins such as Bax to the outer membrane and can account for the selective release of factors from the mitochondrial intermembrane space even in the absence of PTP activation. It is postulated that oligomeric Bax itself may form the pore (Martinez-Caballero et al. 2005).

Changes in inner membrane permeability and consequent depolarization of $\Delta\Psi_m$ have long been associated with cell injury during hypoxic, oxidative or ischemic stress. The two main factors tied to loss of $\Delta\Psi_m$ are Ca^{2+} overload of the mitochondrial matrix and excessive accumulation of ROS. The collapse of $\Delta\Psi_m$ can result in uncoupling of oxidative phosphorylation and reversal of the mitochondrial ATP synthase, leading to rapid consumption of cytosolic ATP and necrotic or apoptotic cell death. Although mitochondrial depolarization is commonly observed under various forms of metabolic stress, the mechanism of $\Delta\Psi_m$ loss is often poorly studied and frequently misinterpreted. How and when specific mitochondrial ion channels contribute to the depolarization of $\Delta\Psi_m$ is not well understood.

10.9. PTP or Not PTP?

Permeabilization of the mitochondrial inner membrane upon treatment with a variety of effectors has been known for more that 30 years (see, for example, Hunter et al. 1976). Perhaps the most physiologically relevant trigger is excessive mitochondrial Ca^{2+} accumulation, which can occur after ischemia and reperfusion. The opening of a specific permeability transition pore (PTP) is thought to underlie this response, allowing ions and metabolites up to ~1500 KDa in mass to exit the matrix compartment. This channel has been characterized electrophysiologically (Bernardi et al. 1992; Szabo et al. 1992) and its opening is promoted by $\Delta\Psi_m$ depolarization, P_i, ROS, and thiol modification, among other factors. The pore can be inhibited by adenine nucleotides, Mg^{2+}, or matrix protons, and also by compounds such as bongkrekic acid (an inhibitor of the adenine nucleotide translocase, ANT) or cyclosporin A (CsA, which binds to the mitochondrial protein cyclophilin D). Although the structure of the PTP is widely portrayed to consist of a multiprotein complex prominently featuring the ANT, voltage dependent anion channel (VDAC), cyclophilin, the F_1F_o ATPase, and other modulatory proteins, in truth, the structure of the pore is presently unknown, as discussed in a recent review (DiLisa et al., 2006). The two proteins central to prior models of PTP structure (Halestrap et al., 2004), ANT (Kokoszka et al. 2006) and cyclophilin D (Baines et al. 2005; Basso et al. 2005; Nakagawa et al. 2005; Schinzel et al. 2005) have been knocked out in recent transgenic mouse studies, and the results support only a modulatory rather than an obligatory role for these proteins in PTP-mediated transport. Multiple lines of evidence support the idea that in tissues such as the heart, PTP opening occurs only during reperfusion after ischemia (Halestrap et al. 2004; Akar et al. 2005) and is a major checkpoint on the route towards cell injury and death.

Ca^{2+} and ROS can undoubtedly induce PTP opening in isolated mitochondria, and conditions which favor Ca^{2+} overload and/or oxidative stress can readily depolarize $\Delta\Psi_m$ in intact cells. However, recent studies have demonstrated that mitochondrial depolarization is not always synonymous with PTP opening (Aon et al., 2003), so one must employ multiple tools to determine whether a permeability transition has occurred in a given situation. These tests include sensitivity of the observed depolarization to the PTP inhibitors CsA or sanglifehrin, and the direct demonstration that small molecules can permeate the mitochondrial inner membrane. Often, PTP activation is invoked to account for $\Delta\Psi_m$ loss without any (or with weak) confirming evidence, so one must be careful to avoid trying to fit data into a preconceived notion about the permeability transition pore.

10.10. IMAC

In order to meet the high energy demands of cardiac muscle, energy supply must be finely regulated to respond quickly to match increases in workload. Accordingly, the control of the mitochondrial oxidative phosphorylation pathway involves a number of nonlinear positive and negative feedback loops utilized to maintain ATP at nearly constant levels at steady-state. The nonlinear dynamics of the bioenergetic system, however, may contribute to unstable or oscillatory behavior under stress. A case in point is the observation that both $\Delta\Psi_m$ and the mitochondrial redox potential can oscillate under metabolic or oxidative stress (O'Rourke et al. 1994). Mitochondrial criticality (Aon et al. 2004; Aon et al. 2006) refers to the cellular conditions leading up to a breakpoint between stable and unstable $\Delta\Psi_m$ in the mitochondrial network of the cardiac cell. The approach to the critical state, in which a small perturbation can induce cell-wide, synchronized and self-sustaining oscillations in $\Delta\Psi_m$ throughout the cell, depends on mitochondrial ROS production exceeding a threshold level in a significant fraction of mitochondria ($\sim$60%) in the network. At this point, the weakly coupled fluctuations of individual mitochondria transition into an emergent spatiotemporal pattern of synchronized limit cycle oscillations (Cortassa et al. 2004). These metabolic oscillations are strongly coupled to the cardiomyocyte's electrophysiological response through energy sensitive sarcK_{ATP} channels on the membrane.

Our current understanding of the underlying mechanism of mitochondrial criticality incorporates the concept of mitochondrial ROS-induced ROS release, a term coined by Zorov et al. (2000) to describe how laser-induced oxidative stress leads to $\Delta\Psi_m$ depolarization. Although PTP opening was implicated in this prior study, we have found that whole-cell oscillations of $\Delta\Psi_m$ in the mitochondrial network, triggered by substrate deprivation or local ROS generation in a tiny fraction of the mitochondrial network, involve a more subtle and selective change in ion permeability, through the activation of the inner membrane anion channel (IMAC). We have examined the mechanism of the mitochondrial oscillator in both experimental and computational studies (Aon et al. 2003; Cortassa et al. 2003; Aon et al. 2004; Cortassa et al. 2004). IMAC, which we currently hypothesize is related to the centum pS channel described in mitoplast patch-clamp studies, is blocked by antagonists of the mitochondrial benzodiazepine receptor (mBzR), a mitochondrial membrane protein whose molecular structure is known but whose function is poorly understood (Gavrish et al. 1999). MBzR has been shown to co-immunoprecipitate with other proteins found at the contact sites between the mitochondrial outer and inner membranes (e.g. VDAC, ANT, etc.), but it is presently unclear what role it plays in modulating mitochondrial function. These questions notwithstanding, inhibition of this receptor-channel (?) complex can acutely stabilize $\Delta\Psi_m$ in the polarized state and prevent the activation of sarcK_{ATP} in cells undergoing metabolic oscillations (Akar et al. 2005). Furthermore, an agonist of the mBzR (FGIN1-27) has the opposite effect, destabilizing $\Delta\Psi_m$ and promoting cellular electrical inexcitability (Aon et al. 2003; Akar et al. 2005).

The close coupling of the energetic and electrical functions of the cell has major consequences in the scenario of ischemia and reperfusion of the whole heart. In a recent study employing optical mapping of electrical activity in isolated perfused hearts subjected to 30 minutes of global ischemia and reperfusion, we demonstrated how failure at the level of the mitochondria could scale to produce post-ischemic arrhythmias in the intact organ. We hypothesize that depolarization of the mitochondrial network in clusters of cells can create "metabolic sinks" of current in heterogeneous regions of the myocardium, resulting in slowed propagation and block of the excitation wave, promoting reentry and ventricular fibrillation. Importantly, the same interventions that stabilize $\Delta\Psi_m$ in single cell experiments can blunt action potential shortening during ischemia and prevent arrhythmias upon reperfusion in intact hearts (Akar et al. 2005).

10.11. Molecular Targets

Based on a long history of investigating the ion permeability of mitochondria using a variety of techniques including swelling assays, fluorescent indicators, measurements of redox changes or respiration, single channel patch-clamp recordings, and reconstitution into lipid bilayers, there is ample evidence that specific ion channels are present. Although the structure of VDAC of the outer membrane is known, it is a disappointing fact that almost nothing is known about the molecular structure of the pore of most mitochondrial inner membrane channels. This is probably because mitochondrial inner membrane channels must be present in low copy numbers to preserve the ability of the mitochondrion to produce ATP.

Provocative analogies with the pharmacological or toxin sensitivities of surface membrane (Liu et al. 2001; Xu et al. 2002) or intracellular (e.g. the ryanodine receptor (RyR) - Beutner et al. 2001, 2005) ion channels have intensified the search for mitochondrial congeners in the inner membrane. Thus far, immunologic data showing reactivity of mitochondrial membranes with a variety of ion channel antibodies is the main evidence supporting the idea that mitochondrial ion channels are similar in structure to surface membrane channels. This approach has not led to definitive proof that a known protein is responsible for the observed ion fluxes. So far, there has been no confirmation that a particular immunoreactive band seen in a purified mitochondrial membrane preparation matches the sequence of the primary target of the antibody. Moreover, proteomic evidence confirming that putative channels of the Kir, K_{Ca}, or RyR subtypes are present in mitochondrial membranes is absent. These deficiencies will likely be overcome as more attention is focused on getting biochemical and structural data confirming that a particular ion flux is associated with a particular protein fraction.

Meanwhile, clues about modulatory interactions between known mitochondrial proteins such as the ANT, the mitochondrial ATP synthase, the mBzR and VDAC may help in the quest for ion channel pores. In this regard, it has been reported that succinate dehydrogenase may regulate mitoK_{ATP} activity in reconstituted proteoliposomes as part of a multiprotein complex (Ardehali et al. 2004), perhaps offering a handle for further purification of mitochondrial K^+ channel activity. Similarly, the structure of IMAC may be revealed through its interaction with the mBzR.

10.12. Conclusions

Mitochondrial ion channels for Ca^{2+}, K^+, or anions have been functionally and pharmacologically characterized at many levels spanning from single molecules to intact cell and organ function. The challenge ahead lies in defining the molecular structures responsible for forming the ion selective pores mediating these important ion transport processes. Achieving these goals will undoubtedly spur the development of novel and specific therapeutic agents targeted to the mitochondria. As the organelle responsible for integrating and responding to environmental challenges, mitochondria are the hub of all cellular functions and they play a central role as a determinant of cell life and death in a variety of pathologies including acute coronary syndrome, neurodegeneration, cancer, and aging.

References

Akar FG, Aon MA, Tomaselli GF, O'Rourke B (2005) The mitochondrial origin of postischemic arrhythmias. J Clin Invest 115:3527–35

Aon MA, Cortassa S, Akar FG, O'Rourke B (2006) Mitochondrial criticality: a new concept at the turning point of life or death. Biochim Biophys Acta 1762:232–40

Aon MA, Cortassa S, Marban E, O'Rourke B (2003) Synchronized whole cell oscillations in mitochondrial metabolism triggered by a local release of reactive oxygen species in cardiac myocytes. J Biol Chem 278:44735–44

Aon MA, Cortassa S, O'Rourke B (2004) Percolation and criticality in a mitochondrial network. Proc Natl Acad Sci U S A 101: 4447–52

Ardehali H, Chen Z, Ko Y, Mejia-Alvarez R, Marban E (2004) Multiprotein complex containing succinate dehydrogenase confers mitochondrial ATP-sensitive K+ channel activity. Proc Natl Acad Sci USA 101:11880–5

Baines CP, Kaiser RA, Purcell NH, Blair NS, Osinska H, Hambleton MA, Brunskill EW, Sayen MR, Gottlieb RA, Dorn GW, Robbins J, Molkentin JD (2005) Loss of cyclophilin D reveals a critical role for mitochondrial permeability transition in cell death. Nature 434:658–62

Basso E, Fante L, Fowlkes J, Petronilli V, Forte MA, Bernardi P (2005) Properties of the permeability transition pore in mitochondria devoid of Cyclophilin D. J Biol Chem 280: 18558–61

Beavis AD (1992) Properties of the inner membrane anion channel in intact mitochondria. J Bioenerg Biomembr 24:77–90

Beavis AD, Powers M (2004) Temperature dependence of the mitochondrial inner membrane anion channel: the relationship between temperature and inhibition by magnesium. J Biol Chem 279:4045–50

Bender E, Kadenbach B (2000) The allosteric ATP-inhibition of cytochrome c oxidase activity is reversibly switched on by cAMP-dependent phosphorylation. FEBS Lett 466:130–4

Bernardi P, Vassanelli S, Veronese P, Colonna R, Szabo I, Zoratti M (1992) Modulation of the mitochondrial permeability transition pore. Effect of protons and divalent cations. J Biol Chem 267:2934–9

Beutner G, Sharma VK, Giovannucci DR, Yule DI, Sheu SS (2001) Identification of a ryanodine receptor in rat heart mitochondria. J Biol Chem 276:21482–8

Beutner G, Sharma VK, Lin L, Ryu SY, Dirksen RT, Sheu SS (2005) Type 1 ryanodine receptor in cardiac mitochondria: transducer of excitation-metabolism coupling. Biochim Biophys Acta 1717:1–10

Billman GE, Englert HC, Scholkens BA (1998) HMR 1883, a novel cardioselective inhibitor of the ATP-sensitive potassium channel. Part II: effects on susceptibility to ventricular fibrillation induced by myocardial ischemia in conscious dogs. J Pharmacol Exp Ther 286:1465–73

Borecky J, Jezek P, D. Siemen D (1997) 108-pS channel in brown fat mitochondria might be identical to the inner membrane anion channel. J Biol Chem 272:19282–9

Bragadin M, Pozzan T, Azzone GF (1979) Kinetics of Ca2+ carrier in rat liver mitochondria. Biochemistry 18:5972–8

Brierley GP, Jurkowitz M, K. Scott KM, Merola AJ (1971) Ion transport by heart mitochondria. XXII. Spontaneous, energy-linked accumulation of acetate and phosphate salts of monovalent cations. Arch Biochem Biophys 147:545–56

Bryan J, Vila-Carriles WH, Zhao G, Babenko AP, Aguilar-Bryan L (2004) Toward linking structure with function in ATP-sensitive K+ channels. Diabetes 53 Suppl 3:S104–12

Cao CM, Chen M, Wong TM (2005a) The K(Ca) channel as a trigger for the cardio-protection induced by kappa-opioid receptor stimulation – its relationship with protein kinase C. Br J Pharmacol 145:984–91

Cao CM, Xia Q, Gao Q, Chen M, Wong TM (2005b) Calcium-activated potassium channel triggers cardioprotection of ischemic preconditioning. J Pharmacol Exp Ther 312:644–50

Chutkow WA, Samuel V, Hansen PA, Pu J, Valdivia CR, Makielski JC, Burant CF (2001) Disruption of Sur2-containing K(ATP) channels enhances insulin-stimulated glucose uptake in skeletal muscle. Proc Natl Acad Sci USA 98:11760–4

Cohen MV, Baines CP, Downey JM (2000) Ischemic preconditioning: from adenosine receptor to KATP channel. Annu Rev Physiol 62:79–109

Cortassa S, Aon MA, Marban E, Winslow RL, O'Rourke B (2003) An integrated model of cardiac mitochondrial energy metabolism and calcium dynamics. Biophys J 84: 2734–55

Cortassa S, Aon MA, Winslow RL, O'Rourke B (2004) A mitochondrial oscillator dependent on reactive oxygen species. Biophys J 87:2060–73

Costa AD, Garlid KD, West IC, Lincoln TM, Downey JM, M. Cohen MV, Critz SD (2005) Protein kinase G transmits the cardioprotective signal from cytosol to mitochondria. Circ Res 97:329–36

Cuong DV, Kim N, Youm JB, Joo H, Warda M, Lee JW, Park WS, Kim T, Kang S, Kim H, Han J (2006) Nitric oxide-cGMP-protein kinase G signaling pathway induces anoxic preconditioning through activation of ATP-sensitive K+ channels in rat hearts. Am J Physiol Heart Circ Physiol 290: H1808–H1817

Demin OV, Kholodenko BN, Skulachev VP (1998) A model of O2.-generation in the complex III of the electron transport chain. Mol Cell Biochem 184:21–33

Denton RM, McCormack JG (1990) Ca2+ as a second messenger within mitochondria of the heart and other tissues. Ann Rev Physiol 52:451–66

Di Lisa F, Bernardi P (2006) Mitochondria and ischemia-reperfusion injury of the heart: Fixing a hole. Cardiovasc Res 70: 191–199

Diwan JJ (1985) Ba2+ uptake and the inhibition by Ba2+ of K+ flux into rat liver mitochondria. J Membr Biol 84:165–71

Douglas MG, Cockrell RS (1974) Mitochondrial cation-hydrogen ion exchange. Sodium selective transport by mitochondria and submitochondrial particles. J Biol Chem 249:5464–71

Ferreira GC, Pedersen PL (1993) Phosphate transport in mitochondria: past accomplishments, present problems, and future challenges. J Bioenerg Biomembr 25:483–92

Gao Q, Zhang SZ, Cao CM, Bruce IC, Xia Q (2005) The mitochondrial permeability transition pore and the Ca2+-activated K+ channel contribute to the cardioprotection conferred by tumor necrosis factor-alpha. Cytokine 32:199–205

Garlid KD (1996) Cation transport in mitochondria–the potassium cycle. Biochim Biophys Acta 1275:123–6

Garlid KD, Beavis AD (1986) Evidence for the existence of an inner membrane anion channel in mitochondria. Biochim Biophys Acta 853:187–204

Garlid KD, Dos Santos P, Xie ZJ, Costa AD, Paucek P (2003) Mitochondrial potassium transport: the role of the mitochondrial ATP-sensitive K(+) channel in cardiac function and cardioprotection. Biochim Biophys Acta 1606:1–21

Garlid KD, Paucek P, Yarov-Yarovoy V, Murray MH, Darbenzio RB, D'Alonzo AJ, Lodge NJ, Smith MA, Grover GJ (1997) Cardioprotective effect of diazoxide and its interaction with mitochondrial ATP-sensitive K+ channels. Possible mechanism of cardioprotection. Circ Res 81:1072–82

Garlid KD, Paucek P, Yarov-Yarovoy V, Sun X, Schindler PA (1996) The mitochondrial KATP channel as a receptor for potassium channel openers. J Biol Chem 271:8796–9

Gavish M, Bachman I, Shoukrun R, Katz Y, Veenman L, Weisinger G, Weizman A (1999) Enigma of the peripheral benzodiazepine receptor. Pharmacol Rev 51:629–50

Ghosh S, Standen NB, Galinanes M (2000) Evidence for mitochondrial K ATP channels as effectors of human myocardial preconditioning. Cardiovasc Res 45:934–40

Grover GJ, Garlid KD (2000) ATP-Sensitive potassium channels: a review of their cardioprotective pharmacology. J Mol Cell Cardiol 32:677–95

Grover GJ, D'Alonzo AJ, Garlid KD, Bajgar R, Lodge NJ, Sleph PG, Darbenzio RB, Hess TA, Smith MA, Paucek P. Atwal KS (2001) Pharmacologic characterization of BMS-191095, a mitochondrial K(ATP) opener with no peripheral vasodilator or cardiac action potential shortening activity. J Pharmacol Exp Ther 297:1184–92

Grover GJ, McCullough JR, Henry DE, Conder ML, Sleph PG (1989) Anti-ischemic effects of the potassium channel activators pinacidil and cromakalim and the reversal of these effects with the potassium channel blocker glyburide. J Pharmacol Exp Ther 251:98–104

Gumina RJ, Pucar D, Bast P, Hodgson DM, Kurtz CE, Dzeja PP, Miki T, Seino S, Terzic A (2003) Knockout of Kir6.2 negates ischemic preconditioning-induced protection of myocardial energetics. Am J Physiol Heart Circ Physiol 284:H2106–13

Gunter KK, Gunter TE (1994) Transport of calcium by mitochondria. J Bioenerg Biomembr 26:471–85

Gunter TE, Pfeiffer DR (1990) Mechanisms by which mitochondria transport calcium. Am J Physiol 258:C755–86

Guzy RD, Hoyos B, Robin E, Chen H, Liu L, Mansfield KD, Simon MC, Hammerling U, Schumacker PT (2005) Mitochondrial complex III is required for hypoxia-induced ROS production and cellular oxygen sensing. Cell Metab 1:401–8

Halestrap AP (1987) The regulation of the oxidation of fatty acids and other substrates in rat heart mitochondria by changes in the matrix volume induced by osmotic strength, valinomycin and Ca^{2+}. Biochem J 244:159–64

Halestrap AP, Clarke SJ, Javadov SA (2004) Mitochondrial permeability transition pore opening during myocardial reperfusion–a target for cardioprotection. Cardiovasc Res 61:372–85

Hansford RG, Lehninger AL (1972) The effect of the coupled oxidation of substrate on the permeability of blowfly flight-muscle mitochondria to potassium and other cations. Biochem J 126:689–700

Hunter DR, Haworth RA, Southard JH (1976) Relationship between configuration, function, and permeability in calcium-treated mitochondria. J Biol Chem 251:5069–77

Inoue I, Nagase H, Kishi K, Higuti T (1991) ATP-sensitive K+ channel in the mitochondrial inner membrane. Nature 352:244–7

Jezek P, Jezek J (2003) Sequence anatomy of mitochondrial anion carriers. FEBS Lett 534:15–25

Juhaszova M, Zorov DB, Kim SH, Pepe S, Fu Q, Fishbein KW, Ziman DB, Wang S, Ytrehus K, Antos CL, Olson EN, Sollott SJ (2004) Glycogen synthase kinase-3beta mediates convergence of protection signaling to inhibit the mitochondrial permeability transition pore. J Clin Invest 113:1535–49

Jung DW, Baysal K, Brierley GP (1995) The sodium-calcium antiport of heart mitochondria is not electroneutral. J Biol Chem 270:672–8

Jung DW, Chavez E, Brierley GP (1977) Energy-dependent exchange of K+ in heart mitochondria. K+ influx. Arch Biochem Biophys 183:452–9

Jung DW, Farooqui T, Utz E, Brierley GP (1984) Effects of quinine on K+ transport in heart mitochondria. J Bioenerg Biomembr 16:379–90

Kinnally KW, Zorov DB, Antonenko YN, Snyder SH, McEnery MW, Tedeschi H (1993) Mitochondrial benzodiazepine receptor linked to inner membrane ion channels by nanomolar actions of ligands. Proc Natl Acad Sci USA 90:1374–8

Kirichok Y, Krapivinsky G, Clapham DE (2004) The mitochondrial calcium uniporter is a highly selective ion channel. Nature 427:360–4

Kokoszka JE, Waymire KG, Levy SE, Sligh JE, Cai J, Jones DP, MacGregor GR, Wallace DC (2004) The ADP/ATP translocator is not essential for the mitochondrial permeability transition pore. Nature 427:461–5

Kopustinskiene DM, Pollesello P, Saris NE (2001) Levosimendan is a mitochondrial K(ATP) channel opener. Eur J Pharmacol 428:311–4

Lamping KA, Gross GJ (1985) Improved recovery of myocardial segment function following a short coronary occlusion in dogs by nicorandil, a potential new antianginal agent, and nifedipine. J Cardiovasc Pharmacol 7:158–66

Liu Y, Ren G, O'Rourke B, Marban E, Seharaseyon J (2001) Pharmacological comparison of native mitochondrial K(ATP) channels with molecularly defined surface K(ATP) channels. Mol Pharmacol 59:225–30

Liu Y, Sato T, O'Rourke B, Marban E (1998) Mitochondrial ATP-dependent potassium channels: novel effectors of cardioprotection? Circulation 97:2463–9

Martinez-Caballero S, Dejean LM, Jonas EA, Kinnally KW (2005) The role of the mitochondrial apoptosis induced channel MAC in cytochrome c release. J Bioenerg Biomembr 37:155–64

Nakagawa T, Shimizu S, Watanabe T, Yamaguchi O, Otsu K, Yamagata H, Inohara H, Kubo T, Tsujimoto Y (2005) Cyclophilin D-dependent mitochondrial permeability transition regulates some necrotic but not apoptotic cell death. Nature 434:652–8

Nicholls DG, Ferguson SJ (2002) Bioenergetics3, Third ed., Academic Press, London

Noma A (1983) ATP-regulated K+ channels in cardiac muscle. Nature 305:147–8

Ockaili R, Emani VR, Okubo S, Brown M, Krottapalli K, Kukreja RC (1999) Opening of mitochondrial KATP channel induces early and delayed cardioprotective effect: role of nitric oxide. Am J Physiol 277:H2425–34

Ockaili R, Salloum F, Hawkins J, Kukreja RC (2002) Sildenafil (Viagra) induces powerful cardioprotective effect via opening of mitochondrial K(ATP) channels in rabbits. Am J Physiol Heart Circ Physiol 283:H1263–9

Ohya S, Kuwata Y, Sakamoto K, Muraki K, Imaizumi Y (2005) Cardioprotective effects of estradiol include the activation of large-conductance Ca(2+)-activated K(+) channels in cardiac mitochondria. Am J Physiol Heart Circ Physiol 289:H1635–42

Oldenburg O, Cohen MV, Downey JM (2003) Mitochondrial K(ATP) channels in preconditioning. J Mol Cell Cardiol 35: 569–75

O'Rourke B (2000) Pathophysiological and protective roles of mitochondrial ion channels. J Physiol 529 Pt 1:23–36

O'Rourke B (2004) Evidence for mitochondrial K+ channels and their role in cardioprotection. Circ Res 94:420–32

O'Rourke B, B. Ramza BM, Marban E (1994) Oscillations of membrane current and excitability driven by metabolic oscillations in heart cells. Science 265:962–6

Otani H (2004) Reactive oxygen species as mediators of signal transduction in ischemic preconditioning. Antioxid Redox Signal 6:449–69

Pain T, X. Yang XM, Critz SD, Yue Y, Nakano A, Liu GS, Heusch G, Cohen MV, Downey JM (2000) Opening of mitochondrial K(ATP) channels triggers the preconditioned state by generating free radicals. Circ Res 87:460–6

Paucek P, Mironova G, Mahdi F, Beavis AD, Woldegiorgis G, Garlid KD (1992) Reconstitution and partial purification of the glibenclamide-sensitive, ATP-dependent K+ channel from rat liver and beef heart mitochondria. J Biol Chem 267:26062–9

Pavlov EV, Priault M, Pietkiewicz D, Cheng EH, Antonsson B, Manon S, Korsmeyer SJ, Mannella CA, Kinnally KW (2001) A novel, high conductance channel of mitochondria linked to apoptosis in mammalian cells and Bax expression in yeast. J Cell Biol 155:725–31

Rousset S, Alves-Guerra MC, Mozo Miroux B, Cassard-Doulcier AM, Bouillaud F, Ricquier D (2004) The biology of mitochondrial uncoupling proteins. Diabetes 53 Suppl 1:S130–5

Sasaki N, Murata M, Guo Y, Jo SH, Ohler A, Akao M, O'Rourke B, Xiao RP, Bolli R, Marban E (2003) MCC-134, a single pharmacophore, opens surface ATP-sensitive potassium channels, blocks mitochondrial ATP-sensitive potassium channels, and suppresses preconditioning. Circulation 107:1183–8

Sasaki N, Sato T, Ohler A, O'Rourke B, Marban E (2000) Activation of mitochondrial ATP-dependent potassium channels by nitric oxide. Circulation 101:439–45

Sato T, Costa AD, Saito T, Ogura T, Ishida H, Garlid KD, Nakaya H (2006) Bepridil, an antiarrhythmic drug, opens mitochondrial KATP channels, blocks sarcolemmal KATP channels, and confers cardioprotection. J Pharmacol Exp Ther 316:182–8

Sato T, O'Rourke B, Marban E (1998) Modulation of mitochondrial ATP-dependent K+ channels by protein kinase C. Circ Res 83:110–4

Sato T, Saito T, Saegusa N, Nakaya H (2005) Mitochondrial Ca2+-activated K+ channels in cardiac myocytes: a mechanism of the cardioprotective effect and modulation by protein kinase A. Circulation 111:198–203

Sato T, Sasaki N, Seharaseyon J, O'Rourke B, Marban E (2000) Selective pharmacological agents implicate mitochondrial but not sarcolemmal K(ATP) channels in ischemic cardioprotection. Circulation 101:2418–23

Schinzel AC, Takeuchi O, Huang Z, Fisher JK, Zhou Z, Rubens J, Hetz C, Danial NN, Moskowitz MA, Korsmeyer SJ (2005) Cyclophilin D is a component of mitochondrial permeability transition and mediates neuronal cell death after focal cerebral ischemia. Proc Natl Acad Sci USA 102:12005–10

Seino S, Miki T (2003) Physiological and pathophysiological roles of ATP-sensitive K+ channels. Prog Biophys Mol Biol 81:133–76

Shintani Y, Node K, Asanuma H, Sanada S, Takashima S, Asano Y, Liao Y, Fujita M, A Hirata A, Shinozaki Y, Fukushima T, Nagamachi Y, Okuda H, Kim J, Tomoike H, Hori M, Kitakaze M (2004) Opening of Ca2+-activated K+ channels is involved in ischemic preconditioning in canine hearts. J Mol Cell Cardiol 37:1213–8

Sorgato MC, Keller BU, Stuhmer W (1987) Patch-clamping of the inner mitochondrial membrane reveals a voltage-dependent ion channel. Nature 330:498–500

Stowe DF, Aldakkak M, Camara AK, Riess ML, Heinen A, Varadarajan SG, Jiang MT (2006) Cardiac mitochondrial preconditioning by Big Ca2+-sensitive K+ channel opening requires superoxide radical generation Am J Physiol Heart Circ Physiol 290:H434–40

Stucki JW (1980) The optimal efficiency and the economic degrees of coupling of oxidative phosphorylation. Eur J Biochem 109:269–83

Suzuki M, Saito T, Sato T, Tamagawa M, Miki T, Seino S, Nakaya H (2003) Cardioprotective effect of diazoxide is mediated by activation of sarcolemmal but not mitochondrial ATP-sensitive potassium channels in mice. Circulation 107:682–5

Suzuki M, Sasaki N, Miki T, Sakamoto N, Ohmoto-Sekine Y, Tamagawa M, Seino S, Marban E, Nakaya H (2002) Role of sarcolemmal K(ATP) channels in cardioprotection against ischemia/reperfusion injury in mice. J Clin Invest 109:509–16

Szabo I, Bernardi P, Zoratti M (1992) Modulation of the mitochondrial megachannel by divalent cations and protons. J Biol Chem 267:2940–6

Szewczyk A, Mikolajek B, Pikula S, Nalecz MJ (1993) Potassium channel openers induce mitochondrial matrix volume changes via activation of ATP-sensitive K+ channel. Pol J Pharmacol 45:437–43

Tanno M, Miura T, Tsuchida A, Miki T, Nishino Y, Ohnuma Y, and Shimamoto K (2001) Contribution of both the sarcolemmal K(ATP) and mitochondrial K(ATP) channels to infarct size limitation by K(ATP) channel openers: differences from preconditioning in the role of sarcolemmal K(ATP) channels. Naunyn Schmiedebergs Arch Pharmacol 364:226–32

Tanonaka K, Taguchi T, Koshimizu M, Ando T, Morinaka T, Yogo T, Konishi F, Takeo S (1999) Role of an ATP-sensitive potassium channel opener, YM934, in mitochondrial energy production in ischemic/reperfused heart. J Pharmacol Exp Ther 291:710–6

Territo PR, Mootha VK, French SA, Balaban RS (2000) Ca(2+) activation of heart mitochondrial oxidative phosphorylation: role of the F(0)/F(1)-ATPase. Am J Physiol Cell Physiol 278:C423–35

Uchiyama Y, Otani H, Wakeno M, Okada T, Uchiyama T, Sumida T, Kido M, Imamura H, Nakao S, Shingu K (2003) Role of mitochondrial KATP channels and protein kinase C in ischaemic preconditioning. Clin Exp Pharmacol Physiol 30:426–36

Wang X, Yin C, Xi L, Kukreja RC (2004) Opening of Ca2+-activated K+ channels triggers early and delayed preconditioning against I/R injury independent of NOS in mice. Am J Physiol Heart Circ Physiol 287:H2070–7

Xu W, Liu Y, Wang S, McDonald T, Van Eyk JE, Sidor A, O'Rourke B (2002) Cytoprotective role of Ca2+- activated K+ channels in the cardiac inner mitochondrial membrane. Science 298:1029–33

Xu Z, Ji X, and Boysen PG (2004) Exogenous nitric oxide generates ROS and induces cardioprotection: involvement of PKG, mitochondrial KATP channels, and ERK. Am J Physiol Heart Circ Physiol 286:H1433–40

Zingman LV, Hodgson DM, Bast PH, Kane GC, Perez-Terzic C, Gumina RJ, Pucar D, Bienengraeber M, Dzeja PP, Miki T, Seino S, Alekseev AE, Terzic A (2002) Kir6.2 is required for adaptation to stress. Proc Natl Acad Sci USA 99:13278–83

Zorov DB, Filburn CR, Klotz LO, Zweier JL, Sollott SJ (2000) Reactive oxygen species (ROS)-induced ROS release: a new phenomenon accompanying induction of the mitochondrial permeability transition in cardiac myocytes. J Exp Med 192:1001–14

Part 4
Mitochondria as Initiators of Cell Death

11
The Mitochondrial Permeability Transition Pore – from Molecular Mechanism to Reperfusion Injury and Cardioprotection

Andrew P. Halestrap, Samantha J. Clarke and Igor Khalilin

11.1. Introduction

In most cells, the primary role played by the mitochondria is the provision of ATP through oxidative phosphorylation to support the numerous energy requiring processes. This is especially so in tissues such as the beating heart where the provision of ATP to meet the demands of muscle contraction and the maintenance of ionic homeostasis are especially heavy. Indeed, even under resting conditions, the heart cannot survive on glycolytic ATP alone and rapidly ceases to beat when oxidative phosphorylation is impaired by anoxia or ischemia. It comes as something of a surprise, therefore, to discover that within the mitochondria there exists a latent mechanism that, once activated, converts them from organelles that energise the cell to those that actively kill the cell via apoptosis or necrosis. This transition, reminiscent of the fictional Dr Jeckyll who turns into the murderous Mr Hyde, is mediated within the mitochondria by the opening of a non-specific pore in the mitochondrial inner membrane, known as the mitochondrial permeability transition pore (MPTP).

In this chapter we will first summarise what is known of the mechanism of the MPTP and how it is controlled. We will then describe the techniques that have been used to measure MPTP opening in isolated cells and intact tissues subject to a variety of stresses. Particular attention will be paid to the ischemic/reperfused heart where the extent of pore opening is a critical determinant of reperfusion injury. Finally, we will explain how inhibiting opening of the MPTP is an effective strategy for protecting the heart from ischemia/reperfusion and other toxic insults.

11.2. The Discovery of the MPTP

A critical property of the inner mitochondrial membrane is that it is impermeable to all but a few selected metabolites and ions. This is essential in order to maintain the membrane potential and pH gradient across the inner membrane that is essential for oxidative phosphorylation. However, it has been known for more than fifty years that mitochondria become leaky, uncoupled and massively swollen if they are exposed to high calcium concentrations, especially in the presence of phosphate and when accompanied by oxidative stress. The phenomenon became known as the permeability transition and was originally thought to reflect activation of endogenous phospholipase A_2 leading to phospholipid breakdown within the inner membrane (Gunter and Pfeiffer 1990). However, seminal studies in the late seventies by Haworth and Hunter (1979) convincingly demonstrated that this could not be the case. The increase in permeability could be rapidly reversed by removal of calcium and exhibited a defined molecular weight cut-off of about 1.5 kDa. These data were consistent with the opening of a non-specific protein channel but little further progress was made until the discovery, by Martin Crompton and colleagues, that the pore could be specifically blocked by sub-micromolar concentrations of the immunosuppressant drug, cyclosporin A (CsA) (Crompton et al. 1988). This critical observation led to pioneering work in this and other laboratories that identified the key components of the MPTP (see below). Parallel studies characterised the properties of the MPTP and the consequences of its opening. The overall features of the MPTP and the consequences of its opening are summarised in Figure 11.1.

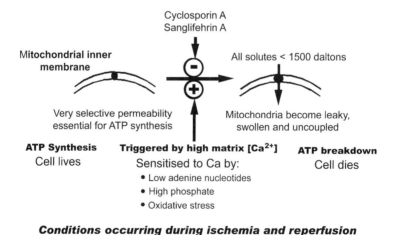

FIGURE 11.1. An overview of the causes and consequences of MPTP opening.

11.3. The Consequences of MPTP Opening (Reviewed in Halestrap et al. 2004; Halestrap et al. 2002)

When the MPTP opens, the inner mitochondrial membrane no longer exerts a permeability barrier to any small molecular weight solute. Consequently, such molecules move freely across the membrane and solute osmotic gradients are lost. However, proteins remain impermeable and as a result exert a colloidal osmotic pressure that causes mitochondria to swell. Unfolding of the cristae allows matrix expansion without compromising the integrity of the inner membrane, but the outer membrane may rupture. This leads to the release of proteins in the intermembrane space such as cytochrome c and other factors that play a critical role in apoptotic cell death (see Chapter 12). Another consequence of MPTP opening is that the inner membrane becomes freely permeable to protons and so mitochondria cannot maintain their pH gradient or membrane potential and become uncoupled. As a result, not only are they unable to synthesise ATP by oxidative phosphorylation, but they actively hydrolyse ATP synthesised by glycolysis as the proton-translocating ATPase reverses direction. If too many mitochondria within a cell are in this condition, intracellular ATP concentrations rapidly decline, leading to the disruption of ionic and metabolic homeostasis and activation of degradative enzymes such as phospholipases, nucleases and proteases. Unless pore closure occurs, these changes will cause irreversible damage to the cell resulting in necrotic death. Even if closure does occur, the mitochondrial swelling and outer membrane rupture may be sufficient to set the apoptotic cascade in motion. Thus it is hardly surprising that the MPTP is kept firmly closed under normal physiological conditions and only opens under pathological conditions.

11.4. Factors that Regulate the MPTP (Reviewed in Halestrap and Brenner 2003; Halestrap et al. 2004; Halestrap et al. 2002)

A matrix facing calcium binding site appears to be essential to trigger MPTP opening, and any factor that influences calcium loading (e.g. Na^+ or Mg^{2+}) will have an effect on MPTP opening. The reader is referred to Chapter 9 by Elinor Griffiths for more information as to how mitochondrial calcium loading is controlled. Interestingly, unlike most mitochondrial calcium-sensitive processes such as the calcium-activated dehydrogenases, Sr^{2+} cannot substitute for Ca^{2+} as a trigger for MPTP opening. Indeed, the calcium trigger site can be inhibited by Sr^{2+} and other divalent cations such as Mg^{2+}, and also by H^+ which accounts for the potent inhibition of MPTP opening by low pH (Halestrap 1991; Szabo et al. 1992). There is an additional divalent cation regulatory site on the MPTP that faces the cytosolic side of the inner membrane and inhibits MPTP opening. This has a broader specificity than the Ca^{2+}-trigger site, and inhibition is observed with many divalent cations including Ca^{2+} and Mg^{2+}

(Bernardi et al. 1993). Although the key factor leading to the opening of the MPTP is mitochondrial calcium overload (i.e. when mitochondrial matrix [Ca^{2+}] is greatly increased), an increase in matrix calcium alone is often inadequate to elicit MPTP opening. Additional factors are required such as oxidative stress, adenine nucleotide depletion, elevated phosphate concentrations and mitochondrial depolarisation (Crompton 1999; Halestrap et al. 2002). These conditions are exactly those that the heart experiences in response to reperfusion after a period of ischemia (reperfusion injury) as will be discussed below. The pore is also sensitive to the ligand induced conformational changes of the adenine nucleotide translocase (ANT), an integral inner membrane protein responsible for ATP and ADP translocation across the membrane. It is also strongly inhibited by a high mitochondrial membrane potential ($\Delta\Psi m$) and when the matrix pH falls below pH 7.0 (Halestrap 1991; Szabo et al. 1992; Bernardi et al. 1993; Bernardi 1999; Crompton, 1999; Halestrap and Brenner, 2003; Halestrap et al., 2002; Szabo et al. 1992). However, many other factors have been described that modulate MPTP (see Table 11.1) and these have helped to define the molecular mechanism of the MPTP as described in Section 11.5 below where their mode of action will be described in more detail.

11.5. The Molecular Mechanism of the MPTP

11.5.1. The Role of Cyclophilin D (Reviewed in Waldmeier et al. 2003)

The discovery in 1988 by Crompton and colleagues that MPTP opening could be inhibited specifically by sub-micromolar concentrations of the immunosuppressive drug, cyclosporin A (CsA) (Crompton et al. 1988) provided a major clue as to the molecular composition of the MPTP. Work in this laboratory identified the matrix protein to which CsA binds as a peptidyl-prolyl cis-trans isomerase (PPIase), now known as cyclophilin D (Cyp-D). PPIases catalyse conformational changes of proteins around proline peptide bonds, and thus suggest that such a conformational change is responsible for pore formation. Indeed, we found that the PPIase activity of CyP-D was blocked by CsA with the same affinity as the drug inhibits MPTP opening (Griffiths and Halestrap 1991; Griffiths and Halestrap 1995; Halestrap and Davidson 1990). CyP-D is a nuclear encoded protein that is translocated into the matrix prior to cleavage of its mitochondrial targeting presequence (Connern and Halestrap 1992; Woodfield et al. 1997; Johnson et al. 1999; Tanveer et al. 1996). It is a distinct gene product from the cytosolic cyclophylin A, whose complex with CsA mediates the immuno-suppressant activity of CsA (Schreiber and Crabtree 1992) as well as other intracellular functions (Crabtree 2001; Rusnak and Mertz 2000). Indeed, CsA analogues such as 6-methyl-ala-CsA, 4-methyl-val-CsA and N-methyl-4-isoleucine-CsA (NIM811) bind to CyP-D and inhibit MPTP opening, but their complexes with CyP-A do not inhibit calcineurin (Griffiths and Halestrap 1995;

TABLE 11.1. Probable sites of action of known effectors of the mitochondrial permeability transition Further details may be found in the text and published reviews (Halestrap et al. 2002; Halestrap and Brenner 2003)

Effect CyP-D binding to ANT[1]	Effect on nucleotide binding to ANT[1]	Effect on Ca^{2+} binding to ANT	Other mode of action
Activatory	**Activatory**	**Activatory**	**Activatory**
• Oxidative stress (e.g. reperfusion, t-butyl hydroperoxide or diamide) or vicinal thiol reagents (e.g. phenylarsine oxide) to cross-link Cys_{160} with Cys_{257} of the ANT	• Thiol reagents attacking Cys_{160} of ANT (e.g. eosine maleimide, phenylarsine oxide)	• High pH	• Some ubiquinone analogues (e.g. decyl-ubiquinone, ubiquinone 10) that may work by binding to Complex 1 or VDAC
• Increased matrix volume	• Oxidative stress (e.g. reperfusion, t-butyl hydroperoxide or diamide) that cross-links Cys_{160} with Cys_{257} of the ANT	**Inhibitory**	**Inhibitory**
• Chaotropic agents	• "c" Conformation of ANT as induced by carboxyatractyloside	• Low pH	
		• Mg^{2+}, Mn^{2+}, Sr^{2+}, Ba^{2+}	
Inhibitory	**Inhibitory**		**Activatory**
• CsA and some analogues (e.g. cyclosporin G, 6-methyl-ala-CsA, 4-methyl-val-CsA and NIM811)	• Adenine nucleotide depletion		• Some ubiquinone analogues (e.g. ubiquinone0,2,5-dihydroxy-6-undecyl-1,4-benzoquinone) an Ro 68-3400 (may work by binding to VDAC)
• SfA (inhibits PPIase activity of CyP-D but not binding)	• High matrix [Pi] and [PPi]		• Trifluoperazine (may work via membrane surface charge)
	• Depolarisation		
	Inhibitory		**Inhibitory**
	• Increase membrane potential		• Phenylglyoxal - modifies arginines of an MPTP component
	• "m" Conformation of ANT as induced by bongkrekic acid		

[1] Note that both CyP-D binding and ADP binding exert their effects through changes in the sensitivity of the MPT to $[Ca^{2+}]$

DiLisa et al. 2001; Waldmeier et al. 2002). We have discovered another extremely potent inhibitor of the MPTP, Sanglifehrin A (SfA), that is unrelated to CsA. Although this drug binds to CyP-D and CyP-A and inhibits their PPIase activity with a $K_{0.5}$ of < 5 nM (Clarke et al. 2002), the SfA-CyP-A complex is without effect on calcineurin activity (Zenke et al. 2001).

Very recently, the role of CyP-D in MPTP opening has been put beyond doubt by studies in four laboratories on MPTP opening in mitochondria from CyP-D knockout mice (Baines et al. 2005; Basso et al. 2005; Schinzel et al. 2003; Nakagawa et al. 2005). MPTP-opening in these mitochondria requires much greater calcium loading and is insensitive to CsA. Indeed, as predicted, the behaviour of these mitochondria exactly mirrors that of normal mitochondria treated with CsA. One important feature to emerge from these experiments is that it confirms that MPTP opening can occur without the involvement of CyP-D, but it requires much greater calcium concentrations (Halestrap 2005). Several groups had previously concluded this because inhibition of MPTP opening by CsA or SfA can be overcome in both heart mitochondria and liver mitochondria if the stimulus (e.g. elevated Ca^{2+}, adenine nucleotide depletion or oxidative stress) was sufficiently great (Clarke et al. 2002; Connern and Halestrap 1994; Connern and Halestrap 1996; Griffiths and Halestrap 1995; Halestrap et al. 1997). Yet under the same conditions, CsA is able to prevent almost totally the binding of CyP-D to the inner mitochondrial membrane (Connern and Halestrap 1994; Halestrap et al. 1997). Thus it would seem that opening of the MPTP involves a protein conformational change that is facilitated by CyP-D rather than dependent upon it.

11.5.2. The Role of the Adenine Nucleotide Translocase (Reviewed in Halestrap and Brenner 2003)

An involvement of the adenine nucleotide translocase (ANT) in the formation or regulation of the MPTP was first suggested more than two decades ago when it was shown that MPTP opening is inhibited by ATP and ADP, but not by their complexes with Mg^{2+} or by other nucleotides such as AMP, GDP or GTP, none of which are transported by the ANT (Hunter and Haworth 1979). It was also demonstrated that carboxyatractyloside (CAT), an inhibitor of the ANT that stabilises it in the "c" conformation (substrate binding site facing the cytosol), greatly sensitises MPTP opening to $[Ca^{2+}]$ whereas bongkrekic acid (BKA) that stabilises the ANT in the "m" conformation, has the opposite (protective) effect (Haworth and Hunter 1979; Hunter and Haworth 1979; LeQuoc and LeQuoc 1988; Halestrap and Davidson 1990; Halestrap et al. 1997). These early observations led us to propose in 1990 that CyP-D might bind to the ANT and induce a conformational change when triggered by Ca^{2+} (Halestrap and Davidson 1990). The adenine nucleotide and BKA binding to the ANT would induce the "m" conformation that is resistant to this change whereas the "c" conformation is susceptible. In order to confirm this hypothesis, we initially demonstrated that conditions such as oxidative stress, that enhanced pore opening, increase the binding of CyP-D to the inner mitochondrial membrane in a CsA-sensitive manner (Connern and Halestrap 1994; Connern and Halestrap 1996). Subsequently, using an immobilised glutathione-S-transferase-CyP-D fusion protein, we and others were able to demonstrate directly that the ANT binds to CyP-D (Crompton et al. 1998; Woodfield et al. 1998). Binding was enhanced by oxidative stress and inhibited by CsA but not SfA (Clarke et al. 2002; Woodfield et al. 1998). The latter observation is intriguing and suggests that CyP-D binding to the ANT occurs as a distinct step from the conformational change that induces pore formation. CsA inhibits the binding whereas SfA still allows binding but prevents the conformational change.

We have demonstrated that critical thiols on the ANT are responsible for the effect of oxidative stress to sensitise the MPTP to [Ca^{2+}]. A disulphide cross-link is formed between two cysteine residues on the matrix facing loops, Cys 160 (in the loop between transmembrane helices (TMs) 3 and 4) and Cys 257 (in the loop between TMs 5 and 6) (McStay et al. 2002). This oxidative cross-link can be mimicked by phenylarsine oxide that is itself a very potent sensitiser of MPTP opening (Halestrap et al. 1997). Very recently, similar cross-linking of cysteines in the mitochondrial carnitine/acylcarnitine transporter has also been demonstrated (Tonazzi et al. 2005). As well as enhancing CyP-D binding to the ANT, this modification greatly reduces adenine nucleotide binding to the ANT and thus prevents the potent inhibition of MPTP opening induced by nucleotide binding. Chemically modifying Cys160 alone using eosine maleimide exerts a similar effect (McStay et al. 2002). It is likely that decreased binding of adenine nucleotides to this site is also the mechanism by which adenine nucleotide depletion, depolarisation and CAT enhance MPTP opening (Halestrap and Brenner 2003; Halestrap et al. 1997).

More direct evidence for a central role of the ANT in MPTP formation has come from experiments with purified ANT from heart mitochondria reconstituted into proteoliposomes (Brustovetsky and Klingenberg 1996; Ruck et al. 1998; Vieira et al. 2000). High (millimolar) concentrations of calcium added to the reconstituted ANT were found to induce the formation of non-specific pores with similar conductance to the MPTP. Pore formation in these reconstituted systems could be inhibited by ADP and BKA and activated by CAT and oxidative stress. However, the sensitivity of the purified ANT to [Ca^{2+}] was far lower than that of the MPTP, although it could be increased by addition of CyP-D whose presence also introduced CsA-sensitivity (Brustovetsky et al. 2002; Crompton et al. 1998; Halestrap et al. 2000).

Although the reconstitution studies described above provide further evidence that CyP-D and the ANT represent the central components of the MPTP, criticisms have been directed against many of these experiments that need to be addressed if further progress is to be made. Problems include the purity of the reconstituted proteins, whether the ANT remains in its native state during purification and reconstitution or represents a denatured form, and the extent to which incomplete detergent removal may distort the results (Halestrap and Brenner 2003). One approach to overcome these problems has been the use of reconstituted recombinant ANT from Neurospora crassa which is both more stable and easier to purify than mammalian ANT (Brustovetsky et al. 2002).

When reconstituted into proteoliposomes this behaved much as the bovine heart ANT and exhibited ADP-inhibitable pore opening at high (1mM) [Ca^{2+}]. When Neurospora cyclophilin was added to the system, the ADP inhibition was abolished in a CsA-sensitive manner. Promising though these data are, no evidence was presented that intact Neurospora crassa mitochondria can undergo a calcium-dependent CsA-inhibitable permeability transition and, if they behave like Saccharomyces cerevisiae mitochondria; it is unlikely that they do (Jung et al. 1997; Manon et al. 1998; Woodfield et al. 1998). Very recently, bovine ANT1 has been crystallised as the CAT complex and a high-resolution structure obtained (PebayPeyroula et al. 2003). It is hoped that this will provide new avenues for the exploration of how the ANT might form a non-specific pore. Indeed, inspection of the structure reveals that it has a large cavity reaching deep into the membrane from the cytosolic side, with only a narrow "gate" preventing the formation of a pore. Were this gate to be wedged open through a conformational change facilitated by CyP-D, it might account for MPTP formation. Consistent with this hypothesis, it is known that modification of specific thiol groups on mitochondrial carrier proteins can convert them from obligate antiporters, to substrate specific uniporters and non-specific channels (Dierks et al. 1990a; Dierks et al. 1990b).

Despite the strong evidence for the ANT playing a critical role in MPTP formation, doubt has been expressed as a result of experiments performed on liver mitochondria from rats in which both ANT1 and ANT2 had been knocked out. These mitochondria still demonstrated a CsA-sensitive MPTP but opening required much higher calcium loading and was insensitive to ligands of the ANT (Kokoszka et al. 2004). It was argued that these data do not support the hypothesis that the ANT is essential for MPTP formation, although it may play a regulatory role. However, this is an oversimplification since, although the data show that a CsA-sensitive MPTP can be formed in the absence of the ANT, it does not rule out the possibility that the ANT is the normal pore forming entity as implied by the accumulating evidence from many laboratories discussed above. The ANT is by far the most abundant member of a family of transporters with a conserved structure (Palmieri 2004), and several members of this family have been demonstrated to form channels under specific conditions (Brustovetsky et al. 2002; Dierks et al. 1990b; Halestrap and Brenner 2003). We have proposed (Halestrap 2004) that in the ANT-knockout mice, another less abundant member of the carrier family becomes the target of CyP-D binding leading to MPTP formation, but with less sensitivity to calcium and no sensitivity to ligands of the ANT. Nevertheless, in the wild-type mitochondria, the dominance of the ANT over the other members of the mitochondrial carrier family makes this the normal target of CyP-D mediated MPTP formation. In passing, it should be noted that the metabolic consequences of a liver in which the mitochondria have no ANT would be expected to be severe and yet no major effects were observed (Halestrap 2004). There is now evidence for an additional ANT isoform in human liver with an orthologue in mouse (Dolce et al. 2005) and it is possible that this enables the liver to continue functioning near normally in the ANT1/ANT2 knockout mice. Of course it would also mean that the MPTP observed in the knockout mice might still involve an ANT isoform! An alternative hypothesis proposed by He & Lemasters (2002) is that the MPTP may be formed from aggregates of denatured membrane proteins, and that these are normally stabilised by a chaperone, but opening into a pore when acted upon by CyP-D in the presence of calcium. However, such a hypothesis does not easily explain how ligands of the ANT can be profound modulators of MPTP opening.

11.5.3. Other Possible Components of the MPTP

There are several other proteins that have been proposed to be components of the MPTP, of which the strongest candidate is the voltage activated anion channel (VDAC, also known as porin). This was originally proposed by Zoratti & Szabo in their seminal review (Zoratti and Szabo 1994) and was based on the observation that VDAC co-purifies as a complex with the ANT and the peripheral benzodiazepine receptor under some conditions (Mcenery et al. 1992). The same proteins are also thought to interact at contact sites, points of intimate contact between the inner and outer mitochondrial membranes (Crompton, 2000). Indeed, Crompton and colleagues demonstrated that a complex of ANT and VDAC from detergent solubilised heart mitochondria bound to a GST-CyP-D affinity column and following elution with glutathione, could be reconstituted into proteoliposomes to form a calcium activated pore that was inhibited by CsA (Crompton et al. 1998). However, our own data suggest that only the ANT binds to CyP-D (Woodfield et al. 1998) and that the complex of ANT and CyP-D in the absence of VDAC is sufficient for the formation of a CsA-sensitive pore. This would imply a regulatory role for VDAC rather than it being an essential structural component. Further evidence for VDAC playing such a role comes from the observation that ubiquinone analogues that act as potent MPTP inhibitors bind tightly to VDAC (Cesura et al. 2003).

In addition to a role for VDAC, the groups of Kroemer and Brdiczka have suggested that several other proteins associated with the contact sites, such hexokinase, creatine kinase, the peripheral benzodiazepine receptor and the anti- and pro-apoptotic proteins Bcl-2 and Bax may also associate with the MPTP (Marzo et al. 1998; Brenner et al. 2000). However, which if any of these components plays an essential in pore formation (as opposed to a potential regulatory role) is unclear. In conclusion, there remains uncertainty over the minimum configuration of the MPTP and clarification will be important if the MPTP is to become a future pharmaceutical target, for example in the protection of the heart from reperfusion injury.

11.5.4. A Model of the MPTP that Explains the Mode of Action of Different Modulators

In Figure 11.2 we summarise our current understanding of the mechanism of the MPTP, which represents a development of the model originally presented in 1990 (Halestrap and Davidson 1990), and is becoming widely accepted. A fuller

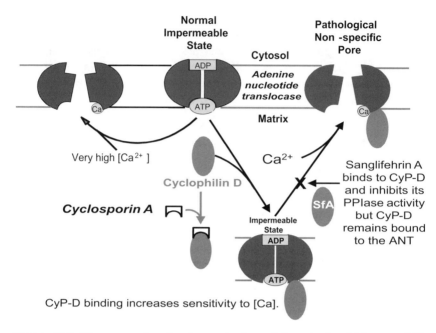

FIGURE 11.2. The proposed molecular mechanism of the mitochondrial permeability transition pore. Supporting data are summarised in the text and extensively reviewed elsewhere (Halestrap et. al. 2002; Halestrap and Brenner 2003). The probable sites of action of known effectors of the MPTP are shown in Table 11.1.

description may be found elsewhere (Halestrap and Brenner 2003; Halestrap et al. 2004; Halestrap et al. 2002). We propose that cyclophilin binds to the ANT on Pro62 and that this binding is greatly enhanced when Cys 160 is cross-linked (e.g. by oxidative stress) to Cys 257. Yeast mitochondria, that lack both these residues, do not possess a CsA-sensitive MPTP (Halestrap et al. 1997; Manon et al. 1998). We suggest that Ca^{2+} binds to a site on the ANT to trigger the conformational change required to induce pore formation probably involving a cis-trans isomerisation of the peptide bond adjacent to Pro62. Binding of adenine nucleotides to the substrate binding site of the ANT greatly reduces the sensitivity of the pore to $[Ca^{2+}]$ and molecular modelling studies suggest that there may be a suitable arrangement of aspartate and glutamate residues in this region that might function as the calcium binding site (Halestrap and Brenner 2003), although this is not immediately apparent in the crystal structure of the ANT-CAT complex (PebayPeyroula et al. 2003). An alternative suggestion is that calcium may bind to the cardiolipin that is tightly associated with the ANT and essential for its activity (Brustovetsky and Klingenberg 1996). However, this seems unlikely because the binding site is totally selective for Ca^{2+}, since not even Sr^{2+} is able to substitute, yet it is strongly inhibited by low pH and by other divalent cations such as Sr^{2+}, Mn^{2+}, Ba^{2+} and Mg^{2+} (Bernardi et al. 1992; Halestrap 1991; Haworth and Hunter 1979; Novgorodov et al. 1994). As noted in Section 11.5.1, the conformational change induced by calcium is facilitated rather than totally dependent on CyP-D.

The model presented in Figure 11.2 is able to provide a plausible explanation of the effects of most known regulators of the pore that may act to modulate binding of adenine nucleotides, calcium or CyP-D to their respective sites as summarised in Table 11.1. However, the modes of action of two other modulators of the MPTP are not so clear. Ubiquinone analogues can act either as activators or inhibitors of the MPTP (Fontaine and Bernardi 1999; Walter et al. 2000). In the light of these data it has been suggested that components of Complex 1 may be involved in the formation and/or regulation of the MPTP (Fontaine and Bernardi, 1999; Walter et al., 2000), but as noted in Section 11.5.3, ubiquinone analogues may also target VDAC (Cesura et al. 2003). However, it is known that uncoupling proteins UCP1, UCP2 and UCP3, may bind oxidised ubiquinone (Echtay et al. 2001; Echtay et al. 2000) and since UCPs are close relatives of ANT (Klingenberg et al. 1995) it would seem plausible that a ubiquinone binding site may also exist on the ANT. Trifluoperazine is a potent inhibitor of the MPTP under energised but not de-energised conditions (Halestrap et al. 1997) and may act through an effect on surface membrane charge that changes the voltage sensitivity of the MPTP (Broekemeier and Pfeiffer 1995).

11.6. The Role of the Mitochondrial Permeability Transition in Reperfusion Injury

In this chapter we will focus on the role of the MPTP in reperfusion injury of the heart. The reader is directed elsewhere for reviews on the critical role of the MPTP in necrotic cell death following chemical toxins or ischemia / reperfusion on other tissues, such as the liver and brain (Friberg and Wieloch 2002; Kim et al. 2003).

11.6.1. The Conditions During Ischemia and Reperfusion Favour MPTP Opening

Ischemia is accompanied by a decrease in total adenine nucleotides, and an increase in both intracellular $[Ca^{2+}]$ and phosphate (Halestrap et al. 2004; Halestrap et al. 1998; Suleiman et al. 2001). There is also evidence for an increase in ROS production (Zweier et al. 1987). All of these factors might be expected to cause MPTP opening, but the low pH that accompanies ischemia will prevent this. Upon reperfusion not only will the renewed supply of oxygen lead to mitochondrial energisation and calcium uptake, but also a burst of oxygen free radical formation as the oxygen reacts with accumulated ubisemiquinone (Zweier et al. 1987). The threat of MPTP opening is therefore increased considerably, but may still be restrained by the low pH. However, as reperfusion continues the pH will return to normal which would be expected to lead to MPTP opening (Halestrap et al. 2004; Halestrap et al. 1998) and this is exactly what is observed as is described below. It may be significant that the commonly prescribed anti-cancer drug, doxorubicin (adriamycin), is a potent inducer of the MPTP (AlNasser 1997; Sokolove 1990) and a major risk factor associated with the use of this drug is the development of a cardiomyopathy (Singal et al. 1997).

11.6.2. Experimental Demonstration that the MPTP Opens During Reperfusion

The most widely used technique for measuring MPTP opening in isolated cells employs fluorescent dyes such as tetramethylrhodamine (red fluorescence) to detect dissipation of the mitochondrial membrane potential as a surrogate indicator of MPTP opening. Such experiments have been performed with cardiac myocytes subject to simulated ischemia and reperfusion and confirm MPTP opening occurs at reperfusion (Duchen et al. 1993). These measurements may also be performed in conjunction with a green fluorescent dye, calcein, that only enters the mitochondria upon MPTP opening. Depending on the method of dye loading, it can either be entrapped within the matrix and then released upon pore opening or be excluded from the mitochondria, but enter when the pore opens (Bernardi et al. 1999; Lemasters et al. 1999). Although this technique has been applied successfully to heart cells (Katoh et al. 2002) only limited data are available. In any case, isolated cardiac myocytes are not an ideal model of the whole ischemic reperfused heart and in order to measure MPTP opening in the perfused heart two alternative techniques have been developed. One, devised by Di Lisa and colleagues (2001), is to determine the loss of mitochondrial NAD^+ that accompanies reperfusion as a surrogate indicator of pore opening. Inhibition by CsA is taken as verification that this loss is through the MPTP. A more direct approach, devised in this laboratory, is to measure the mitochondrial entrapment of a radioactive marker, [^{3}H]-2-deoxyglucose (^{3}H-DOG). In this technique Langendorff perfused hearts are first loaded with ^{3}H-DOG that accumulates within the cytosol as ^{3}H-DOG-6-phosphate (^{3}H-DOG-6P) but can only enter the mitochondria when the MPTP opens (Griffiths and Halestrap 1995). The extent to which the ^{3}H-DOG enters the mitochondria can be determined by their rapid isolation in the presence of EGTA to reseal the pores and so entrap the ^{3}H-DOG within them. Measurement of the ^{3}H content of the mitochondria gives a quantitative value for the extent of pore opening, provided that suitable controls and corrections are performed to account for variations in mitochondrial recovery and loading of the heart with ^{3}H-DOG. Both techniques confirm that the MPTP stays closed during ischemia but opens after about 2 minutes of reperfusion. This is the point at which the intracellular pH has returned from its ischemic value of <6.5 (inhibitory to the MPTP (Halestrap 1991)) to pre-ischemic values and the heart enters hypercontracture (Griffiths and Halestrap 1995; Halestrap et al. 1997).

One limitation of the ^{3}H-DOG entrapment technique is that it only determines the extent of MPTP opening at the point of reperfusion and does not take into account the possibility that some mitochondria might "open" and then "close" again and recover their bioenergetic integrity. Such mitochondria will not lose the ^{3}H-DOG that they have taken up and so will register as still "open". To circumvent this problem, we introduced a "post-loading" technique in which the hearts were loaded with ^{3}H-DOG after they had been through ischemia and reperfusion, and then the extent of mitochondrial ^{3}H-DOG entrapment determined. In this case, mitochondria that were open but are now closed will not take up the ^{3}H-DOG and by comparing data for pre-loaded and post-loaded hearts we were able to show that about half of the mitochondria that initially opened on reperfusion subsequently resealed (Javadov et al. 2003; Kerr et al. 1999). If sufficient mitochondria reseal within a cardiac myocyte, it will not become necrotic and will regain its function. However, there will remain the risk of the cell entering apoptosis as will be described below (Section 11.8).

Another feature of the ^{3}H-DOG entrapment technique poses a severe limitation in its ability to determine the extent of necrosis when this is extensive. The permeability barrier imposed by the plasma membrane of necrotic tissue is compromised and the mitochondria are totally disrupted. Hence all the ^{3}H-DOG that might have been entrapped in the cytosol and mitochondria will be lost and thus not detected. This will lead to a major underestimate of pore opening (Javadov et al. 2003). Thus the ^{3}H-DOG entrapment technique is most valuable for the detection of cells that have not yet progressed to full necrosis but are on their way there, and already exhibit compromised mitochondrial function through opening of the MPTP. The extent to which myocytes within the heart have already undergone necrosis with loss of plasma membrane integrity can be detected by classical techniques such as enzyme release.

11.7. The MPTP as A Target for Protecting Hearts from Reperfusion Injury.

It would be predicted that if opening of the MPTP is a critical factor in the transition from reversible to irreversible reperfusion injury of the heart, inhibitors of pore opening would offer protection (Suleiman et al. 2001). There is now increasing evidence that this is the case, and the resistance of hearts and brains from CyP-D knockout mice to ischemia/reperfusion provide the ultimate validation of this conclusion (Baines et al. 2005; Halestrap 2005; Nakagawa et al. 2005; Schinzel et al. 2005). Indeed, it now seems likely that almost any procedure that reduces reperfusion injury, including ischemic preconditioning (IPC), will be associated with a decrease in MPTP opening or an increase in subsequent pore closure. However, this effect may be mediated either through direct inhibition of the pore with agents such as CsA or SfA, or through an indirect effect associated with a decrease in the factors responsible for MPTP opening such as oxidative stress or mitochondrial calcium overload. Specific examples are considered below.

11.7.1. Targeting CyP-D with Cyclosporin A and Sanglifehrin A

In view of their potency as inhibitors of the MPTP (Ki of about 5nM (Clarke et al. 2002; Griffiths and Halestrap 1991; Halestrap and Davidson 1990)) CsA and SfA would be predicted to be effective inhibitors of reperfusion injury and this has been demonstrated in a variety of models. Protection by CsA of isolated rat cardiac myocytes from cell death induced by anoxia and re-oxygenation has been demonstrated in several studies (Nazareth et al. 1991; Griffiths et al. 2000; Xu et al. 2001), including in human atrial myocytes (Shanmuganathan et al. 2005). Interestingly, in rat cardiomyocytes it has been shown that there is a correlation between mitochondrial [Ca^{2+}] content and subsequent cell death (Griffiths et al. 1998; Miyata et al. 1992). Work from this laboratory and that of Di Lisa has confirmed that CsA and its non-immunosuppressive analogues protect the Langendorff perfused heart from reperfusion injury, and this is associated with a decrease in MPTP opening (Clarke et al. 2002; DiLisa et al. 2001; Griffiths and Halestrap 1993; Griffiths and Halestrap 1995; Halestrap et al. 1997; Stowe and Riess 2004). When CsA was present prior to ischemia and then also in the reperfusion medium, hearts were better able to re-establish a regular beat and left ventricular developed pressure (LVDP). This protective effect was accompanied by a return of the tissue ATP/ADP ratios and AMP levels to control values, greatly reduced release of intracellular lactate dehydrogenase (an indicator of necrosis) and lower end diastolic pressure (EDP) (an indicator of contracture whose elevation reflects elevated [Ca^{2+}]). We have subsequently observed similar protective effects with SfA (Clarke et al. 2002). Other laboratories have used measurement of infarct size relative to area at risk following coronary occlusion and reflow to confirm protection by CsA and SfA (Argaud et al. 2004; Hausenloy et al. 2003; Hausenloy et al. 2002). In this model of reperfusion injury CsA and SfA significantly reduce infarct size even when added only at reperfusion (Hausenloy et al. 2003).

Taken together, these data strongly support direct inhibition of the MPTP by CsA and SfA as being an effective means of inhibiting reperfusion injury. However, two major drawbacks to the use of CsA should be noted. First, it can potentially exert additional undesirable effects on the heart through inhibition of calcineurin-mediated processes (Periasamy 2002; Rusnak and Mertz 2000). This problem is not shared by SfA or by some CsA analogues such as 6-methyl-ala-CsA, 4-methyl-val-CsA and N-methyl-4-isoleucine-CsA (NIM811) (Argaud et al. 2005; Clarke et al. 2002; DiLisa et al. 2001; Griffiths and Halestrap 1995). The second problem associated with the use of CsA is that it only protects within a narrow concentration range (DiLisa et al. 2001; Griffiths and Halestrap 1993). In both isolated cardiac myocytes and Langendorff perfused hearts, the optimal concentration of CsA for protection is about 0.2μM with the effect being lost at higher concentrations (DiLisa et al. 2001; Griffiths and Halestrap 1993; Griffiths and Halestrap 1995). This may partly reflect inhibition of calcineurin-dependent processes in the heart, but may also involve the emerging role of cyclophilins in the response of cells to oxidative stress (Doyle et al. 1999; Halestrap et al. 2004).

11.7.2. Targeting the MPTP with Bongkrekic Acid and Ubiquinone-Derivatives

BKA is a potent inhibitor of the MPTP in isolated mitochondria (see section 11.5.2) and it can inhibit apoptosis in cultured cells (Halestrap and Brenner 2003; Halestrap et al. 2002). However, it is inappropriate for use in the perfused heart since its primary action is to inhibit export of ATP derived from oxidative phosphorylation from the mitochondria to the cytosol where it is essential to drive contraction. The same considerations apply when using atractyloside to demonstrate a role for MPTP opening in reperfusion injury and studies where this has been attempted should be treated with caution (Hausenloy et al. 2002). The use of ubiquinone derivatives that inhibit the MPTP (see Sections 11.2 and 11.3) has also proved inappropriate since these too have detrimental effects on the heart (Halestrap et al. 2004).

11.7.3. Indirect Mechanisms of Inhibiting the MPTP

In addition to inhibiting pore opening through the use of drugs that target components of the MPTP directly, it would be predicted that inhibition of pore opening could be induced indirectly. Thus reducing calcium overload, ROS production and adenine nucleotide depletion, the provision of ROS scavengers or maintaining a lower pH longer during reperfusion would all be expected to offer protection through reducing MPTP opening. In many cases direct measurement of MPTP opening has not been made, but data are consistent with this mode of action. Ruthenium red, which blocks mitochondrial calcium uptake, or calcium antagonists that block plasma membrane Ca channels, do offer protection from mitochondrial Ca-overload and reperfusion injury in the perfused heart although they are unlikely to prove clinically useful (Suleiman et al. 2001). Low pH on reperfusion has been shown to protect a variety of cells and tissues to reperfusion injury (Lemasters 1999) and this may well be a major factor in the protective effects of Na^+/H^+ antiporter inhibitors such as cariporide as described below (Section 11.7.3.1). There is also extensive evidence that free radical scavengers can protect hearts from reperfusion injury (Dhalla et al. 2000; Halestrap et al. 1993) and for two of these, pyruvate and the anesthetic propofol, direct evidence for inhibition of pore opening has been obtained (see Sections 11.7.3.2 and 11.7.3.3). Magnesium is well known to protect hearts from ischemia and reperfusion injury and it is generally accepted that it exerts its protective effects on the heart by inhibiting L-type calcium channels and the Na^+/Ca^{2+} antiporter, thus decreasing calcium overload (Lareau et al. 1995; Suleiman 1994). There are also data to suggest that the presence of supra-physiological [Mg^{2+}] prior to ischemia exerts an antioxidant effect during reperfusion (Maulik et al., 1999), but this could be secondary to inhibition of MPTP opening (Batandier et al. 2004).

11.7.3.1. Na$^+$/ H$^+$ Antiporter Inhibitors

It has been well established that Na$^+$/H$^+$ antiporter inhibitors such as amiloride and cariporide, which decrease the rate at which pH returns to normal during reperfusion (Vandenberg et al. 1993), can offer protection against reperfusion injury (Allen and Xiao 2003; Karmazyn et al. 2001). These drugs have the added benefit of decreasing sodium loading and hence calcium loading during ischemia, which may also play a role in their protective effects. Recent data using the ^{3}H-DOG entrapment technique has confirmed that this protection is associated with inhibition of MPTP opening as predicted (Javadov et al. 2005).

11.7.3.2. Pyruvate

The ability of pyruvate to protect hearts and other tissues, including gut, liver, kidney and brain, against ischemia-reperfusion (anoxia-reoxygenation) injury has been known for many years and may partially reflect its ability to act as a free radical scavenger and as an excellent respiratory substrate (Halestrap et al. 2004; Kerr et al. 1999; Mallet 2000). In addition, it slows the return of the intracellular pH to normal during reperfusion, perhaps by increasing the accumulation of intracellular lactate (Cross et al. 1995; Kerr et al. 1999). Our own data demonstrated that 10 mM pyruvate given before ischemia and during reperfusion after 40 min global ischemia enabled 100% recovery of hemodynamic function compared to 36% recovery in control hearts. Using the ^{3}H-DOG entrapment technique, we demonstrated that this was associated with a substantial decrease in MPTP opening on reperfusion. Furthermore, when the post-loading ^{3}H-DOG technique was employed, we were able to show that in the presence of pyruvate, even those mitochondria that still opened upon reperfusion subsequently closed again, consistent with the 100% recovery in hemodynamic performance (Kerr et al. 1999).

11.7.3.3. Propofol

The anesthetic propofol is frequently used during cardiac surgery and in post-operative sedation and has been demonstrated to act as a free radical scavenger that protects the Langendorff perfused heart against reperfusion injury and damage caused by hydrogen peroxide-induced oxidative stress (Javadov et al. 2000; Suleiman et al. 2001). It may also inhibit plasma membrane calcium channels (Li et al. 1997) and, at concentrations higher than those used clinically, it may inhibit the MPTP directly (Sztark et al. 1995). We have used the ^{3}H-DOG-entrapment technique to demonstrate directly that in addition to improving hemodynamic recovery from ischemia, propofol, at clinically relevant concentrations, inhibits MPTP opening in situ. Furthermore, mitochondria isolated from the propofol-treated hearts exhibited less pore opening than control mitochondria exposed to the same [Ca^{2+}] (Javadov et al. 2000). We have also demonstrated a cardioprotective effect of propofol on the functional recovery of the working rat heart following cold cardioplegic ischemic arrest, a model that is closer to the situation experienced in open heart surgery (Javadov et al. 2000). Most recently we have extended these studies to an in vivo pig model of cardiopulmonary bypass with warm blood cardioplegia that closely matches current clinical practice. Normal clinical concentrations of propofol not only improved functional recovery of the heart, but also reduced troponin-I release and maintained higher tissue ATP levels and lower lactate output (Lim et al. 2005). These data suggest that propofol and pyruvate, alone or together, may be a useful adjunct to the cardioplegic solutions used in cardiac surgery.

11.7.3.4. Ischemic Preconditioning

One of the most effective ways to protect the heart (and indeed other tissues) against reperfusion injury is to subject them to brief ischemic periods with intervening recovery periods before the prolonged period of ischemia is initiated. This phenomenon, known as "ischemic preconditioning" (IPC), is associated with two phases of protection; an immediate effect that lasts for an hour or two and a "second window" that occurs 24-48 hours later (Murphy 2004; Yellon and Downey 2003). The exact mechanisms involved in preconditioning are still debated but several processes have been implicated. The longer-term effects are probably caused by stimulation of the transcription of specific genes, perhaps through a mechanism activated by free radicals and stress-activated protein kinases. The mechanisms responsible for the short-term effects are even less clear, but activation of protein kinase C (PKC) has been strongly implicated either by reactive oxygen species (ROS) released during the short intervening reperfusion periods or by factors released during the brief ischemic periods such as adenosine, bradykinin, noradrenaline and opioids. Other studies, many dependent on pharmacological agents whose specificity is debatable, have implicated sulphonylurea-sensitive K_{ATP} channels, either plasma membrane or mitochondrial, and also other protein kinases including cyclic GMP dependent protein kinase, protein kinase B (Akt) and glycogen synthase kinase 3 (Costa et al. 2005; Juhaszova et al. 2004; Murphy 2004; Yellon and Downey 2003).

Whatever the exact mechanism of preconditioning, several studies have now confirmed that, as might be predicted, it is associated with a decrease in MPTP opening (Argaud et al. 2004; Hausenloy et al. 2004; Javadov et al. 2003; Juhaszova et al. 2004). Our own studies have used the ^{3}H-DOG entrapment technique to demonstrate inhibition of the initial MPTP opening at reperfusion and greater subsequent closure (Javadov et al. 2003). Others have shown that IPC increases the time after laser induced ROS production at which MPTP opening occurs, as assessed by mitochondrial depolarisation (Hausenloy et al. 2004; Juhaszova et al. 2004). This is taken to reflect a direct effect of the preconditioning protocol on the MPTP, perhaps through a direct modification of some component of the MPTP machinery as suggested by some workers (Baines et al. 2003). However, data from us and others have shown that mitochondria isolated after the preconditioning protocol, but before hearts went into ischemia, showed no change in their sensitivity to MPTP opening (Javadov et al. 2003; Khaliulin et al. 2004). In contrast, mitochondria isolated from preconditioned hearts after ischemia and reperfusion have been reported to be less sensitive to

MPTP opening (Argaud et al. 2004; Khaliulin et al. 2004). Our own studies initially failed to show this difference after 30 min reperfusion (Javadov et al. 2003), but recently we have repeated these experiments at 5 min reperfusion and confirmed inhibition of MPTP opening in the mitochondria from preconditioned hearts. Data are illustrated in Figure 11.3. All these data would be consistent with the protection against MPTP opening mediated by IPC being secondary to the decreased calcium over-load and free radical production at reperfusion that has been reported in response to preconditioning (Javadov et al. 2003). Indeed, Khaliulin et al. (2004) have shown that mitochondria from preconditioned hearts subject to ischemia and reperfusion exhibit less protein carbonylation, an indicator of ROS-induced oxidation, and our own data confirm this (Figure 11.3).

How preconditioning protocols exert their effects on the MPTP, be it a direct or an indirect effect, remains unclear. Hausenloy et al. have suggested that transient opening of the MPTP during the preconditioning stimulus may be important (Hausenloy et al. 2004) but our own data do not support this (Javadov et al. 2003) and it is hard to envisage an appropriate mechanism. There has been considerable attention given to the potential role of mitochondrial potassium channels and especially sulphonylurea-sensitive K_{ATP} channels in preconditioning (Garlid et al. 2003; Oldenburg et al. 2003; O'Rourke 2004). This is discussed by Brian O'Rourke in greater detail in Chapter 10. However, our own studies provide no evidence for this (Das et al. 2003; Lim et al. 2002) and there is no consensus as to how opening of such channels might lead to inhibition of the MPTP (Garlid et al. 2003; O'Rourke 2004). Furthermore, there is increasing doubt over the specificity of the pharmacological agents used to support the hypothesis and we have even questioned the existence of mitochondrial K_{ATP} channels (Das et al. 2003; Hanley et al. 2005; Hanley et al. 2002). Data from the laboratories of Downey and Yellon have suggested that there are kinase pathways, including PI-3-kinase mediated Akt activation, that are activated by preconditioning, whether by ischemia or pharmacological agents, and protect the mitochondria during reperfusion (Hausenloy et al. 2005a; Hausenloy and Yellon 2004; Yellon and Downey 2003). However, the ultimate target of these kinases is unknown and the protection against MPTP opening may still be indirect involving reductions in ROS and calcium overload.

11.7.3.5. Postconditioning

Vinter Johansen and colleagues have recently identified a novel way of protecting hearts from reperfusion injury that they have called postconditioning (Kin et al. 2004; Zhao et al. 2003). This protocol involves interspersing the initial phase of reperfusion with several very brief (10s) ischemic periods. Postconditioning, like preconditioning, has been shown to be accompanied by activation of PI-3-kinase and Akt (Tsang et al. 2004) and to involve decreased ROS production and mitochondrial calcium overload (Sun et al. 2005) together with inhibition of MPTP opening (Argaud et al. 2005). Thus there appear to be many similarities between preconditioning and postconditioning (Tsang et al. 2005). However, it

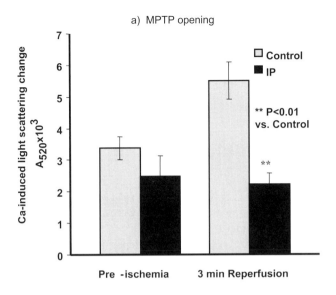

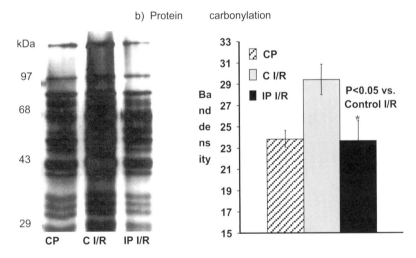

FIGURE 11.3. Mitochondria from preconditioned hearts subject to ischemia / reperfusion show less protein carbonylation and are less sensitive to MPTP opening. The opening of the MPTP was determined under de-energised conditions as described in Javadov et al. (2003) and data are shown as the change in light scattering induced by addition of 80μM buffered free [Ca^{2+}]. Data are presented as means ± S.E.M. (error bars) of 6 separate experiments on different mitochondrial preparations. Protein carbonylation was measured as described by Khaliulin et al. (2004). A typical Western blot is shown and for each experiment these blots were scanned to give an average track density. Data are presented as means ± S.E.M. (error bars) for the 6 mitochondrial preparations of each condition (CP, control non-ischemic perfusion; C I/R, control ischemic reperfused; IP I/R, preconditioned ischemic reperfused).

would be predicted that another mechanism involved in post-conditioning would be the maintenance of an acid pH longer during the reperfusion phase. This would lead to inhibition of MPTP opening in its own right (Halestrap 1991). Indeed, it is quite possible that the inhibition of ROS production is actually a consequence rather than a cause of the decreased MPTP opening (Batandier et al. 2004; Zorov et al. 2000).

11.8. The Mitochondrial Permeability Transition Pore and Apoptosis

As noted under Section 11.6.2, by comparing the ^{3}H-DOG entrapment obtained using the pre- and post-loading techniques, we were able to show that some transient MPTP opening occurs during reperfusion followed by closure. Even hearts that showed 100% recovery following reperfusion gave some initial opening of the MPTP, but these mitochondria subsequently sealed allowing total functional recovery. Indeed, the extent of MPTP opening and mitochondrial dysfunction. In contrast, at the periphery where the insult is less, transient MPTP opening would cause apoptosis. This is illustrated in Figure 11.4.

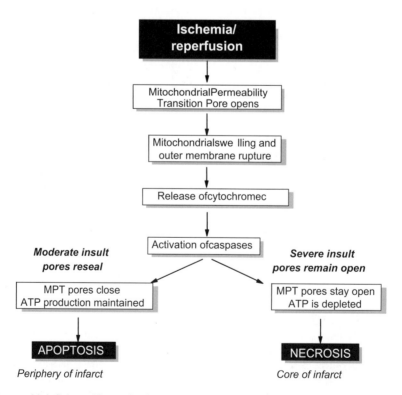

FIGURE 11.4. Scheme illustrating how the extent of MPTP opening and subsequent closure may determine whether cell death is necrotic or apoptotic. Further details may be found in the text.

11.9. Conclusions

When the MPTP opens, mitochondria undergo a Jekyll to Hyde conversion that turns them from the providers of ATP to sustain normal cell function into executioners that initiate cell death. Even if the pores open only transiently, allowing maintenance of tissue ATP levels, the swelling that accompanies pore opening can lead to outer membrane rupture and the release of pro-apoptotic proteins that commit the cell to apoptosis. Whatever the final pathway of cell death, pharmacological interventions that can inhibit MPTP opening and enhance pore closure, either directly (CsA and SfA) or indirectly (propofol, pyruvate and pre- or post- conditioning protocols), provide protection from reperfusion injury. In cardiac surgery, where it is possible to administer pharmacological agents prior to ischemia, there is good reason to think that targeting the MPTP will be an effective therapeutic strategy. It may also be possible to inhibit the MPTP only upon reperfusion and give significant protection (Hausenloy et al. 2003; Hausenloy et al. 2005b), suggesting that this strategy could be efficacious in the treatment of a coronary thrombosis using angioplasty or clot busters. It appears that there is about a 15 minute window into reperfusion within which the MPTP inhibition must be initiated (Hausenloy et al. 2003; Hausenloy et al. 2005b). This observation implies that after the initial MPTP opening there is ongoing and progressive MPTP opening that occurs throughout the early phase of reperfusion and that after 15 minutes this has reached the point of no return in sending the heart towards necrosis. One explanation for this is that MPTP opening itself produces ROS (Batandier et al. 2004; Zorov et al. 2000) that will in turn sensitise more mitochondria to undergo the permeability transition. Inhibiting the MPTP at any point in this early phase may thus prevent sufficient pore opening to avoid the point of no return at which necrosis becomes inevitable. The narrow therapeutic window of CsA makes this drug of limited use and a major goal is the development of novel, specific and potent inhibitors of the MPTP that can enter the heart rapidly and hopefully reach the ischemic area from the co-lateral circulation. The development of a reconstituted form of the MPTP would be of immense value in screening for such drugs.

Acknowledgements. This work was supported by project and programme grants from the British Heart Foundation

References

Allen DG, Xiao XH (2003) Role of the cardiac Na^+/H^+ exchanger during ischemia and reperfusion. Cardiovasc Res 57: 934–941

Al Nasser IA (1997) Prevention of adriamycin aglycone-induced changes of inner mitochondrial membrane permeability by cyclosporin A. Med Sci Res 25: 249–251

Anversa P, Cheng W, Liu Y, Leri A, Redaelli G, Kajstura J (1998) Apoptosis and myocardial infarction. Basic Res Cardiol 93: 8–12

Argaud L, Gateau Roesch O, Chalabreysse L, Gomez L, Loufouat J, Thivolet Bejui F, Robert D, Ovize M (2004) Preconditioning delays Ca^{2+}-induced mitochondrial permeability transition. Cardiovasc Res 61: 115–122

Argaud L, GateauRoesch O, Raisky O, Loufouat J, Robert D, Ovize M (2005) Postconditioning inhibits mitochondrial permeability transition. Circulation 111: 194–197

Baines CP, Kaiser RA, Purcell NH, Blair NS, Osinska H, Hambleton MA, Brunskill EW, Sayen MR, Gottlieb RA, Dorn GW, Robbins J, Molkentin JD (2005) Loss of cyclophilin D reveals a critical role for mitochondrial permeability transition in cell death. Nature 434: 658–662

Baines CP, Song CX, Zheng YT, Wang GW, Zhang J, Wang OL, Guo Y, Bolli R, Cardwell EM, Ping PP (2003) Protein kinase C epsilon interacts with and inhibits the permeability transition pore in cardiac mitochondria. Circ Res 92: 873–880

Basso E, Fante L, Fowlkes J, Petronilli V, Forte MA, Bernardi P (2005) Properties of the permeability transition pore in mitochondria devoid of cyclophilin D. J Biol Chem 280: 18558–18561

Batandier C, Leverve X, Fontaine E (2004) Opening of the mitochondrial permeability transition pore induces reactive oxygen species production at the level of the respiratory chain complex I. J Biol Chem 279: 17197–17204

Bernardi P (1999) Mitochondrial transport of cations: Channels, exchangers, and permeability transition. Physiol Rev 79: 1127–1155

Bernardi P, Scorrano L, Colonna R, Petronilli V, Di.Lisa F (1999) Mitochondria and cell death - Mechanistic aspects and methodological issues. Eur J Biochem 264: 687–701

Bernardi P, Vassanelli S, Veronese P, Colonna R, Szabo I, Zoratti M (1992) Modulation of the mitochondrial permeability transition pore - effect of protons and divalent cations. J Biol Chem 267: 2934–2939

Bernardi P, Veronese P, Petronilli V (1993) Modulation of the Mitochondrial Cyclosporin A-Sensitive Permeability Transition Pore .1. Evidence for 2 separate Me^{2+} binding sites with opposing effects on the pore open probability. J Biol Chem 268: 1005–1010

Brenner C, Cadiou H, Vieira HLA, Zamzami N, Marzo I, Xie ZH, Leber B, Andrews D, Duclohier H, Reed JC, Kroemer G (2000) Bcl-2 and Bax regulate the channel activity of the mitochondrial adenine nucleotide translocator Oncogene 19: 329–336

Broekemeier KM, Pfeiffer DR (1995) Inhibition of the mitochondrial permeability transition by cyclosporin a during long time frame experiments: Relationship between pore opening and the activity of mitochondrial phospholipases. Biochemistry 34: 16440–16449

Brustovetsky N, Klingenberg M (1996) Mitochondrial ADP/ATP carrier can be reversibly converted into a large channel by Ca2+. Biochemistry 35: 8483–8488

Brustovetsky N, Tropschug M, Heimpel S, Heidkamper D, Klingenberg M (2002) A large Ca^{2+}-dependent channel formed by recombinant ADP/ATP carrier from Neurospora crassa resembles the mitochondrial permeability transition pore. Biochemistry 41: 11804–11811

Cesura AM, Pinard E, Schubenel R, Goetschy V, Friedlein A, Langen H, Polcic P, Forte MA, Bernardi P, Kemp JA (2003) The voltage-dependent anion channel is the target for a new class of inhibitors of the mitochondrial permeability transition pore. J Biol Chem 278: 49812–49818

Clarke SJ, McStay GP, Halestrap AP (2002) Sanglifehrin A acts as a potent inhibitor of the mitochondrial permeability transition and reperfusion injury of the heart by binding to cyclophilin-D at a different site from cyclosporin A. J Biol Chem 277: 34793–34799

Connern CP, Halestrap AP (1992) Purification and N-terminal sequencing of peptidyl-prolyl cis-trans-isomerase from rat liver mitochondrial matrix reveals the existence of a distinct mitochondrial cyclophilin. Biochem J 284: 381–385

Connern CP, Halestrap AP (1994) Recruitment of mitochondrial cyclophilin to the mitochondrial inner membrane under conditions of oxidative stress that enhance the opening of a calcium-sensitive non-specific channel. Biochem J 302: 321–324

Connern CP, Halestrap AP (1996) Chaotropic agents and increased matrix volume enhance binding of mitochondrial cyclophilin to the inner mitochondrial membrane and sensitize the mitochondrial permeability transition to [Ca^{2+}]. Biochemistry 35: 8172–8180

Costa AD, Garlid KD, West IC, Lincoln TM, Downey JM, Cohen MV, Critz SD (2005) Protein kinase G transmits the cardioprotective signal from cytosol to mitochondria. Circ Res 97: 329–336

Crabtree GR (2001) Calcium, calcineurin, and the control of transcription. J Biol Chem, 276: 2313–2316

Crompton M (1999) The mitochondrial permeability transition pore and its role in cell death. Biochem J 341: 233–249

Crompton M (2000) Mitochondrial intermembrane junctional complexes and their role in cell death. J Physiol 529: 11–21

Crompton M, Ellinger H, Costi A (1988) Inhibition by cyclosporin A of a Ca^{2+}-dependent pore in heart mitochondria activated by inorganic phosphate and oxidative stress. Biochem J 255: 357–360

Crompton M, Virji S, Ward JM (1998) Cyclophilin-D binds strongly to complexes of the voltage-dependent anion channel and the adenine nucleotide translocase to form the permeability transition pore. Eur J Biochem 258: 729–735

Cross HR, Clarke K, Opie LH, Radda GK (1995) Is lactate-induced myocardial ischemic injury mediated by decreased pH or increased intracellular lactate? J Mol Cell Cardiol 27: 1369–1381

Das M, Parker JE, Halestrap AP (2003) Matrix volume measurements challenge the existence of diazoxide/glibencamide-sensitive K-ATP channels in rat mitochondria. J Physiol 547: 893–902

Dhalla NS, Elmoselhi AB, Hata T, Makino N (2000) Status of myocardial antioxidants in ischemia-reperfusion injury. Cardiovasc Res 47: 446–456

Dierks T, Salentin A, Heberger C, Kramer R (1990a) The Mitochondrial Aspartate/Glutamate and ADP/ATP Carrier switch from obligate counterexchange to unidirectional transport after modification by SH-reagents. Biochim Biophys Acta 1028: 268–280

Dierks T, Salentin A, Kramer R (1990b) Pore-like and carrier-like properties of the mitochondrial aspartate/glutamate carrier after modification by SH-reagents - Evidence for a preformed channel as a structural requirement of carrier-mediated transport. Biochim Biophys Acta 1028: 281–288

Di Lisa F, Menabo R, Canton M, Barile M, Bernardi P (2001) Opening of the mitochondrial permeability transition pore causes depletion of mitochondrial and cytosolic NAD(+) and is a causative event in the death of myocytes in postischemic reperfusion of the heart. J Biol Chem 276: 2571–2575

Dolce V, Scarcia P, Iacopetta D, Palmieri F (2005) A fourth ADP/ATP carrier isoform in man: identification, bacterial expression, functional characterization and tissue distribution. FEBS Lett 579: 633–637

Doyle V, Virji S, Crompton M (1999) Evidence that cyclophilin-A protects cells against oxidative stress. Biochem J 341: 127–132

Duchen MR, Mcguinness O, Brown LA, Crompton M (1993) On the Involvement of a Cyclosporin-A Sensitive Mitochondrial Pore in Myocardial Reperfusion Injury. Cardiovasc Res 27: 1790–1794

Echtay KS, Winkler E, Frischmuth K, Klingenberg M (2001) Uncoupling proteins 2 and 3 are highly active H+ transporters and highly nucleotide sensitive when activated by coenzyme Q (Ubiquinone) . Proc Natl Acad Sci USA 98: 1416–1421

Echtay KS, Winkler E, Klingenberg M (2000) Coenzyme Q is an obligatory cofactor for uncoupling protein function. Nature 408: 609–613

Fliss H, Gattinger D (1996) Apoptosis in ischemic and reperfused rat myocardium. Circ Res 79: 949–956

Fontaine E, Bernardi P (1999) Progress on the mitochondrial permeability transition pore: Regulation by complex I and ubiquinone analogs. J Bioenerg Biomembr 31: 335–345

Friberg H, Wieloch T (2002) Mitochondrial permeability transition in acute neurodegeneration. Biochimie 84: 241–250

Garlid KD, Dos Santos P, Xie ZJ, Costa ADT, Paucek P (2003) Mitochondrial potassium transport: the role of the mitochondrial ATP-sensitive K+ channel in cardiac function and cardioprotection. Biochim Biophys Acta 1606: 1–21

Griffiths EJ, Halestrap AP (1991) Further evidence that cyclosporin-A protects mitochondria from calcium overload by inhibiting a matrix peptidyl-prolyl cis-trans isomerase - implications for the immunosuppressive and toxic effects of cyclosporin. Biochem J 274: 611–614

Griffiths EJ, Halestrap AP (1993) Protection by cyclosporin A of ischemia reperfusion-induced damage in isolated rat hearts. J Mol Cell Cardiol 25: 1461–1469

Griffiths E.J, Halestrap AP (1995) Mitochondrial non-specific pores remain closed during cardiac ischemia, but open upon reperfusion. Biochem J 307: 93–98

Griffiths EJ, Ocampo CJ, Savage JS, Rutter GA, Hansford RG, Stern MD, Silverman HS (1998) Mitochondrial calcium transporting pathways during hypoxia and reoxygenation in single rat cardiomyocytes. Cardiovasc Res 39: 423–433

Griffiths EJ, Ocampo CJ, Savage JS, Stern MD, Silverman HS (2000) Protective effects of low and high doses of cyclosporin A against reoxygenation injury in isolated rat cardiomyocytes are associated with differential effects on mitochondrial calcium levels. Cell Calcium 27: 87–95

Gunter TE, Pfeiffer DR (1990) Mechanisms by Which Mitochondria Transport Calcium. Am J Physiol 258: C755–C786

Halestrap A (2005) A pore way to die. Nature 434: 578–579

Halestrap AP (1991) Calcium-dependent opening of a non-specific pore in the mitochondrial inner membrane is inhibited at pH values below 7 - implications for the protective effect of low pH against chemical and hypoxic cell damage. Biochem J 278: 715–719

Halestrap AP (2004) Dual role for the ADP/ATP translocator? Nature 430: U1

Halestrap AP, Brenner C (2003) The adenine nucleotide translocase: A central component of the mitochondrial permeability transition pore and key player in cell death. Curr Med Chem 10: 1507–1525

Halestrap AP, Clarke SJ, Javadov SA (2004) Mitochondrial permeability transition pore opening during myocardial reperfusion - a target for cardioprotection. Cardiovasc Res 61: 372–385

Halestrap AP, Connern CP, Griffiths EJ, Kerr PM (1997) Cyclosporin A binding to mitochondrial cyclophilin inhibits the permeability transition pore and protects hearts from ischemia/reperfusion injury. Mol Cell Biochem 174: 167–172

Halestrap AP, Davidson AM (1990) Inhibition of Ca^{2+}-induced large amplitude swelling of liver and heart mitochondria by Cyclosporin A is probably caused by the inhibitor binding to mitochondrial matrix peptidyl-prolyl cis-trans isomerase and preventing it interacting with the adenine nucleotide translocase. Biochem J 268: 153–160

Halestrap AP, Doran E, Gillespie JP, O'Toole A (2000) Mitochondria and cell death. Biochem Soc Trans 28: 170–177

Halestrap AP, Griffiths EJ, Connern CP (1993) Mitochondrial calcium handling and oxidative stress. Biochem Soc Trans 21: 353–358

Halestrap AP, Kerr PM, Javadov S, Woodfield KY (1998) Elucidating the molecular mechanism of the permeability transition pore and its role in reperfusion injury of the heart. Biochim Biophys Acta 1366: 79–94

Halestrap AP, McStay GP, Clarke SJ (2002) The permeability transition pore complex: another view. Biochimie 84: 153–166

Halestrap AP, Woodfield KY, Connern CP (1997) Oxidative stress, thiol reagents, and membrane potential modulate the mitochondrial permeability transition by affecting nucleotide binding to the adenine nucleotide translocase. J Biol Chem 272: 3346–3354

Hanley PJ, Drose S, Brandt U, Lareau RA, Banerjee AL, Srivastava DK, Banaszak LJ, Barycki JJ, VanVeldhoven PP, Daut J (2005) 5-Hydroxydecanoate is metabolised in mitochondria and creates a rate-limiting bottleneck for beta-oxidation of fatty acids. J Physiol 562: 307–318

Hanley PJ, Mickel M, Loffler M, Brandt U, Daut J (2002) K-ATP channel-independent targets of diazoxide and 5-hydroxydecanoate in the heart. J Physiol 542: 735–741

Hausenloy D, Wynne A, Duchen M, Yellon D (2004) Transient mitochondrial permeability transition pore opening mediates preconditioning-induced protection. Circulation 109: 1714–1717

Hausenloy DJ, Duchen MR, Yellon DM (2003) Inhibiting mitochondrial permeability transition pore opening at reperfusion protects against ischemia-reperfusion injury. Cardiovasc Res 60: 617–625

Hausenloy DJ, Maddock HL, Baxter GF, Yellon DM (2002) Inhibiting mitochondrial permeability transition pore opening: a new paradigm for myocardial preconditioning? Cardiovasc Res 55: 534–543

Hausenloy DJ, Tsang A, Mocanu MM, Yellon DM (2005a) Ischemic preconditioning protects by activating prosurvival kinases at reperfusion. Am J Physiol 288: H971-H976

Hausenloy DJ, Tsang A, Yellon DM (2005b) The reperfusion injury salvage kinase pathway: A common target for both ischemic preconditioning and postconditioning. Trends Cardiovasc Med 15: 69–75

Hausenloy DJ, Yellon DM (2004a) New directions for protecting the heart against ischemia-reperfusion injury: targeting the Reperfusion Injury Salvage Kinase (RISK)-pathway. Cardiovasc Res 61: 448–460

Hausenloy DJ, Yellon DM, Mani-Babu S, Duchen MR (2004b) Preconditioning protects by inhibiting the mitochondrial permeability transition. Am J Physiol 287: H841–H849

Haworth RA, Hunter DR (1979) The Ca^{2+}-induced membrane transition in mitochondria. II. Nature of the Ca^{2+} trigger site. Arch Biochem Biophys 195: 460–467

He L, Lemasters JJ (2002) Regulated and unregulated mitochondrial permeability transition pores: a new paradigm of pore structure and function? FEBS Lett 512: 1–7

Hunter DR, Haworth RA (1979) The Ca^{2+}-induced membrane transition in mitochondria. I. The protective mechanisms. Arch Biochem Biophys 195: 453–459

Javadov S, Huang C, Kirshenbaum L, Karmazyn M (2005) NHE-1 inhibition improves impaired mitochondrial permeability transition and respiratory function during postinfarction remodelling in the rat. J Mol Cell Cardiol 38: 135–143

Javadov SA, Clarke S, Das M, Griffiths EJ, Lim KHH, Halestrap AP (2003) Ischemic preconditioning inhibits opening of mitochondrial permeability transition poRes. in the reperfused rat heart. J Physiol 549: 513–524

Javadov SA, Lim KHH, Kerr PM, Suleiman MS, Angelini GD, Halestrap AP (2000) Protection of hearts from reperfusion injury by propofol is associated with inhibition of the mitochondrial permeability transition. Cardiovasc Res 45: 360–369

Johnson N, Khan A, Virji S, Ward JM, Crompton M (1999) Import and processing of heart mitochondrial cyclophilin D. Eur J Biochem 263: 353–359

Juhaszova M, Zorov DB, Kim SH, Pepe S, Fu Q, Fishbein KW, Ziman BD, Wang S, Ytrehus K, Antos CL, Olson EN, Sollott SJ (2004) Glycogen synthase kinase-3 beta mediates convergence of protection signaling to inhibit the mitochondrial permeability transition pore. J Clin Invest 113: 1535–1549

Jung DW, Bradshaw PC, Pfeiffer DR (1997) Properties of a cyclosporin-insensitive permeability transition pore in yeast mitochondria. J Biol Chem 272: 21104–21112

Karmazyn M, Sostaric JV, Gan XT (2001) The myocardial Na+/H+ exchanger - A potential therapeutic target for the prevention of myocardial ischemic and reperfusion injury and attenuation of postinfarction heart failure. Drugs 61: 375–389

Katoh H, Nishigaki N, Hayashi H (2002) Diazoxide opens the mitochondrial permeability transition pore and alters Ca2+ transients in rat ventricular myocytes. Circulation 105: 2666–2671

Kerr PM, Suleiman MS, Halestrap AP (1999) Reversal of permeability transition during recovery of hearts from ischemia and its enhancement by pyruvate. Am J Physiol 276: H496–H502

Khaliulin I, Schwalb H, Wang P, Houminer E, Grinberg L, Katzeff H, Borman JB, Powell SR (2004) Preconditioning improves postischemic mitochondrial function and diminishes oxidation of mitochondrial proteins. Free Radic Biol Med 37: 1–9

Kim JS, He L, Qian T, Lemasters JJ (2003) Role of the mitochondrial permeability transition in apoptotic and necrotic death after ischemia/reperfusion injury to hepatocytes. Curr Mol Med 3: 527–535

Kin H, Zhao ZQ, Sun HY, Wang NP, Corvera JS, Halkos ME, Kerendi F, Guyton RA, Vinten Johansen J (2004) Postconditioning attenuates myocardial ischemia-reperfusion injury by inhibiting events in the early minutes of reperfusion. Cardiovasc Res 62: 74–85

Klingenberg M, Winkler E, Huang SG (1995) ADP/ATP carrier and uncoupling protein. Methods Enzymol 260: 369–389

Kokoszka JE, Waymire KG, Levy SE, Sligh JE, Cal JY, Jones DP, MacGregor GR, Wallace DC (2004) The ADP/ATP translocator is not essential for the mitochondrial permeability transition pore. Nature 427: 461–465

Kroemer G, Reed JC (2000) Mitochondrial control of cell death. Nat Med.6: 513–519

Lareau S, Boyle AJ, Stewart LC, Deslauriers R, Hendry P, Keon WJ, Labow RS (1995) The role of magnesium in myocardial preservation. Magnes Res 8: 85–97

Lemasters JJ (1999) The mitochondrial permeability transition and the calcium, oxygen and pH paradoxes: one paradox after another. Cardiovasc Res 44: 470–473

Lemasters JJ, Trollinger DR, Qian T, Cascio WE, Ohata H (1999) Confocal imaging of Ca^{2+}, pH, electrical potential, and membrane permeability in single living cells. Methods Enzymol 302: 341–358

LeQuoc K, LeQuoc D (1988) Involvement of the ADP/ATP carrier in calcium-induced perturbations of the mitochondrial inner membrane permeability: importance of the orientation of the nucleotide binding site. Arch Biochem Biophys 265: 249–257

Li YC, Ridefelt P, Wiklund L, Bjerneroth G (1997) Propofol induces a lowering of free cytosolic calcium in myocardial cells. Acta Anaesthesiol Scand 41: 633–638

Lim KHH, Halestrap AP, Angelini GD, Suleiman MS (2005) Propofol is cardioprotective in a clinically relevant model of normothermic blood cardioplegic arrest and cardiopulmonary bypass. Exp Biol Med 230: 413–420

Lim KHH, Javadov SA, Das M, Clarke SJ, Suleiman MS, Halestrap AP (2002) The effects of ischemic preconditioning, diazoxide and 5-hydroxydecanoate on rat heart mitochondrial volume and respiration. J Physiol 545: 961–974

Mallet RT (2000) Pyruvate: Metabolic protector of cardiac performance. Proc Soc Exp Biol Med 223: 136–148

Manon S, Roucou X, Guerin M, Rigoulet M, Guerin B (1998) Minireview: Characterization of the yeast mitochondria unselective channel: A counterpart to the mammalian permeability transition pore? J Bioenerg Biomembr 30: 419–429

Martinou JC, Green DR (2001) Breaking the mitochondrial barrier. Nat Rev Mol Cell Bio 2: 63–67

Marzo I, Brenner C, Zamzami N, Susin SA, Beutner G, Brdiczka D, Remy R, Xie ZH, Reed JC, Kroemer G (1998) The permeability transition pore complex: A target for apoptosis regulation by caspases and Bcl-2-related proteins. J Exp Med 187: 1261–1271

Maulik M, Maulik SK, Kumari R (1999) Importance of timing of magnesium administration: a study on the isolated ischemic-reperfused rat heart. Magnes Res 12: 37–42

McEnery MW, Snowman AM, Trifiletti RR, Snyder SH (1992) Isolation of the mitochondrial benzodiazepine receptor - association with the voltage-dependent anion channel and the adenine nucleotide carrier. Proc Natl Acad Sci USA 89: 3170–3174

McStay GP, Clarke SJ, Halestrap AP (2002) Role of critical thiol groups on the matrix surface of the adenine nucleotide translocase in the mechanism of the mitochondrial permeability transition pore. Biochem J 367: 541–548

Miyata H, Lakatta EG, Stern MD, Silverman HS (1992) Relation of mitochondrial and cytosolic free calcium to cardiac myocyte recovery after exposure to anoxia. Circ Res 71: 605–613

Murphy E (2004) Primary and secondary signaling pathways in early preconditioning that converge on the mitochondria to produce cardioprotection. Circ Res 94: 7–16

Nakagawa T, Shimizu S, Watanabe T, Yamaguchi O, Otsu K, Yamagata H, Inohara H, Kubo T, Tsujimoto Y (2005) Cyclophilin D-dependent mitochondrial permeability transition regulates some necrotic but not apoptotic cell death. Nature 434: 652–658

Nazareth W, Yafei N, Crompton M (1991) Inhibition of Anoxia-Induced Injury in Heart Myocytes by Cyclosporin-A. J Mol Cell Cardiol 23: 1351–1354

Nicotera P, Leist M (1997) Mitochondrial signals and energy requirement in cell death. Cell Death Differ 4: 516–516

Novgorodov SA, Gudz TI, Brierley GP, Pfeiffer DR (1994) Magnesium ion modulates the sensitivity of the mitochondrial permeability transition pore to cyclosporin A and ADP. Arch Biochem Biophys 311: 219–228

Oldenburg O, Cohen MV, Downey JM (2003) Mitochondrial K-ATP channels in preconditioning. J Mol Cell Cardiol 35: 569–575

O'Rourke B (2004) Evidence for mitochondrial K+ channels and their role in cardioprotection. Circ Res 94: 420–432

Palmieri F (2004) The mitochondrial transporter family (SLC25): physiological and pathological implications. Pflugers Arch 447: 689–709

Pebay.Peyroula, E, Dahout.Gonzalez C, Kahn R, Trezeguet V, Lauquin GJM, Brandolin R (2003) Structure of mitochondrial ADP/ATP carrier in complex with carboxyatractyloside. Nature 426: 39–44

Periasamy M (2002) Calcineurin and the heartbeat, an evolving story. J Mol Cell Cardiol 34: 259–262

Rück A, Dolder M, Wallimann T, Brdiczka D (1998) Reconstituted adenine nucleotide translocase forms a channel for small molecules comparable to the mitochondrial permeability transition pore. FEBS Lett 426: 97–101

Rusnak F, Mertz P (2000) Calcineurin: Form and function. Physiol Rev 80: 1483–1521

Schinzel AC, Takeuchi O, Huang Z, Fisher JK, Zhou Z, Rubens J, Hetz C, Danial NN, Moskowitz MA, Korsmeyer SJ (2005) Cyclophilin D is a component of mitochondrial permeability transition and mediates neuronal cell death after focal cerebral ischemia. Proc Natl Acad Sci USA 102: 12005–12010

Schreiber SL, Crabtree GR (1992) The mechanism of action of cyclosporin-A and FK506. Immunol Today 13: 136–142

Shanmuganathan S, Hausenloy DJ, Duchen MR, Yellon DM (2005) Mitochondrial permeability transition pore as a target for cardioprotection in the human heart. Am J Physiol 289: H237–H242

Singal PK, Iliskovic N, Li TM, Kumar D (1997) Adriamycin cardiomyopathy: pathophysiology and prevention. FASEB J 11: 931–936

Sokolove PM (1990) Inhibition by cyclosporin A and butylated hydroxytoluene of the inner mitochondrial membrane permeability transition induced by Adriamycin aglycones. Biochem Pharmacol 40: 2733–2736

Suleiman MS (1994) New concepts in the cardioprotective action of magnesium and taurine during the calcium paradox and ischemia of the heart. Magnes Res 7:295–312

Suleiman MS, Halestrap AP, Griffiths EJ (2001) Mitochondria: a target for myocardial protection. Pharmacol Therapeut 89: 29–46

Sun HY, Wang NP, Kerendi F, Halkos M, Kin H, Guyton RA, Johansen JV, Zhao ZQ (2005) Hypoxic postconditioning reduces cardiomyocyte loss by inhibiting ROS generation and intracellular Ca2(+) overload. Am J Physiol 288: H1900–H1908

Szabo I, Bernardi P, Zoratti M (1992) Modulation of the Mitochondrial Megachannel by Divalent Cations and Protons. J Biol Chem 267: 2940–2946

Sztark F, Ichas F, Ouhabi R, Dabadie P, Mazat JP (1995) Effects of the anaesthetic propofol on the calcium-induced permeability transition of rat heart mitochondria: Direct pore inhibition and shift of the gating potential. FEBS Lett 368: 101–104

Tanveer A, Virji S, Andreeva L, Totty NF, Hsuan JJ, Ward JM, Crompton M (1996) Involvement of cyclophilin D in the activation of a mitochondrial pore by Ca^{2+} and oxidant stress. Eur J Biochem 238: 166–172

Tonazzi A, Giangregorio N, Indiveri C, Palmieri F (2005) Identification by site-directed mutagenesis and chemical modification of three vicinal cysteine residues in rat mitochondrial carnitine/acylcarnitine transporter. J Biol Chem 280: 19607–19612

Tsang A, Hausenloy DJ, Mocanu MM, Yellon DM (2004) Postconditioning: A form of "modified reperfusion" protects the myocardium by activating the phosphatidylinositol 3-kinase-Akt pathway. Circ Res 95: 230–232

Tsang A, Hausenloy DJ, Yellon DM (2005) Myocardial postconditioning: reperfusion injury revisited. Am J Physiol 289: H2-H7

Vandenberg JI, Metcalfe JC, Grace AA (1993) Mechanisms of intracellular pH recovery following global ischemia in the perfused heart. Circ Res 72: 993–1003

Vieira HLA, Haouzi D, ElHamel C, Jacotot E, Belzacq AS, Brenner C, Kroemer G (2000) Permeabilization of the mitochondrial inner membrane during apoptosis: impact of the adenine nucleotide translocator. Cell Death Differ 7: 1146–1154

Waldmeier PC, Feldtrauer JJ, Qian T, Lemasters JJ (2002) Inhibition of the mitochondrial permeability transition by the nonimmunosuppressive cyclosporin derivative NIM811. Mol Pharmacol 62: 22–29

Waldmeier PC, Zimmermann K, Qian T, Tintelnot Blomley M, Lemasters JJ (2003) Cyclophilin D as a drug target. Curr Med Chem 10: 1485–1506

Walter L, Nogueira V, Leverve X, Heitz MP, Bernardi P, Fontaine E (2000) Three classes of ubiquinone analogs regulate the mitochondrial permeability transition pore through a common site. J Biol Chem 275: 29521–29527

Woodfield K, Rück A, Brdiczka D, Halestrap AP (1998) Direct demonstration of a specific interaction between cyclophilin-D and the adenine nucleotide translocase confirms their role in the mitochondrial permeability transition. Biochem J 336: 287–290

Woodfield KY, Price NT, Halestrap AP (1997) cDNA cloning of rat mitochondrial cyclophilin. Biochim Biophys Acta 1351: 27–30

Xu MF, Wang YG, Hirai K, Ayub A, Ashraf A (2001) Calcium preconditioning inhibits mitochondrial permeability transition and apoptosis. Am J Physiol 280: H899–H908

Yellon DM, Downey JM (2003) Preconditioning the myocardium: From cellular physiology to clinical cardiology. Physiol Rev 83: 1113–1151

Zenke G, Strittmatter U, Fuchs S, Quesniaux VF, Brinkmann V, Schuler W, Zurini M, Enz A, Billich A, Sanglier JJ, Fehr T (2001) Sanglifehrin A, a novel cyclophilin-binding compound showing immunosuppressive activity with a new mechanism of action. J Immunol 166: 7165–7171

Zhao ZQ, Corvera JS, Halkos ME, Kerendi F, Wang NP, Guyton RA, Vinter Johansen J (2003) Inhibition of myocardial injury by ischemic postconditioning during reperfusion: comparison with ischemic preconditioning. Am J Physiol 285: H579–H588

Zoratti M, Szabo I (1994) Electrophysiology of the inner mitochondrial membrane. J Bioenerg Biomembr 26: 543–553

Zorov DB, Filburn CR, Klotz LO, Zweier JL, Sollott SJ (2000) Reactive oxygen species (ROS)-induced ROS release: A new phenomenon accompanying induction of the mitochondrial permeability transition in cardiac myocytes. J Exp Med 192: 1001–1014

Zweier JL, Flaherty JT, Weisfeldt ML (1987) Direct measurement of free radical generation following reperfusion of ischemic myocardium. Proc Natl Acad Sci USA 84: 1404–1407

12
The Apoptotic Mitochondrial Pathway – Modulators, Interventions and Clinical Implications

M-Saadeh Suleiman and Stephen W. Schaffer

12.1. An Overall View of Cardiac Apoptotic Cell Death

The orderly, energy-dependent death by apoptosis is an essential process that ensures the functional and structural integrity of a variety of adult tissues and organs. This process is associated with morphological and biochemical changes, culminating in nuclear DNA fragmentation and cell shrinkage but without the dramatic disruption of the sarcolemma that is the main feature of death by necrosis (Hetts 1998). However, to ensure the safe and effective physiological role of apoptosis, a number of regulators and controllers are involved. Loss of control over apoptosis can lead to pathological conditions (e.g. proliferative disorders and degenerative conditions (reviewed in (Hetts 1998)). Although there is evidence suggesting that apoptosis may also play an important role in the physiology and pathophysiology of the cardiovascular system, this issue remains controversial. Whereas a physiological role for apoptosis is expected in the vascular tissue of this system (Namiki et al. 1995; Diez et al. 1998;

Abbreviations:
ACE: angiotensin converting enzyme; AIF: apoptosis-inducing factor; AP-1: activator protein-1; Apaf: apoptosis-activating factor; ARC: apoptosis repressor with caspase recruitment domain; BAD: Bcl-2 antagonist of cell death; BAX: Bcl-2 associated X protein; BAK: Bcl-2 antagonist/killer protein; Bcl-2: B cell leukemia/lymphoma-2; BNIP-3: Bcl-2/adenovirus E1B 19 kDa interacting protein-3; HtrA2/omi:high temperature requirement A endoprotease; Caspase: cysteine-dependent aspartate-directed protease; cIAP: cellular inhibitor of apoptosis protein; CrmA: cytokine response modifier protein; dATP: deoxy adenosine triphosphate; DD: death domain; Diablo: direct inhibitor of apoptosis protein-binding protein with low pI; DISC: death-inducing signaling complex; DNase: Deoxyribonucleic acidase; EndoG: Endonuclease G; eNOS: endothelial nitric oxide synthase; FADD: Fas-associated death domain protein; HSP60: heat shock protein; IAP: inhibitor of apoptosis protein; IGF-1: insulin-like growth factor-1; iNOS: inducible nitric oxide synthase; NO: nitric oxide; PARP: poly-ADP ribose polymerase; ROS: reactive oxygen species; XIAP: X-chromosome-linked inhibitor of apoptosis protein; VDAC: voltage-dependent anion channel; 2-VAD-fmk: benzyloxycarbonyl-Val-Ala-Asp-fluoromethylketone

Rossig et al. 2001; Doonan and Cotter 2004; Li and Shah 2004), the potential role of apoptotic death of cardiac myocytes is not clear. This is because adult cardiomyocytes are terminally differentiated and the myocardium is known for its poor ability to regenerate after an injury. Subsequently and in contrast to other tissues that are capable of self-renewal, apoptosis-induced cardiomyocyte loss is likely to dramatically alter cardiac structure and function. In support of this are reports suggesting that apoptotic death is relevant in a number of cardiac pathologies and aging (Kajstura et al. 1996; Narula et al. 1996; Sharov et al. 1997; Heinrich and Holz 1998; Knowlton et al. 1998; Narula et al. 1999; Rayment et al. 1999; Qin et al. 2003). In fact, apoptosis has been detected in myocardial samples obtained from patients with end-stage congestive heart failure, diabetic cardiomyopathy, cardiac allograft rejection, diseases of the conduction system, arrhythmogenic right ventricular dysplasia, and myocardial infarction (Mallat et al. 1996; Olivetti et al. 1997; Saraste et al. 1997a; Saraste et al. 1997b; Saraste et al. 1997c). Furthermore, apoptosis has been detected in cardiac myocytes in a number of experimental models including hypoxia/reoxygenation (Tanaka et al. 1994; Webster et al. 1999; Wang et al. 2001; Tatsumi et al. 2003; Sun et al. 2004), acidosis (Webster et al. 1999), mechanical stretch (Cheng et al. 1995; Leri et al. 1998; Liao et al. 2004), ischemia/reperfusion injury (Gottlieb et al. 1994; Yue et al. 1998; Yaoita et al. 2000; Zuurbier et al. 2005), oxidative stress (von Harsdorf et al. 1999), serum and glucose deprivation and metabolic inhibition (Bialik et al. 1999; Malhotra and Brosius 1999), β-adrenergic agonists (Communal, et al. 1998; Shizukuda et al. 1998) and angiotensin II (Kajstura et al., 1997). Therefore, a better understanding of molecular events and of the cellular factors and signaling pathways that regulate apoptotic cell death in the heart is essential for formulating strategies to assist the heart following a cardiac insult or during either remodeling or increased work conditions that may promote apoptosis. In this chapter, we review aspects relating to cardiac apoptosis, with particular emphasis on the mitochondrial apoptotic pathway in the heart and suggest potential anti-apoptotic interventions in clinical settings.

12.2. The Mitochondrial Apoptotic Pathway

Two major pathways of apoptosis exist in mammalian cells, known as extrinsic and intrinsic (Crow et al. 2004; Fischer and Schulze-Osthoff 2005). The extrinsic cell death pathway is mediated by the death receptor and is initiated by the recruitment of adapter proteins, like FADD, via the DD, which then bind to the death effector domain-containing caspase-8 or -10. Formation of this DISC results in the activation of caspase-8. Activated caspase-8 then cleaves downstream procaspase-3, which breaks down intracellular substrates killing the cell.

The intrinsic pathway of apoptosis is mitochondrial mediated (Richter et al. 1996; Li et al. 1997; Hu et al. 1999; Borutaite and Brown 2003). The mitochondria are the source of ATP necessary for execution of apoptosis.

However, they also contain cytochrome c and apoptogens, which enter the cytoplasm and activate caspases, which in turn cause cellular damage and induce nuclear fragmentation. Cells undergo apoptotic death when they are stressed by one or more stimulators (extra or intracellular). These include radiation, genotoxic and cytotoxic drugs, reactive oxygen species, metabolic poisons, cytokines, growth/survival factor-deprivation, endoplasmic reticular stress agents, toxins disrupting the actin cytoskeleton, detachment of cells from the extracellular matrix, etc. (Fumarola and Guidotti 2004). The signalling pathway elicited by these stimuli is likely to be caused by a perturbation of the mitochondrial membrane integrity in response to physiological or pathogenic stimuli.

The intracellular and extracellular stress signals are transmitted to the mitochondria by pro-apoptotic Bcl-2 family members; Bax and Bak, which translocate to the mitochondria (Figure 12.1). The BH3-only protein Bid activates Bax and Bak to mediate the release of cytochrome c and other apoptogens that results in the formation of an apoptosome, which in turn activates procaspase-9. Bid, a direct substrate of caspase-8, connects the extrinsic and intrinsic pathways. After cleavage, its C-terminus translocates to and inserts into the outer

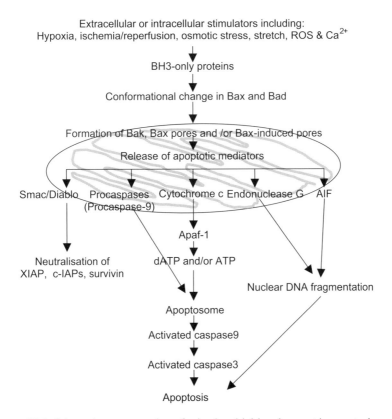

FIGURE 12.1. Schematic representation of mitochondrial involvement in apoptosis.

mitochondrial membrane, triggering activation of Bax and Bak and promoting cytochrome c release.

12.2.1. Bax, Bad Translocation and Leaky Outer Mitochondrial Membrane

The anti-apoptotic (Bcl-2, Bcl-xL) and pro-apoptotic (Bad, Bax) Bcl-2 family of proteins play a central role in regulating the mitochondrial apoptotic pathway. Many of the apoptotic stimulators alter the interaction, the expression and the localization of these proteins. For example, oxidative stress-induced apoptosis is associated with an increase in the protein content of Bad and the translocation of Bax and Bad from the cytosol to the mitochondria, where these factors form heterodimers with Bcl-2, which is followed by the release of cytochrome c (von Harsdorf et al. 1999).

Stretch-induced apoptosis in myocytes also increases the expression of the pro-apoptotic Bcl-2 family of proteins and causes the accumulation of Bax in the mitochondria (Liao et al. 2004). Bax has also been shown to translocate from the cytosol to the mitochondria in neonatal lambs following cardiopulmonary bypass and cardioplegic arrest (Caldarone et al. 2004). It is worth noting that the expression and localization of anti-apoptotic and pro-apoptotic proteins changes during development (Cook et al. 1999). The anti-apoptotic proteins, Bcl-2 and Bcl-xL, are expressed at high levels in neonatal cardiomyocytes, where their expression is sustained during development. In contrast, the pro-apoptotic proteins, Bad and Bax, are present at high levels in neonatal hearts yet barely detectable in adult hearts. In neonatal cardiomyocytes, Bcl-2 and Bcl-xL are associated with the mitochondria, but Bad and Bax are predominantly present in the cytosol.

12.2.2. Release of Mitochondrial Apoptogens and Apoptosome Formation

Mitochondrial outer-membrane permeabilization by pro-apoptotic Bcl-2 family members is an essential step in the induction of apoptosis (van Loo et al. 2002). This permeabilization step appears to facilitate apoptosis by directly causing the release of different mitochondrial apoptogenic factors. Work in HeLa cells or with isolated mitochondria has shown that pro-apoptotic Bcl-2 proteins cause the release of cytochrome c, Smac/Diablo and HtrA2/Omi but not endonuclease G (EndoG) and apoptosis-inducing factor (AIF) (Arnoult et al. 2003). The selective mitochondrial efflux of cytochrome c, Smac/Diablo and HtrA2/Omi into the cytosol causes caspase activation. This results in the opening of an EndoG- and/or AIF-conducting pore. Once in the cytosol, both factors can precipitate apoptotic death. Cardiomyocytes have also been shown to undergo mitochondria-dependent apoptosis, in which cytochrome c and Smac/DIABLO are released from the mitochondria into the cytosol (Liao et al. 2004).

A major pathway of apoptosis involves the activation of caspase-9 through apoptotic signals that induce the release of cytochrome c from the mitochondrial intermembrane space into the cytosol (Czerski and Nunez 2004). In the presence of dATP/ATP, cytochrome c triggers the assembly of a protein complex named the 'apoptosome'. This complex contains the apoptotic protease-activating factor 1 (Apaf-1) and caspase-9. Apaf-1 is a 130 kDa cytosolic protein that was originally shown to play a role in cytochrome c-dependent caspase activation (Zou et al. 1997). Evidence for the formation of the apoptosome complex has come from studies with purified components, revealing that Apaf-1, pro-caspase-9, cytochrome c, and dATP are necessary and sufficient for the formation of an active apoptosome complex (Bratton et al. 2001). Using a cell-free model, (Bratton et al. 2001) also showed that XIAP, as well as binding active caspases-9 and -3, also associate with Apaf-1. Therefore, XIAP appears not only to influence the activation of caspase-3 by caspase-9, but also to inhibit the release of active caspase-3 from the complex.

Several of the apoptotic stimulators trigger the formation of an apoptosome. For example, simulated ischemia facilitates the formation of the oligomeric Apaf-1/caspase-9 apoptosome (Takatani et al. 2004). On the other hand, certain regulators are capable of either delaying or altering apoptosome formation. Most of these regulators prevent apoptosome formation by inhibiting cytochrome c release from the mitochondria. However, apoptosome formation can also be blocked by either disruption of Apaf-1 oligomerization or modulation of the interaction between caspase 9 with Apaf-1 (Takatani et al. 2004).

12.2.3. Activation of Caspases and DNA Fragmentation

The morphological changes associated with apoptosis (e.g. membrane blebbing, chromatin condensation, nuclear condensation and cell shrinkage) result from the activation of caspases which in turn cleave substrates such as cytoskeletal proteins, DNA repair enzymes and protein kinases (reviewed in Logue et al. 2005).

Caspases, known as the executioners of apoptosis (Czerski and Nunez 2004), are synthesized as inactive pro-enzymes and are identified as either initiators or effectors of apoptosis (Riedl and Shi, 2004). Activation of the effector caspases involves proteolytic cleavage by initiator caspases, whose activation is complex but probably involves protein-protein interactions. For example, effector caspase 9 is activated by its interaction with the proteins of the apoptosome, although a detailed mechanism of the activation process awaits further study. Following the interaction of procaspase-9 within the apoptosome complex, the disassociation of active caspase 9 from the apoptosome complex, makes it available to cleave and activate downstream caspases, such as effector caspase-3 (Zou et al. 1999). The effector caspases cleave a broad spectrum of proteins, causing the distinct cellular features of apoptosis, including DNA fragmentation, nuclear condensation, cell shrinkage, blebbing and phosphatidylserine externalization (Thornberry and Lazebnik 1998).

Mitochondrial apoptogens can trigger either caspase-dependent or caspase-independent death pathways (Arnoult et al. 2003). A protein involved in a caspase-dependent pathway is cytochrome c, which activates caspase-9 through its association with apoptosis protease activating factor-1 (Apaf-1). Smac/Diablo and HtrA2/Omi are also mitochondrial apoptogens that lead to the activation of caspases, however, their actions depend upon their neutralization of inhibitor of apoptosis proteins (IAPs). The caspase-independent death proteins include apoptosis-inducing factor (AIF) and endonuclease G (EndoG). AIF induces chromatin condensation and DNA fragmentation when released into the cytosol (Susin et al. 1999). During apoptosis both EndoG and AIF translocate to the nucleus where EndoG causes oligonucleosomal DNA fragmentation (Li et al. 2001). EndoG can catalyze both high molecular weight DNA cleavage and oligonucleosomal DNA breakdown in a sequential fashion; it cooperates with exonuclease and DNase I to facilitate DNA processing.

12.3. Stimulators of Apoptosis

Both external and internal stimulators have been shown to trigger mitochondrial apoptotic death of cardiomyocytes via activation of caspases. Examples of such stimulators include hypoxia, ischemia/reperfusion, osmotic stress, stretch, diabetes, reactive oxygen species (of mitochondrial or cytoplasmic origin), Ca^{2+}, excessive nitric oxide, cytotoxic drugs, angiotensin II and catecholamines (reviewed in Chen et al., 2002; Chen and Tu, 2002).

12.3.1. Hypoxia/Ischemia

Work on neonatal cardiac myocytes has shown that hypoxia (in the absence of substrates) induces apoptosis, as revealed by the presence of characteristic features of apoptosis as well as by the translocation of cytochrome c from the mitochondria to the cytosol (Malhotra and Brosius 1999). When such myocytes are exposed to simulated ischaemia, they undergo the mitochondrial pathway of apoptotic death (Takatani et al. 2004). A model of neonatal cardiac myocytes in culture is important for understanding mechanisms underlying a variety of effectors. However, these myocytes are neonatal, enzymatically dissociated and incubated in culture media. It is likely that under these conditions, the response of heart cells to a cardiac insult will be different than the response of freshly isolated adult cells. Because adult cardiomyocytes are terminally differentiated, apoptosis is of major importance. Furthermore, the expression of pro-apoptotic and anti-apoptotic proteins appears to be age related (Cook et al. 1999). Therefore studies investigating the role of the mitochondria-mediated pathway of apoptosis during hypoxia and/or ischaemia should be confirmed using the intact adult heart, preferably in situ. Apoptosis has been shown to occur in such an *in vivo* model after only 30 min of occluding the left anterior descending coronary artery (Lundberg and Szweda 2004). Furthermore, the authors have found that the cytosolic content of cytochrome c increases and pro-caspase-9 content disappears only after reperfusion, a period in which Bax translocates to the mitochondria. Apoptosis is also seen in failing, infarcted, and hibernating human hearts, and during open heart surgery (Valen 2003). In all of these conditions, impaired control over intracellular Ca^{2+} handling, as well as a rise in reactive oxygen species (ROS), are major changes that result in necrotic and apoptotic death (Suleiman et al. 2001).

12.3.2. Reactive Oxygen Species

The role of ROS as inducers of apoptotic death in isolated myocytes is well established (Tanaka et al. 1994; Tanaka et al. 1998; Cook et al. 1999; von Harsdorf et al. 1999). More importantly, the signaling pathways responsible for ROS-induced apoptosis have also been extensively investigated. For example, H_2O_2-induced apoptotic cell death in cardiomyocytes has been shown to involve specific pathways (von Harsdorf et al. 1999). H_2O_2 was found to induce translocation of Bax and Bad from the cytosol to the mitochondria, where they would form heterodimers with Bcl-2. This was followed by the release of cytochrome c, activation of CPP32/caspase 3, and cleavage of poly (ADP-ribose) polymerase. The signaling cascade appears to occur over a relatively short period of H_2O_2 exposure (Cook et al. 1999), as exemplified by the translocation of cytochrome c from the mitochondria to the cytosol within 15 to 30 minutes of H_2O_2 exposure. There is also evidence that endogenous generation of ROS also triggers apoptosis. For example, high glucose levels in diabetic animals generate ROS that are involved in both death-receptor- and mitochondrion-dependent apoptosis in the heart in vivo (Bojunga et al. 2004). Elevated lipid content is also a source of ROS that can trigger apoptosis (Schaffer et al., 2005). This highlights the potential therapeutic role for preventing cardiovascular damage in diabetes mellitus and obesity.

In addition to experimental models, ROS have also been shown to play an important role in the induction of apoptosis in clinical settings of cardiac pathologies. This is partly due to the fact that heart cells have a large energy demand and so contain a very high density of mitochondria. Additionally, mitochondria provide the energy essential for the completion of apoptosis, a source of ROS, release pro-apoptotic factors and are a site of action of the apoptosis regulatory proteins of the Bcl-2 family, as well as being a source of ROS.

Chronic congestive heart failure carries a poor prognosis and remains a leading cause of cardiovascular death. Reactive oxygen species are required for the normal, physiologic activity of cardiac cells. However, accumulating evidence suggests that ROS play an important role in the development and progression of heart failure. ROS have been implicated in the development of agonist-induced cardiac hypertrophy, cardiomyocyte apoptosis and remodeling of the failing myocardium (reviewed in Sorescu and Griendling 2002; Byrne and Grieve 2003). Indeed, heart remodeling itself is associated with the loss of cardiomyocytes and an increase in fibrous tissue owing to the abnormal mechanical load in a number of heart disease conditions (Liao et al. 2004).

In vitro and in vivo studies suggest the importance of mitochondria and the activation of caspases in cell death occurring in failing hearts. Oxidants, excessive nitric oxide, angiotensin II and catecholamines have been shown to trigger apoptotic death of cardiomyocytes. Eliminating these inducers reduces apoptosis and reverses the loss of contractile function in many cases, indicating the feasibility of interventions with pharmacological agents, such as antioxidants, nitric oxide synthetase inhibitors, ACE inhibitors, angiotensin II receptor antagonists and adrenergic receptor antagonists (Chen and Tu 2002). Apoptosis also plays a role in post-infarction left ventricular remodeling (Baldi et al. 2002).

ROS induced by high glucose (diabetes) are involved in both death-receptor- and mitochondrion-dependent apoptosis in the heart *in vivo*. This suggests that antioxidants may be a therapeutic option for preventing cardiovascular damage in diabetes mellitus in humans (Bojunga et al. 2004).

12.3.3. Osmotic Stress

Another stimulus that has been shown to trigger apoptosis is severe osmotic stress. Sorbitol-induced osmotic stress initiates apoptosis in cardiac fibroblasts and in cardiac myocytes by impairment of mitochondrial function, activation of caspase-3, elevation in poly-[ADP-ribose] polymerase (PARP) degradation and DNA fragmentation (Mockridge et al. 2000; Morales et al. 2000). A link between apoptosis and cellular osmotic balance is underscored by the requirement of cell shrinkage in the apoptotic cascade. Blocking cell shrinkage terminates the apoptotic cascade at the cell shrinkage step (Bortner and Cidlowski, 1998).

12.3.4. Stretch

Stretch also triggers apoptotic death of the cardiomyocytes (Liao et al. 2004). It does so by increasing the expression of the pro-apoptotic Bcl-2 family of proteins (e.g. Bax and Bad) and facilitating the association of Bax with the mitochondria. Stretch also activates the angiotensin II receptor and stimulates ion channels that elevate $[Ca^{2+}]_i$ (Leri et al. 1998; Liao et al. 2004). Significantly, Bax also accumulates in the mitochondria following stretch. These observations highlight the importance of Bcl-2 family proteins, Ca^{2+} and angiotensin II in the coupling of stretch signaling to the mitochondrial apoptotic pathway.

12.3.5. Nitric Oxide

Nitric oxide (NO) has various actions on the cardiovascular system, but its action on cardiac myocytes remains controversial, as highlighted in recent reviews (Andreka et al. 2004). NO has been shown to induce apoptosis (Kawaguchi et al. 1997) and has been implicated in the pathophysiology of heart failure (Andreka et al. 2004). The role of NO in cardiac apoptosis is somewhat puzzling (Razavi et al. 2005), as high levels of NO produced by inducible nitric oxide synthase (iNOS) promote apoptosis while basal levels of NO production from endothelial nitric oxide synthase (eNOS) protect cardiomyocytes from apoptosis. A clear cut effect of NO on the heart *in vivo* is likely to be complicated by the fact that NO targets both myocytes and the vasculature (Liu et al. 2002). However, NO remains a strong candidate for potential therapeutic intervention in cardiac pathologies.

12.3.6. Beta-Adrenergic Agonists

Beta-adrenergic activation has been shown to stimulate apoptosis in adult cardiomyocytes, an effect that has been implicated in the progression of myocardial failure (Communal et al. 1998). The underlying mechanism for this action is a rise in intracellular Ca^{2+}, the source of which appears to be extracellular, as inhibitors of the L-type calcium channels antagonize apoptosis (Rabkin and Kong 2000)

12.3.7. Calcium

Although Ca^{2+} signals play a major role in triggering the mitochondria-dependent pathway of apoptosis, the link between apoptosis and other Ca^{2+} events within the cell, such as the control of mitochondrial energy metabolism, is poorly understood (Pacher et al. 2001). In addition to its direct effects on the mitochondria, Ca^{2+} might promote apoptosis by activating DNase I, an effect that is independent from the expression of Bax and Bcl-2 (Nitahara et al. 1998). This scenario has been proposed to explain apoptotic death of aging myocytes, in which there is an increase in diastolic calcium.

12.3.8. Angiotensin II

The renin-angiotensin system is activated in response to impaired renal function, hypotension, hyperglycemia and increased sympathetic stimulation. Although elevations in angiotensin II are usually compensatory, they also trigger apoptosis. Unlike many cell types, angiotensin II-mediated apoptosis in the cardiomyocyte is initiated by the AT_1 receptor but not by the AT_2 receptor (Kajstura et al. 1997; Sugino et al. 2001). The AT_1 receptor appears to act in part by stimulating Ca^{2+}-dependent DNase I activity (Kajstura et al. 1997). However, angiotensin II-mediated apoptosis is also associated with an elevation in the Bax/Bcl-2 ratio (Ravassa et al. 2000; Grishko et al. 2003). Recently, Ricci et al. (2005) proposed that a key factor directing the angiotensin II treated cardiomyocyte into apoptosis is oxidative damage to mitochondrial DNA, leading to impaired electron transport. In the proposed scenario, electrons are diverted from the electron transport chain to oxygen, producing excessive superoxide that overwhelms the antioxidant defenses. The cells die of apoptosis after activating the mitochondrial permeability transition.

12.3.9. Doxorubicin

Doxorubicin initiates apoptosis in the cardiomyocyte through the activation of several pathways. One of the pathways involves the generation of ROS by the mitochondria, leading to the release of cytochrome c and the activation of caspase 3 (Wang et al., 2001; Green and Leeuwenburg 2002). Besides promoting the mitochondrial permeability transition, ROS also impacts other pro-apoptotic events, such as the activation of caspase 12 (Jang et al. 2004), upregulation of Bax (Lou et al. 2005), sensitization to Fas-induced apoptosis (Nitobe et al. 2003), stimulation of p53 (Liu et al. 2004) and stimulation of mitogen-activated protein kinases (Lou et al. 2005).

The other mitochondrial-linked pathway stimulated by doxorubicin involves the slow accumulation of ceramides (Delpy et al. 1999). Although ceramides may contribute to cell shrinkage in the doxorubicin-treated cardiomyocyte (d'Anglemont de Tussigny et al. 2004), the permeabilization of the mitochondrial membrane is considered the dominant factor causing apoptosis (Novgorodov et al. 2005).

12.4. Strategies for Preventing Cardiac Apoptotic Cell Death

In view of the fact that adult cardiomyocytes are terminally differentiated, cardiac insults (e.g. hypoxia, hemodynamic and oxidative stress, ischaemia) leading to apoptosis pose an acute and long term disruption to cardiac pump function. This is also the case for cardiac pathologies (e.g. heart failure, coronary disease) that are associated with the loss of myocytes. The prognosis of heart failure is worse than that of most cancers, but new therapeutic interventions using stem and other cell-based therapies are succeeding in the fight against it, and old drugs, with new twists, are making a comeback (Benjamin and Schneider 2005). Therefore maneuvers that are aimed at opposing apoptosis will be of important preventative and therapeutic value. Interventions can either strengthen an already existing anti-apoptotic machinery within the myocyte or target the triggers and executioners of apoptotic death. A number of agents have already been identified which can be used to prevent apoptotic death. These include caspase inhibitors, Bcl-2 family, IGF-1, heat shock proteins, calcium antagonists and antioxidants (Gill et al. 2002). Antioxidants, nitric oxide synthetase inhibitors, ACE inhibitors, angiotensin II receptor antagonists and adrenergic receptor antagonists have been proposed as a means of preventing/reducing apoptosis and thus protect the integrity of the myocardium (Chen and Tu 2002). The β-amino acid taurine has also been shown to exhibit anti-apoptotic activity and may be used for therapeutic intervention (Takatani et al. 2004).

The fact that cardiac myocytes do undergo apoptosis in disease states (e.g. myocardial infarction, heart failure, myocarditis, arrhythmogenic right ventricular dysplasia, and immune rejection after cardiac transplantation) renders this process a suitable target for myocardial protection (Andreka et al. 2004).

12.4.1. Caspase Inhibitors

ARC (apoptosis repressor with caspase recruitment domain), an apoptotic inhibitor that is expressed predominantly in cardiac and skeletal muscle, can protect heart myogenic H9c2 cells from hypoxia-induced apoptosis (Ekhterae et al. 1999). ARC prevents cytochrome c release by acting upstream of caspase activation, perhaps at the mitochondrial level. Another endogenous inhibitor with potential use in the clinical setting is the caspase-9 inhibitor, caspase-9S. This inhibitor is downregulated in terminally failing hearts (Scheubel et al. 2002). The pan-caspase inhibitor z-VAD-fmk has been shown to suppress BNIP3-induced apoptotic cell death of ventricular myocytes in a dose-dependent manner (Regula et al. 2002)

Pretreatment of ventricular myocytes with the peptide-caspase inhibitors suppress the cleavage of PARP and apoptosis, but have no effect on cytochrome c release by the mitochondria (de Moissac et al. 2000). The synthetic serpin caspase inhibitor protein CrmA exhibits anti-apoptotic activity in ventricular myocytes during prolonged durations of hypoxia (Gurevich et al. 2001). Yet caspase inhibitors, while preventing apoptosis, can leave the heart in a compromised state, one in which mitochondrial function remains impaired.

12.4.2. Bcl-2 Family Proteins

Another therapeutic approach could involve the Bcl-2 family of proteins where the anti-apoptotic proteins could be upregulated whilst pro-apoptotic pathways are inhibited. Indeed, overexpression of Bcl-2 attenuates apoptosis and mechanical dysfunction following an ischemia-reperfusion insult (Chen et al. 2001).

12.4.3. Antioxidant and Reducing Oxidative Stress

An interesting way of preventing apoptotic cell death would be to strengthen the cellular antioxidant ability of heart cells. For example, training has been shown to improve cell defense systems and to reduce oxidative stress. Training seems to protect by altering the content and activities of Bax, the Bax-to-Bcl-2 ratio, and tissue caspase-3 activity (Ascensao et al. 2005). Antioxidants have been shown to be anti-apoptotic in cardiac myocytes (Kumar et al. 1999). However, a major drawback to the use of antioxidant therapy is the involvement of ROS in cell signaling, which would be disrupted by the use of ROS scavengers to prevent apoptosis.

12.4.4. Heat Shock Protein

Heat shock protein 60 has been shown to form complexes with Bax in the cytosol that can be dissociated by hypoxia, resulting in the translocation of cytosolic HSP60 to the plasma membrane and Bax to the mitochondria. Thus, hypoxia abolishes the ability of HSP60 to neutralize Bax and prevent apoptosis. This is sufficient to trigger apoptosis. Others have also shown a key anti-apoptotic role for cytosolic HSP60. To our knowledge, this is the first report suggesting that interactions of HSP60 with Bax and/or Bak regulate apoptosis (Gupta and Knowlton 2002; Kirchhoff et al. 2002). Interventions that can alter the dissociation rate of the cytosolic complex are also invariably anti-apoptotic. Additionally, overexpression of heat shock proteins should favor complex formation and render the cell resistant to apoptosis. Overexpression of phospholipid hydroperoxide glutathione peroxidase and heat shock proteins exerts a significant anti-apoptotic protective effect in neonatal cardiomyocytes (Hollander et al. 2003).

Results showing that the neonatal left ventricle is more vulnerable than the right ventricle to apoptotic death indicates a relationship between increased afterload and apoptosis-related mitochondrial dysfunction (Caldarone et al. 2004). Moreover, elevated afterload renders the adult heart more susceptible to ischemic injury (Mozaffari and Schaffer, 2003). Therefore, manipulation of afterload after neonatal cardiac operations might be an important cardioprotective intervention.

12.4.5. ACE Inhibitors and Angiotensin II Antagonists

Based on the CONSENSUS trial study and the SOLVD investigation, the angiotensin II antagonists are one of the few drugs that prevent the development and progression of heart failure after infarction. They also significantly reduce mortality, reinfarction or readmission for heart failure when administered post-infarction (ACE Inhibitor Myocardial Infarction Collaborative Group 1998; Flather et al. 2000). While the effect of angiotensin II antagonists on apoptosis has not been investigated in clinical trials, animal studies suggest that angiotensin II-induced apoptosis is an important determinant of left ventricular remodeling and survivability following either a myocardial infarction (Harada et al. 1999; Leri et al. 2000) or experimental heart failure (Goussev et al. 1998; Weinberg et al. 1994). Angiotensin II antagonists have also been found to reduce the incidence of apoptosis in endomyocardial biopsies of patients with essential hypertension, an effect associated with an elevated Bcl-2/Bax ratio (Gonzalez et al. 2002). These findings are not surprising because the transition from compensated hypertrophy to overt heart failure depends upon cell loss, including angiotensin II-mediated apoptosis.

12.4.6. DNA Repair

DNA damage has been shown to induce apoptosis through either the activation of p53 or initiation of a cascade leading to enhanced superoxide generation by the mitochondria. Because DNA damage is caused by cytotoxic drugs, toxic levels of natural substrates and oxidative stress mediated by ischemia, heart failure, diabetes etc, DNA damage has important pathological significance. If the damage is allowed to induce mutations, cardiomyopathies can develop (Marin-Garcia et al. 2001). However, mitochondrial DNA damage, independent of any mutation, can itself elevate oxidative stress by reducing mitochondrial protein expression and diverting electrons from the electron transport chain to oxygen (Ide et al. 2001; Ricci et al. 2005). Therefore, it is not surprising that repairing menadione-induced DNA damage prevents cytochrome c release from the mitochondria and activation of caspase 9 (Druzhyna et al. 2003). Similarly, overexpression of the DNA repair enzyme, 8-oxoguanine glycosylase, in the mitochondria of neonatal cardiomyocytes prevents palmitate-mediated cytochrome c release (Schaffer et al. 2005). Thus, repair of mitochondrial DNA damage can prevent not only apoptosis, but also mitochondrial dysfunction. Clearly, further studies are warranted to test the value of DNA repair as a therapeutic modality.

12.4.7. Antiapoptotic Clinical Interventions

Insufficient protection during open heart surgery may result in morbidity and long term functional impairments. Although this is likely to be associated with necrotic death, apoptosis also plays an important role (Suleiman et al. 2001). Several methods and interventions can be utilized to prevent or attenuate apoptosis in patients undergoing cardiac surgery (Khoynezhad et al. 2004). These include using beta-blockers, phosphodiesterase inhibitors and afterload reduction with medication or intraaortic balloon. Use of antiapoptotic agents, such as the caspase inhibitors, antioxidants, calcium-channel blockers and IGF-1, add to the arsenal of weapons capable of treating heart failure and preventing apoptosis in patients undergoing cardiac surgery.

12.5. Conclusion

Despite extensive research work on apoptosis, the potential role of apoptotic death of cardiac myocytes remains controversial. This is because adult cardiomyocytes are terminally differentiated and the myocardium is known for its poor ability to regenerate after an injury. Therefore apoptosis-induced cardiomyocyte loss is likely to dramatically alter cardiac structure and function. A better understanding of molecular events and of the cellular factors and signaling pathways that regulate apoptotic cell death in the heart is essential for formulating cardioprotective strategies. In this chapter, we have focused on the mitochondrial apoptotic pathway in the heart highlighting the importance and the role of different signaling pathways. External and internal stimulators of mitochondrial apoptotic pathway are becoming more extensive and expanding, thus supporting recent and mounting evidence that this pathway is indeed an important determinant of cardiomyocyte life and death. Examples of mitochondrial apoptotic pathway stimulators include hypoxia, ischemia/reperfusion, osmotic stress, stretch, diabetes, reactive oxygen species (of mitochondrial or cytoplasmic origin), Ca^{2+}, excessive nitric oxide, cytotoxic drugs, angiotensin II and catecholamines. The increasing number of apoptotic stimulators has triggered work aimed at combating such triggers. A number of agents have already been identified which can be used to prevent apoptotic death. These include caspase inhibitors, Bcl-2 family, IGF-1, heat shock proteins, calcium, antioxidants, nitric oxide synthetase inhibitors, ACE inhibitors, angiotensin II receptor antagonists and adrenergic receptor antagonists. These antiapoptotic agents may be used for therapeutic intervention. What is currently lacking is an improved understanding of cellular signaling in cardiomyocytes undergoing apoptosis in disease states (e.g. myocardial infarction, heart failure, myocarditis, arrhythmogenic right ventricular dysplasia, and immune rejection after cardiac transplantation). This will further improve our existing strategies to protect the heart in clinical settings.

References

Ace inhibitor myocardial infarction collaborative group (1998) Indications for ACE inhibitors in the early treatment of acute myocardial infarction: systematic overview of individual data from 100000 patients in randomized trials. Circulation 97: 2202–2212

Andreka P, Nadhazi Z, Muzes G, Sxantho G, Vandor L, Konya L, Turner MS, TulassayZ, Bishopric NH (2004a) Possible therapeutic targets in cardiac myocyte apoptosis. Curr Pharm Des 10(20): 2445–61

Andreka P, Tran T, Webster KA, Bishopric NH (2004b) Nitric oxide and promotion of cardiac myocyte apoptosis. Mol Cell Biochem 263(1–2): 35–53

Arnoult D, Gaume B, Karbowski M, Sharpe J, Cecconi F, Youle R (2003) Mitochondrial release of AIF and EndoG requires caspase activation downstream of Bax/Bak-mediated permeabilization. EMBO J 22(17): 4385–4399

Ascensao A, Magalhaes J, Soares JM, Ferreira R, Neuparth MJ, Marques F, Oliveira PJ, Duarte JA (2005) Moderate endurance training prevents doxorubicin-induced in vivo mitochondriopathy and reduces the development of cardiac apoptosis. Am J Physiol Heart Circ Physiol 289(2): H722–31

Baldi A, Abbate A, Bussani R, Patti G, Melfi R, Angelini A, Dobrina A, Rossiello R, Silvestri F, Baldi F, DiSciascio G (2002) Apoptosis and post-infarction left ventricular remodeling. J Mol Cell Cardiol 34(2): 165–1

Benjamin IJ, Schneider MD (2005) Learning from failure: congestive heart failure in the postgenomic age. J Clin Invest 115: 495–9

Bialik S, Cryns VL, Drimcic A, Miyata S, Wollowick AL, Srinivasan A, Kitsis RN (1999) The mitochondrial apoptotic pathway is activated by serum and glucose deprivation in cardiac myocytes. Circ Res 85: 403–14

Bojunga J, Nowak D, Mitrou P S, Hoelzer D, Zeuzem S, Cchow KU (2004) Antioxidative treatment prevents activation of death-receptor- and mitochondrion-dependent apoptosis in the hearts of diabetic rats. Diabetologia 47: 2072–80

Bortner CD, Cidlowski JA (1998) A necessary role for cell shrinkage in apoptosis. Biochem Pharmacol 56: 1549–1559

BorutaiteV, Brown GC (2003) Mitochondria in apoptosis of ischemic heart. FEBS Lett 541: 1–5

Bratton S, Walker G, Srinivasula S, Sun X, Butterworth M, Alnemri E, Cohen G (2001) Recruitment, activation and retention of caspases-9 and-3 by Apaf-1 apoptosome and associated XIAP complexes. Embo J 20: 998–1009

Byrne JA, Grieve DJ, Cave AC, Shan AM (2003) Oxidative stress and heart failure. Arch Mal Coeur Vaiss 96: 214–21

Caldarone CA, Barner EW, Wang L, Karimi M, Mascio CE, Hammel JM, Segar JL, Du C, Scholz TD (2004) Apoptosis-related mitochondrial dysfunction in the early postoperative neonatal lamb heart. Ann Thorac Surg 78: 948–955

Chen QM, Tu VC (2002) Apoptosis and heart failure: mechanisms and therapeutic implications. Am J Cardiovasc Drugs 2: 43–57

Chen Z, Chua CC, Ho YS, Hamdy RC, Chua BHL (2001) Overexpression of Bcl-2 attenuates apoptosis and protects against myocardial I/R injury in transgenic mice. Am J Physiol Heart Circ Physiol 280: H2313–H2320

Cheng W, Li B, Kajstura J, Li P, Wolin M, Sonnenblick E, Hintze T, Olivetti G, Anversa P (1995) Stretch-induced programmed myocyte cell-death. J Clin Invest 96: 2247–2259

Communal C, Singh K, Pimentel D, Colucci W (1998). Norepinephrine stimulates apoptosis in adult rat ventricular myocytes by activation of the beta-adrenergic pathway. Circulation 98: 1329–1334

Cook SA, Sugden PH, Clerk A. (1999) Regulation of bcl-2 family proteins during development and in response to oxidative stress in cardiac myocytes: association with changes in mitochondrial membrane potential. Circ Res 85: 940–949

Crow MT, Mani K, Nam YJ, Kitsis RN (2004) The mitochondrial death pathway and cardiac myocyte apoptosis. Circ Res 95: 957–970

Czerski L, Nunez G (2004) Apoptosome formation and caspase activation: is it different in the heart? J Mol Cell Cardiol 37: 643–652

d'Anglemont de Tassigny A, Souktani R, Henry P, Ghaleh B, Berdeaux A (2004) Volume-sensitive chloride channels ($I_{Cl,vol}$) mediated doxorubicin-induced apoptosis through apoptotic volume decrease in cardiomyocytes. Fundam Clin Pharmacol 18: 531–539

de Moissac D, Gurevich RM, Zheng H, Singal PK, Kirshenbaum LA (2000) Caspase activation and mitochondrial cytochrome C release during hypoxia-mediated apoptosis of adult ventricular myocytes. J Mol Cell Cardiol 32: 53–63

Deply E, Hatem SN, Andrieu N, De Vaumas C, Henaff M, Rucker-Martin C, Ajffrezou JP, Laurent G, Levade T, Mercadier JJ (1999) Doxorubicin induces slow ceramide accumulation and late apoptosis in cultured adult rat ventricular myocytes. Cardiovasc Res 43: 398–407

Diez J, Fortuno M, Zalba G, Etayo J, Fortuno A, Ravassa S, Beaumont J (1998) Altered regulation of smooth muscle cell proliferation and apoptosis in small arteries of spontaneously hypertensive rats. Eur Heart J 19: G29–G33

Doonan F, Cotter TG (2004) Apoptosis: A potential therapeutic target for retinal degenerations. Curr Neurovasc Res 1: 41–53

Druzhyna NM, Hollensworth SB, Kelley MR, Wilson GL, Ledoux SP (2003) Targeting human 8-oxoguanine glycosylase to mitochondria of oligodendrocytes protects against menadione-induced oxidative stress. Glia 42: 370–378

Ekhterae D, Lin Z, Lundberg MS, Crow MT, Brosius F C III, Nunez G (1999) ARC inhibits cytochrome c release from mitochondria and protects against hypoxia-induced apoptosis in heart-derived H9c2 cells. Circ Res 85: e70–7

Fischer U, Schulze-Osthoff K (2005) New approaches and therapeutics targeting apoptosis in disease. Pharmacol Rev 57: 187–215

Flather MD, Yusuf S, Kober L, Pfeffer M, Hall A, Murray G, Torppedersen C, Ball S, Pogue J, Moye L, Braunwald E (2000) Long-term ACE inhibitor therapy in patients with heart failure or left-ventricular dysfunction: a systematic overview of dtaa from individual patients. The Lancet 355: 1575–1581

Frustaci A, Kajstura J, Chimenti C, Jakoniuk I, Leri A, Maseri A, Nadal-Ginard B, Anversa P (2000) Myocardial cell death in human diabetes. Circ Res 87: 1123–1132

Fumarola C, Guidotti GG (2004) Stress-induced apoptosis: toward a symmetry with receptor–mediated cell death. Apoptosis 9: 77-82

Gill C, Mestril R, Samali A (2002) Losing heart: the role of apoptosis in heart disease--a novel therapeutic target? FASEB J 16: 135-46

Gonzalez A, Lopez B, Ravassa S, Querejeta R, Larman M, Diez J, Fortuno MA (2002) Stimulation of apoptosis in essential hypertension: potential role of angiotensin II. Hypertension 39: 75–80

Gottlieb R, Burleson K, Kloner R, Babior B, Engler R (1994) Reperfusion injury induces apoptosis in rabbit cardiomyocytes. J Clin Invest 94: 1621–1628

Goussev A, Sharov VG, Shimoyama H, Tanimura M, Lesch M, Goldstein S, Sabbah HN (1998) Effects of ACE inhibition on cardiomyocyte apoptosis in dogs with heart failure. Am J Physiol Heart Circ Physiol 275: H626–H631

Green PS, Leeuwenburg C (2002) Mitochondrial dysfunction is an early indicator of doxorubucin-induced apoptosis. Biochim Biophys Acta 1588: 94–101

Grishko V, Pastukh V, Solodushko V, Gillespie M, Azuma J, Schaffer S (2003) Apoptotic cascade initiated by angiotensin II in neonatal cardiomyocytes: role of DNA damage. Am J Physiol Heart Circ Physiol 285: H2364–H2372

Gupta S, Knowlton AA (2002) Cytosolic heat shock protein 60, hypoxia, and apoptosis. Circulation 106: 2727–33

Gurevich RM, Regula KM, Kirshenbaum LA (2001). Serpin protein CrmA suppresses hypoxia-mediated apoptosis of ventricular myocytes. Circulation 103: 1984–91

Harada K, Sugaya T, Murakami K, Yazaki Y, Komuro I (1999) Angiotensin II type 1A receptor knockout mice display less left ventricular remodeling and improved survival after myocardial infarction. Circulation 100: 2093–2099

Heinrich H, Holz J (1998) Myocardial apoptosis in the overloaded and the aging heart: a critical role of mitochondria? Eur Cytokine Netw 9: 693–5

Hetts SW (1998). To die or not to die: an overview of apoptosis and its role in disease. JAMA 279: 300–7

Hollander JM, Lin KM, Scott BT, Dillmann WH (2003) Overexpression of PHGPx and HSP60/10 protects against ischemia/reoxygenation injury. Free Radic Biol Med 35: 742–51

Hu Y, Benedict M, Ding L, Nunez G (1999) Role of cytochrome c and dATP/ATP hydrolysis in Apaf-1-mediated caspase-9 activation and apoptosis. Embo J 18: 3586–3595

Ide T, Tsutsui H, Hayashidani S, Kang D, Suematsu N, Nakamura KE, Utsumi H, Hamasaki N, Takeshita A (2001) Mitochondrial DNA damage and dysfunction associated with oxidative stress in failing hearts after myocardial infarction. Circ Res 88: 529–555

James TN, St Martin E, Willis P W III, Lohr TO (1996) Apoptosis as a possible cause of gradual development of complete heart block and fatal arrhythmias associated with absence of the AV node, sinus node and internodal pathways. Circulation 93: 1424–1438

Jang YM, Kendaiah S, Drew B, Phillips T, Selman C, Julian D, Leeuwenburg C (2004) Doxorubicin treatment in vivo activates caspase-12 mediated cardiac apoptosis in both male and female rats. FEBS Lett 577: 483–490

Kajstura J, Cheng W, Sarabgarajan R, Li P, Li B, Nitahara J, Chapnick S, Reiss K, Olivetti G, Anversa P (1996) Necrotic and apoptotic myocyte cell death in the aging heart of Fischer 344 rats. Am J Physiol Heart Circ Physiol 40: H1215–H1228

Kajstura J, Cigola E, Malhotra A, Li P, Cheng W, Meggs LG, Anversa P (1997) Angiotensin II induces apoptosis of adult ventricular myocytes in vitro. J Mol Cell Cardiol 29: 859–870

Kawaguchi H, Shin WS, Wang Y, Inukai M, Kato M, Matsuo-Okai Y, Sakamoto A, Uehara Y, Kaneda Y, Toyo-Oka T (1997) In vivo gene transfection of human endothelial cell nitric oxide synthase in cardiomyocytes causes apoptosis-like cell death. Identification using Sendai virus-coated liposomes. Circulation 95: 2441–7

Khoynezhad A, Jalali Z, Tortolani AJ (2004) Apoptosis: pathophysiology and therapeutic implications for the cardiac surgeon. Ann Thorac Surg 78: 1109–18

Kirchhoff SR, Gupta S, Knowlton AA (2002) Cytosolic heat shock protein 60, apoptosis, and myocardial injury. Circulation 105: 2899–904

Knowlton AA, Kapadia S, Torre-Amione G, Durand JB, Bies R, Young J, Mann DL (1998) Differential expression of heat shock proteins in normal and failing human hearts. J Mol Cell Cardiol 30: 811–8

Kumar D, Kirshenbaum L, Li T, Danelisen I, Singal P (1999) Apoptosis in isolated adult cardiomyocytes exposed to adriamycin. Heart in Stress 874: 156–168

Leri A, Claudio P, Li Q, Wang X, Reiss K, Wang S, Malhotra A, Kajstura J, Anversa P (1998) Stretch-mediated release of angiotensin II induces myocyte apoptosis by activating p53 that enhances the local renin-angiotensin system and decreases the Bcl-2-to-Bax protein ratio in the cell. J Clin Invest 101: 1326–1342

Leri A, Liu Y, Li B, Fiordaliso F, Malhotra A, Latini R, Kajstura J, Anversa P (2000) Up-regulation of AT_1 and AT_2 receptors in postinfarcted hypertrophied myocytes and stretch-mediated apoptotic cell death. Am J Pathol 156: 1663–1672

Li JM, Shah AM (2004) Endothelial cell superoxide generation: regulation and relevance for cardiovascular pathophysiology. Am J Physiol Regul Integr Comp Physiol 287: R1014–30

Li L, Luo L, Wang X (2001) Endonuclease G is an apoptotic DNase when released from mitochondria. Nature 412: 95–99

Li P, Nijhawan D, Budihardjo I, Srinivasula S, Ahmad M, Alnemri E, Wang X (1997) Cytochrome c and dATP-dependent formation of Apaf-1/caspase-9 complex initiates an apoptotic protease cascade. Cell 91: 479–489

Liao XD, Wang XH, Jin HJ, Chen LY, Chen Q (2004) Mechanical stretch induces mitochondria-dependent apoptosis in neonatal rat cardiomyocytes and G2/M accumulation in cardiac fibroblasts. Cell Res 14: 16–26

Liu P, Xu B, Forman LJ, Carsia R, Hock CE (2002) L-NAME enhances microcirculatory congestion and cardiomyocyte apoptosis during myocardial ischemia-reperfusion in rats. Shock 17: 185–92

Liu X, Chua CC, Gao J, Chen Z, Landy CLC, Hamdy R, Chua BHL (2004) Pifithrin-a protects against doxorubicin-induced apoptosis and acute cardiotoxicity in mice. Am J Physiol Heart Circ Physiol 286: H933–H939

Logue SE, Gustafsson AB, Samali A, Gottlieb RA (2005) Ischemia/reperfusion injury at the intersection with cell death. J Mol Cell Cardiol 38: 21–33

Lou H, Danielsen I, Singal PK (2005) Involvement of mitogen-activated protein kinases in adriamycin-induced cardiomyopathy. Am J Physiol Heart Circ Physiol 288: H1925-H1930

Lundberg KC, Szweda LI (2004) Initiation of mitochondrial-mediated apoptosis during cardiac reperfusion. Arch Biochem Biophys 432: 50–7

Malhotra R, Brosius FC III (1999) Glucose uptake and glycolysis reduce hypoxia-induced apoptosis in cultured neonatal rat cardiac myocytes. J Biol Chem 274: 12567–75

Mallat Z, Tedgui A, Fontaliran F, Frank R, Durigom M, Fontaine G (1996) Evidence of apoptosis in arrhythmogenic right ventricular dysplasia. Circulation 94: 2493–2493

Marin-Garcia J, Goldenthal MJ, Moe GW (2001) Mitochondrial pathology in cardiac failure. Cardiovasc Res 49: 17–26

Mockridge JW, Benton EC, Andreeva LV, Latchman DS, Marber MS, Heads RJ (2000) IGF-1 regulates cardiac fibroblast apoptosis induced by osmotic stress. Biochem Biophys Res Commun 273: 322–7

Morales MP, Galvez A, Eltit JM, Ocaranza P, Diaz-Araya G, Lavandero S (2000) IGF-1 regulates apoptosis of cardiac myocyte induced by osmotic-stress. Biochem Biophys Res Commun 270: 1029–35

Mozaffari MS, Schaffer SW (2003) Effect of hypertension and hypertension-glucose intolerance on myocardial ischemic injury. Hypertension 42: 1042–1049

Namiki A, Brogi E, Kearney M, Kim E, Wu T, Varticovski L, Isner J (1995) Hypoxia induces vascular endothelial growth-factor nRNA expression and protein-production in human endothelial-cells in-vitro. Circulation 92: 527–527

Narula J, Haider N, Virmani R, DiSalvo T, Kolodgie F, Hajjar R, Schmidt U, Semigran M, Dec G, Khaw B (1996) Apoptosis in myocytes in end-stage heart failure. New England J Med 335: 1182–1189

Narula J, Pandey P, Arbustini E, Haider N, Narula N, Kolodgie FD, Dal Bello B, Semigran MJ, Bielsa-Masdeu A, Dec GW, Israels S, Ballester M, Virmani R, Saxena S, Kharbanda S (1999) Apoptosis in heart failure: release of cytochrome c from mitochondria and activation of caspase-3 in human cardiomyopathy. Proc Natl Acad Sci USA 96: 8144–9

Nitahara J, Cheng W, Liu Y, Li B, Leri A, Li P, Mogul D, Gambert S, Kajstura J, Anversa P (1998) Intracellular calcium, DNase activity and myocyte apoptosis in aging Fischer 344 rats. J Mol Cell Cardiol 30: 519–535

Nitobe J, Yamaguchi S, Okuyama S, Nozaki N, Sata M, Miyamoto T, Takeishi Y, Kubota I, Tomoike H (2003) Reactive oxygen species regulate FLICE inhibitory protein (FLIP) and susceptibility to Fas-mediated apoptosis in cardiac myocytes. Cardiovasc Res 57: 119–128

Novogovodov SA, Szulc ZM, Luberto C, Jones JA, Bielawski J, Bielawska A, Hannun YA, Obeid LM (2005) Positively charged ceramide is a potent induced of mitochondrial permeabilization. J Biol Chem 280: 16096–16015

Olivetti G, Abbi R, Quaini F, Kajstura J, Cheng W, Nitahara J, Quaini E, DiLoreto C, Beltrami C, Krajwski S, Reed J, Anversa P (1997) Apoptosis in the failing human heart. New England J Med 336: 1131–1141

Pacher P, Csordas G, Hajnoczky G (2001) Mitochondrial Ca^{2+} signaling and cardiac apoptosis. Biol Signals Recept 10: 200–23

Qin F, Shite J, Mao W, Liang CS (2003) Selegiline attenuates cardiac oxidative stress and apoptosis in heart failure: association with improvement of cardiac function. Eur J Pharmacol 461: 149–58

Rabkin S, Kong J (2000) Nifedipine does not induce but rather prevents apoptosis in cardiomyocytes. Eur J Pharmacol 388: 209–217

Rayment N, Haven A, Madden B, Murday A, Trickey R, Shipley M, Davies M, Katz D (1999) Myocyte loss in chronic heart failure. J Pathol 188: 213–219

Ravassa S, Fortuno MA, Gonzalez A, Lopez B, Zalba G, Fortuno A, Diez J (2000) Mechanisms of increased susceptibilityof angiotensin II-induced apoptosis in ventricular cardiomyocytes of spontaneously hypertensive rats. Hypertension 36: 1065–1071

Razavi HM, Hamilton JA, Feng, Q (2005) Modulation of apoptosis by nitric oxide: implications in myocardial ischemia and heart failure. Pharmacol Ther 106: 147–62

Regula KM, Ens K, Kirshenbaum LA (2002) Inducible expression of BNIP3 provokes mitochondrial defects and hypoxia-mediated cell death of ventricular myocytes. Circ Res 91: 226–31

Ricci C, Pastukh V, Schaffer SW (2005) Involvement of the mitochondrial permeability transition pore in angiotensin II-mediated apoptosis. Exp Clin Cardiol 10: 160–164

Richter C, Schweizer M, Cossarizza A, Franceschi C (1996) Control of apoptosis by the cellular ATP level. FEBS Lett 378: 107–110

Riedl SJ, Shi Y (2004) Molecular mechanisms of caspase regulation during apoptosis. Nature Rev Mol Cell Biol 5: 897–907

Rossig L, Hoffmann J, Hugel B, Mallat Z, Haase A, Freyssinet JM, Tedgui A, Aicher A, Zeiher AM, Dimmeler S (2001) Vitamin C inhibits endothelial cell apoptosis in congestive heart failure. Circulation 104: 2182–7

Saito S, Hiroi Y, Zou Y, Aikawa R, Toko H, Shibasaki F, Yazaki Y, Nagai R, Komuro I (2000) Beta-adrenergic pathway induces apoptosis through calcineurin activation in cardiac myocytes. J Biol Chem 275: 34528–33

Saraste A, Pulkki K, Kallajoki M, Heikkila P, Laine P, Nieminen M, Mattila S, Parvinem M, VoipioPulkki L (1997) Cardiomyocyte apoptosis is observed in explanted failing human hearts with and without coronary artery disease. Circulation 96: 651–651

Saraste A, Pulkki K, Kallajoki M, Henriksen K, Parvinen M, VoipioPulkki L (1997) Apoptosis in human acute myocardial infarction. Circulation 95: 320–323

Saraste A, VoipioPulkki L, Parvinen M, Pulkki K (1997) Apoptosis in the heart. New England J Med 336: 1025–1026

Schaffer S, Ricci C, Pastukh V, Wilson G (2005) DNA damage-involvement in fatty acid-mediated apoptosis. J Mol Cell Cardiol 38, 859

Scheubel RJ, Bartling B, Simm A, Silber RE, Drogaris K, Darmer D, Holtz J (2002) Apoptotic pathway activation from mitochondria and death receptors without caspase-3 cleavage in failing human myocardium: fragile balance of myocyte survival? J Am Coll Cardiol 39: 481–8

Sharov VG, Sabbah HN, Ali AS, Shimoyama H, Lesch M, Goldstein S (1997) Abnormalities of cardiocytes in regions bordering fibrous scars of dogs with heart failure. Int J Cardiol 60: 273–9

Shizukuda Y, Buttrick P, Geenen D, Borczuk A, Kitsis R, Sonnenblick E (1998) beta-Adrenergic stimulation causes cardiocyte apoptosis: influence of tachycardia and hypertrophy. Am J Physiol Heart Circ Physiol 44: H961–H968

Sorescu D, Griendling KK (2002) Reactive oxygen species, mitochondria, and NAD(P)H oxidases in the development and progression of heart failure. Congest Heart Fail 8: 132–40

Sugino H, Ozono R, Kurisu S, Matsuuura H, Ishida M, Oshima T, Kambe M, Teranishi Y, Masaki H, Matsubara H (2001) Apoptosis is not increased in myocardium overexpressing type 2 angiotensin II receptor in transgenic mice. Hypertension 37: 1394–1398

Suleiman MS, Halestrap AP, Griffiths EJ (2001) Mitochondria: a target for myocardial protection. Pharmacol Ther 89: 29–46

Sun HY, Wang NP, Halkos ME, Kerendi F, Kin H, Wang RX, Guyton RA, Zhao ZQ (2004) Involvement of Na^+/H^+ exchanger in hypoxia/re-oxygenation-induced neonatal rat cardiomyocyte apoptosis. Eur J Pharmacol 486: 121–31

Susin S, Lorenzo H, Zamzami N, Marzo I, Snow B, Brothers G, Mangion J, Jacotot E, Costantini P, Loeffler M, Larochette N, Goodlett D, Aebersold R, Siderovski D, Penninger J, Kroemer G (1999) Molecular characterization of mitochondrial apoptosis-inducing factor. Nature 397: 441–446

Szabolcs M, Michler RE, Yang X, Aji W, Roy D, Athan E, Sciacca RR, Minanova OP, Cannon PJ (1996) Apoptosis of cardiac myocytes during cardiac allograft rejection: relation to induction of nitric oxide synthase. Circulation 94: 1665–1673

Takatani T, Takahashi K, Uozumi Y, Shikata E, Yamamoto Y, Ito T, Matsuda T, Schaffer SW, Fujio Y, Azuma J (2004) Taurine inhibits apoptosis by preventing formation of the Apaf-1/caspase-9 apoptosome. Am J Physiol Cell Physiol 287: C949–53

Tanaka K, Pracyk JB, Takeda K, Yu ZX, Ferrans V J, Deshpande SS, Ozaki M, Hwang PM, Lowenstein CJ, Irani K, Finkel T (1998) Expression of Id1 results in apoptosis of cardiac myocytes through a redox-dependent mechanism. J Biol Chem 273, 25922–8

Tanaka M, Ito H, Adachi S, Akimoto H, Nishikawa T, Kasajima T, Marumo F, Hiroe M (1994) Hypoxia induces apoptosis with enhanced expression of fas antigen messenger-RNA in cultured neonatal rat cardiomyocytes. Circulation 90: 426–426

Tatsumi T, Shiraishi J, Keira N, Akashi K, Mano A, Yamanaka S, Matoba S, Fushiki S, Fliss H, Nakagawa M (2003) Intracellular ATP is required for mitochondrial apoptotic pathways in isolated hypoxic rat cardiac myocytes. Cardiovasc Res 59: 428–40

Thornberry N, Lazebnik Y (1998) Caspases: Enemies within. Science 281: 1312–1316.

Valen G (2003) The basic biology of apoptosis and its implications for cardiac function and viability. Ann Thorac Surg 75: S656–60

Van Loo G, Saelens X, Van Gurp M, MacFarlane M, Martin S, Vandenabelle P (2002) The role of mitochondrial factors in apoptosis: a Russian roulette with more than one bullet. Cell Death and Differentiation 9: 1031–1042

Von Harsdorf R, Li PF, Dietz R (1999) Signaling pathways in reactive oxygen species-induced cardiomyocyte apoptosis. Circulation 99: 2934–41

Wang GW, Klein JB, Kang YJ (2001) Metallothionein inhibits doxorubicin-induced mitochondrial cytochrome c release and caspase-3 activation in cardiomyocytes. J Pharmacol Exp Ther 298: 461–468

Wang GW, Zhou Z, Klein JB, Kang YJ (2001) Inhibition of hypoxia/reoxygenation-induced apoptosis in metallothionein-overexpressing cardiomyocytes. Am J Physiol Heart Circ Physiol 280: H2292–9

Webster K, Discher D, Kaiser S, Hernandez O, Sato B, Bishopric N (1999) Hypoxia-activated apoptosis of cardiac myocytes requires reoxygenation or a pH shift and is independent of p53. J Clin Invest 104: 239–252

Yaoita H, Ogawa K, Maehara K, Maruyama Y (2000) Apoptosis in relevant clinical situations: contribution of apoptosis in myocardial infarction. Cardiovasc Res 45: 630–41

Yue T, Sanjay K, Feng G, Louden C, Wang C, Gu J, Lee J, Feuerstein G, Ma X (1998) Inhibition of P38 MAP kinase decreases cardiomyocyte apoptosis and improves cardiac function after myocardial ischemia and reperfusion. Naunyn-Schiedebergs Arch Pharmacol 358: R621-R621

Zou H, Henzel W, Liu X., Lutschg A, Wang X (1997) Apaf-1, a human protein homologous to C-elegans CED-4, participates in cytochrome c-dependent activation of caspase-3. Cell 90: 405–413

Zou H, Li Y, Liu H, Wang X (1999) An APAF-1 center dot cytochrome c multimeric complex is a functional apoptosome that activates procaspase-9. J Biol Chem 274: 11549–11556

Zuurbier CJ, Eerbeek O, Meijer AJ (2005) Ischemic preconditioning, insulin, and morphine all cause hexokinase redistribution. Am J Physiol Heart Circ Physiol 289: H496–9

13
The Role of Mitochondria in Necrosis Following Myocardial Ischemia-Reperfusion

Elizabeth Murphy and Charles Steenbergen

13.1. Introduction

Using the recently adopted nomenclature of cell death (Levin et al. 1999), necrosis refers to cell death and disintegration, regardless of the pathway; the modifiers apoptotic and oncotic are used to refer to the mechanism of death. Traditionally cell death following ischemia has been thought to be primarily oncotic necrosis. As detailed in numerous reviews, during ischemia when the ATP falls to very low levels, the ion pumps cannot function resulting in a rise in Ca^{2+} which further consumes ATP (Farber and Gerson 1984; Jennings et al. 1990; Jennings and Reimer 1981; Jennings and Steenbergen 1985). The rise in Ca^{2+} during ischemia and reperfusion leads to mitochondrial Ca^{2+} accumulation, particularly during reperfusion when oxygen is reintroduced. Reintroduction of oxygen provides a terminal electron acceptor (oxygen) allowing electron transport to occur; however damage to electron transport chain can lead to increased mitochondrial generation of reactive oxygen species (ROS). The resulting Ca^{2+} overload of mitochondria, and increased ROS further exacerbates mitochondrial dysfunction and can result in opening of the mitochondrial permeability transition pore (mPTP), which further compromises cellular energetics (Halestrap 2005; Kroemer et al. 1998; Schneider 2005). The resultant low ATP and altered ion homeostasis result in rupture of the plasma membrane and irreversible, oncotic (necrotic) cell death (Jennings and Reimer 1981; Jennings and Steenbergen 1985). Mitochondria have long been proposed as central players in oncotic (necrotic) cell death, since the mitochondria are central to synthesis of both ATP and ROS and since mitochondrial and cytosolic Ca^{2+} overload are key components of oncotic (necrotic) cell death. More recently it has been demonstrated that apoptotic cell death can occur during I/R (Anversa et al. 1998; Gottlieb et al. 1994), although the precise percentage of apoptosis versus oncosis has been debated, and may depend on the precise conditions (Rodriguez and Schaper 2005). In addition, it has recently been recognized that oncosis can be a regulated process (Zong and Thompson 2006). For example there are some recent data showing that anti-apoptotic proteins such as Bcl-2 may reduce apoptosis as well as necrosis. This review will cover three topics: 1) it will review the role of mitochondria in oncotic necrosis, 2) it will discuss the relationship between oncosis and apoptosis in ischemia-reperfusion mediated cell death, and 3) it will evaluate the role of mitochondria in reducing oncotic necrosis.

13.2. Mitochondrial and Oncosis

13.2.1. Mitochondrial Changes During Ischemia

Ischemic injury has long been divided into reversible and irreversible phases. In the well characterized in vivo canine model (Jennings and Reimer 1981) the reversible phase (first 15 min of severe ischemia) is characterized by stimulation of anaerobic glycolysis, an increase in lactate, a decrease in glycogen, a decrease in ATP and pH, and an increase in inosine, adenosine and hypoxanthine. If flow is restored during this time period the myocytes will recover and ATP and other metabolites will return to normal levels in a few days. If the ischemia persists longer than 20 minutes, increasing number of myocytes enter the irreversible phase in which glycolysis slows and then stops, ATP falls to levels less than 2 μmol/g, there is clumping of nuclear chromatin, the mitochondria are swollen and contain amorphous matrix densities, and there are defects in the plasmalemmal membrane (Jennings and Reimer 1981). Many of the changes occurring in irreversible, necrotic (or oncotic) cells can be attributed to alteration in mitochondrial function. Cardiomyocytes depend almost entirely on oxidative metabolism for the generation of ATP, and commensurate with their importance, mitochondria comprise greater than 30% of the myocyte volume. During ischemia, in the absence of oxygen, electron transport and mitochondrial ATP generation is inhibited (Figure 13.1). Thus many of the well characterized changes in ischemia, such as the fall in ATP and the increase in anaerobic glycolysis with the resultant generation of lactate and fall in pH, result from the inhibition of mitochondrial electron transport. In addition, as described in the seminal paper by Jennings and coworkers, mitochondria in oncotic cells are swollen and have characteristic amorphous densities (Jennings and Ganote 1976). Furthermore, since ATP is used to fuel contraction and ion pumps, the loss of ATP leads to loss of muscle contraction and dysregulation of ion gradients. There are considerable data showing that the fall in pH during ischemia (increase in [H^+]) stimulates Na-H exchange (NHE), and due to the fall in ATP, the Na^+ that enters via NHE cannot be extruded fully by the Na-K-ATPase, resulting in an increase in cytosolic Na^+ which stimulates reverse mode Na-Ca exchange (NCX) leading to an increase in cytosolic Ca^{2+} (Karmazyn 1988; Murphy et al. 1991). This increase in cytosolic Ca^{2+} is causally associated with irreversible ischemic injury; mechanisms, such as NHE inhibitors, that reduce the rise in cytosolic Ca^{2+} during ischemia reduce oncosis (Murphy et al. 1991; Steenbergen et al. 1987; Steenbergen et al. 1990). During reperfusion, the mitochondria are reenergized and some of the excess calcium in the cytosol can be imported into the mitochondria and cause altered

13. The Role of Mitochondria in Necrosis Following Ischemia-Reperfusion   293

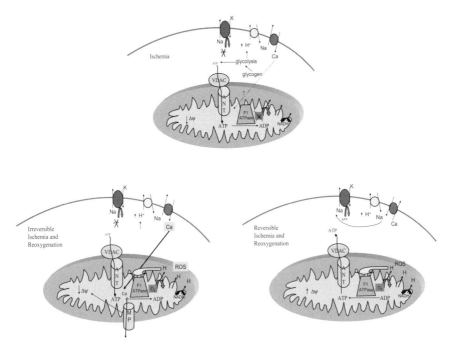

FIGURE 13.1. During ischemia, due to lack of oxygen, electron transport is inhibited thereby decreasing oxidatively generated ATP. ATP is produced by breakdown of glycogen. However this glycolytically generated ATP is consumed by reverse mode of the F_1-F_0-ATPase following entry of ATP into the mitochondria via VDAC and ANT. The breakdown of ATP and generation of lactate results in a decrease in pH that stimulates Na-H exchange leading to an increase in intracellular Na^+ because the decrease in ATP limits activity of the Na-K ATPase. The increase in Na^+ stimulates Ca^{2+} entry via the Na-Ca exchanger. When oxygen is re-introduced during reperfusion, the electron transport resumes and can lead to generation of ROS. Furthermore at the start of reperfusion there is additional Ca^{2+} entry via Na-Ca exchange and with a repolarized mitochondrial membrane potential Ca^{2+} is accumulated by the mitochondria. If the level of Ca^{2+} overload and ROS generation reach a critical threshold they can lead to sustained activation of the mPTP which will depolarize the mitochondrial membrane resulting in further degradation of ATP and irreversible injury.

mitochondrial function. Furthermore mitochondria are a primary site of ROS generation and damage to the mitochondrial electron transport chain can lead to inefficient electron transport and increased generation of ROS during reperfusion. There is considerable data showing defects in complex I during ischemia-reperfusion (Paradies et al. 2004). It is also interesting that defects in complex I have been suggested to play a role in cell death due to chronic diseases such as Parkinson's and some cardiomyopathies (Dawson and Dawson 2003; Maloyan et al. 2005).

13.2.2. Mitochondrial Energetics and Transporters

During ischemia the mitochondrial membrane potential has also been shown to decline (Di Lisa and Bernardi 1998). ATP generated by glycolysis appears to be consumed via reversal of the F_1F_0-ATPase in an attempt to maintain the mitochondrial membrane potential. In slow heart rate animals, an inhibitor of the F_1F_0-ATPase has been shown to translocate to the mitochondria during ischemia which would inhibit ATP consumption by the mitochondria during ischemia (Rouslin and Broge 1996). There is also an extensive literature (Griffiths and Halestrap 1993; Halestrap et al. 2004; Halestrap et al. 1997; Zamzami et al. 2005) reporting that opening of the mPTP occurs during ischemia-reperfusion and may play a major role in the process of cell death. The mPTP is a large conductance mega-channel which is poorly defined, but has been suggested to be composed of the adenine nucleotide translocator (ANT), the voltage dependent anion channel (VDAC) and cyclophilin D. Prolonged opening of the mPTP would collapse the mitochondrial membrane potential, release calcium stored in the mitochondria, and deplete cell ATP. Opening of the mPTP has been reported to play an important role in both oncotic (necrotic) and apoptotic cell death (Kroemer et al. 1998; Schneider 2005). Somewhat surprisingly, transient opening of the mPTP has been suggested to be required for preconditioning (Hausenloy et al. 2004). Opening of mPTP is stimulated by many conditions that occur during ischemia such as high Ca, oxidative stress and high inorganic phosphate. Cyclosporin, which binds to cyclophilin, is a commonly used inhibitor of the mPTP. Griffiths and Halestrap (1993) have shown that inhibition of the mPTP with cyclosporin reduces ischemia-reperfusion injury. As discussed later, there are some intriguing recent data suggesting that loss of cyclophilin protects against oncotic ischemia-reperfusion mediated cell death (Baines et al. 2005; Nakagawa et al. 2005). Interestingly, loss of cyclophilin did not block apoptosis related to overexpression of BAX (Baines et al. 2005). Taken together, there is considerable data to suggest that mitochondrial dysfunction directly leads to oncosis.

Although loss of ATP plays a major role in the events leading to oncosis, the level of ATP has been suggested also to be a key determinant of whether a cell undergoes apoptosis or oncosis during ischemia-reperfusion (Shiraishi et al. 2001). ATP is required for apoptosis. Thus if ATP falls below a critical level necessary for apoptosis, the process may convert to a process which more closely resembles oncosis. Differences in ATP levels may also be important for some of the morphologic differences between cells undergoing oncosis and apoptosis. A feature of oncosis is cell swelling, largely related to failure of ion pumps and electrolyte imbalances, whereas cell shrinkage is a feature of apoptosis, and requires active ion transport. Thus the ability to generate ATP may be necessary for many of the characteristics of classic apoptosis, and although there may be sufficient ATP in the early stages of ischemia to support classic apoptosis, at the time when cell death is occurring during ischemia, there may be insufficient ATP to complete the apoptotic process, and the dying cells may come to resemble oncotic cells. Alternatively, there could be sufficient ATP synthesis during reperfusion to permit apoptosis if the duration of ischemia is short enough that mitochondria are intact and can function to generate ATP in the immediate reperfusion period.

13.3. Oncotic Versus Apoptotic Death During Ischemia and Reperfusion

As mentioned, there are two primary forms of cell death, oncotic necrosis and apoptotic necrosis, both of which seem to occur during myocardial ischemia-reperfusion injury (Anversa et al. 1998; Bartling et al. 1998; Imahashi et al. 2004; Zong and Thompson 2006). The distinction between these pathways can sometimes be blurred and mitochondria are key elements of both modes of cell death. The primary reason for trying to understand the relative contribution of these two forms of cell death is that different interventions may be more beneficial for preventing or minimizing each form of cell death, and ultimately, the goal of any intervention is to minimize the number of dead myocytes because clinical outcome is most closely correlated with infarct size. It is commonly stated that apoptosis is a programmed cell death, whereas oncotic necrosis is accidental and unregulated. Clearly apoptosis that occurs during development is a highly regulated process that appears to be genetically programmed, and in some experimental models, can be prevented by selective mutation. However, the response of a cell or organ to an accidental insult such as ischemia-reperfusion does not appear to be genetically controlled. Cell death as a result of a sustained severe accidental insult cannot be prevented by any known genetic manipulation although it can be delayed by a variety of genetic alterations. Furthermore, there are data suggesting that the response to accidental insults can be both apoptotic and oncotic, and in fact many agents such as ischemia-reperfusion result in both apoptotic and oncotic cell death. Many toxic agents at low doses kill primarily by apoptosis whereas at higher doses they kill by oncosis. Interestingly these same agents or manipulations at even lower doses can induce protection. Furthermore, there are recent data suggesting that oncotic necrosis can be regulated (Zong and Thompson 2006). Thus in the case of accidental insults, the response can be either apoptotic or oncotic and both types of death involve mitochondrial pathways and both can be inhibited by manipulating similar mitochondrial signaling pathways.

Recent studies have suggested that classic inhibitors of apoptosis, such as Bcl-2, can reduce what appears to be oncotic cell death. Imahashi et al. (Imahashi et al. 2004) have shown that cardiac specific overexpression of Bcl-2 significantly reduces necrosis, measured by triphenyl tetrazolium chloride (TTC) staining after two hours of reperfusion. Necrosis was $38 \pm 5\%$ in wild type hearts and was significantly lower at $18 \pm 3\%$ in hearts with overexpression of Bcl-2. Apoptosis as assessed by TUNEL staining was also reduced in hearts with Bcl-2 overexpression ($1.2 \pm 0.9\%$) compared to wild type hearts ($4.3 \pm 1.7\%$). However, the percentage of apoptosis as measured by TUNEL was very small compared to the total cell death. For example in wild type hearts 38% of the heart was dead as measured by TTC, yet only 4.3% of the death was apoptotic. With overexpression of Bcl-2, 18% of the heart was dead as measured by TTC, but this reduction in cell death cannot be accounted for solely by a reduction in apoptosis, since apoptosis was only 4% in wild type hearts. These data suggest that Bcl-2 also reduces what is commonly classified as necrotic or oncotic cell death. These results are consistent with the findings of Chen et al. (2001); although in the study of Chen et al., it is difficult to compare the percentage of total death (TTC staining) to apoptotic death because apoptosis was measured after 30 minutes of ischemia while TTC measurements were made after 50 minutes of ischemia. It has been suggested that Bcl-2 can reduce necrosis because it blocks release of cytochrome c, which will lead to depletion of ATP and necrosis (Kuznetsov et al. 2004). This would suggest that release of cytochrome c is part of the oncotic pathway. It has also been suggested that Bcl-2 can directly alter mitochondrial Ca^{2+} uptake, possibly by altering the activity of mitochondrial NCX (Zhu et al. 2001), which could play a role in reducing oncotic cell death as well as apoptotic cell death. It has also been suggested that Bcl-2 might bind to other mitochondrial proteins such as VDAC and alter consumption of glycolytically generated ATP during ischemia (Imahashi et al. 2004). During ischemia, the mitochondrial membrane potential falls and glycolytically generated ATP is consumed by the mitochondria due to reversal of the mitochondrial F_1F_0-ATPase. DiLisa and coworkers (Di Lisa and Bernardi 1998; Di Lisa et al. 1995) and others (Leyssens et al. 1996) have shown that inhibition of the F_1F_0-ATPase with oligomycin results in a larger and faster decline in the mitochondrial membrane potential, but a slower decline in ATP and reduced generation of lactate. Imahashi et al. (2004) have shown that overexpression of Bcl-2 has an effect that is similar to oligomycin; overexpression of Bcl-2 reduces the rate of decline in ATP and the fall in pH during ischemia. Furthermore the effects of Bcl-2 and oligomycin were not additive. Bcl-2 and VDAC were also shown by immunoprecipitation to have increased association during ischemia. Imahashi et al. (2004) also demonstrated that addition of uncoupler to mitochondria with increased Bcl-2 slowed the rate of uncoupler mediated consumption of ATP. Taken together, these data suggest that during ischemia, there may be increased association of Bcl-2 and VDAC which results in a reduction in ATP entry into the mitochondria via VDAC thereby conserving glycolytically generated ATP. The interaction between VDAC and Bcl-2 may also inhibit cytochrome c release via VDAC or the mPTP, since VDAC is thought to be a component of the mPTP. The role of Bcl-2 in modulating oncotic cell death is an area for future study. It is possible the amount of apoptosis in an in vivo model would be higher than in an isolated perfused heart. However, it is clear that inhibitors of apoptosis can inhibit what has commonly been considered necrosis.

Opening of the mPTP has been proposed to be an important component of both apoptosis and oncosis (Halestrap 2005; Zamzami et al. 2005). However, recent studies suggest that hearts null for cyclophilin are protected against oncotic ischemia-reperfusion injury, but are not protected from apoptosis induced by overexpression of BAX (Baines et al. 2005; Nakagawa et al. 2005). These data suggest an important role for cyclophilin mediated activation of mPTP in oncosis associated with ischemia-reperfusion injury.

13.4. Mitochondria and Cardioprotection

A clear example of the important role of mitochondria in necrotic cell death is illustrated by the important role of mitochondria in cardioprotection against ischemia-reperfusion injury, since oncotic necrosis appears to be the predominant form of death in ischemia-reperfusion. This topic is fully covered in another chapter, but it is worthwhile to reiterate the key points of the role of mitochondria in reducing necrotic cell death. Preconditioning, one of the best studied mechanisms of cardioprotection has a number of mitochondrial targets (Figure 13.2). A study of these targets provides insight into the role of mitochondria in necrosis. Preconditioning is activated by release of agonists such as bradykinin, adenosine, or opioids, resulting in activation of G-protein coupled receptor signaling and activation of a signaling cascade involving PI3-kinase, protein kinase C, nitric oxide synthase (NOS), ERK, and glycogen synthase kinase. There is increasing data suggesting that these signals converge on the mitochondria to activate the mitochondrial K_{ATP} channel (Garlid et al., 1997; O'Rourke, 2004), to modify the mitochondrial voltage dependent anion channel (VDAC) (Baines et al., 2003; Imahashi et al., 2004), to phosphorylate BAD (Baines et al., 2002), and to inhibit the opening of the mPTP (Javadov et al., 2003). Downey, Cohen and coworkers have shown that acetylcholine mediated cardioprotection involves

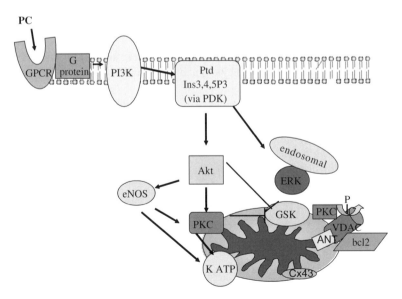

FIGURE 13.2. Activation of mitochondrial signaling pathways reduces necrotic cell death. See text for the details.

activation of PI3-kinase, AKT, NOS and protein kinase G, which then transmits the protective signal to the mitochondria, where it leads to activation of PKC-epsilon and activation of the mitochondrial K-ATP channel (Costa et al. 2005). Ping and coworkers (Baines et al., 2003) have reported that preconditioning leads to mitochondrial localization of PKC-epsilon which results in phosphorylation of VDAC. Baines et al. (2002) have also reported that PKC-epsilon mediated protection involves activation of mitochondrial ERK leading to phosphorylation of mitochondrial BAD. Boengler et al. (2005) have demonstrated that preconditioning results in the translocation of connexin 43 to the mitochondria, and that inhibition of this pathway blocks the protection. Sollott and coworkers (Juhaszova et al. 2004) have demonstrated that a large number of cardioprotective signaling pathways all result in an increase in mitochondrial phospho-GSK. They further show that phosphorylation and the resultant inhibition of GSK mediates protection, at least in part, by inhibition of the mPTP. Imahashi et al. (2004) have also reported that overexpression of Bcl-2 reduces oncotic necrosis by interaction with VDAC resulting in decreased entry of glycolytically generated ATP into the mitochondria, thus limiting consumption of glycolytic ATP by the reverse mode of the F_1F_0-ATPase. It is also possible that interaction of Bcl-2 with VDAC may reduce the probability of opening of the mPTP. Taken together, these data suggest that mitochondria are key modulators of oncotic necrosis. These data suggest an important role for mPTP opening in necrosis and agents that inhibit mPTP opening reduce necrosis.

13.5. Summary

In summary, mitochondria are key regulators of cell survival and death. As discussed mitochondria are key regulators of energetic, ROS and calcium homeostasis and are therefore key regulators of cardioprotection and death. There are interesting data suggesting that the length of opening a mitochondrial channel the mPTP determines whether cells are protected or die by apoptosis or oncosis. Transient opening of the mPTP has been shown to lead to cardioprotection, whereas longer, but reversible opening of mPTP results in apoptosis and prolonged opening results in oncosis. Thus a better understanding of mitochondrial regulation will provide new therapeutic approaches for reducing cell death.

References

Anversa P, Cheng W, Liu Y, Leri A, Redaelli G, Kajstura J (1998) Apoptosis and myocardial infarction. Basic Res Cardiol 93 Suppl 3: 8–12

Baines CP, Kaiser RA, Purcell NH, Blair NS, Osinska H, Hambleton MA, Brunskill EW, Sayen MR, Gottlieb RA, Dorn GW, Robbins J, Molkentin JD (2005) Loss of cyclophilin D reveals a critical role for mitochondrial permeability transition in cell death. Nature 434: 658–662

Baines CP, Song CX, Zheng YT, Wang GW, Zhang J, Wang OL, Guo Y, Bolli R, Cardwell EM, Ping P (2003) Protein kinase Cepsilon interacts with and inhibits the permeability transition pore in cardiac mitochondria. Circ Res 92: 873–880

Baines CP, Zhang J, Wang GW, Zheng YT, Xiu JX, Cardwell EM, Bolli R, Ping P (2002) Mitochondrial PKCepsilon and MAPK form signaling modules in the murine heart: enhanced mitochondrial PKCepsilon-MAPK interactions and differential MAPK activation in PKCepsilon-induced cardioprotection. Circ Res 90: 390–397

Bartling B, Holtz J, Darmer D (1998) Contribution of myocyte apoptosis to myocardial infarction? Basic Res Cardiol 93: 71–84

Boengler K, Dodoni G, Rodriguez-Sinovas A, Cabestrero A, Ruiz-Meana M, Gres P, Konietzka I, Lopez-Iglesias C, Garcia-Dorado D, Di Lisa F, Heusch G, Schulz R (2005) Connexin 43 in cardiomyocyte mitochondria and its increase by ischemic preconditioning. Cardiovasc Res 67: 234–244

Chen Z, Chua CC, Ho YS, Hamdy RC, Chua BH (2001) Overexpression of Bcl-2 attenuates apoptosis and protects against myocardial I/R injury in transgenic mice. Am J Physiol Heart Circ Physiol 280: H2313–2320

Costa AD, Garlid KD, West IC, Lincoln TM, Downey JM, Cohen MV, Critz SD (2005) Protein kinase G transmits the cardioprotective signal from cytosol to mitochondria. Circ Res 97: 329–336

Dawson TM, Dawson VL (2003) Molecular pathways of neurodegeneration in Parkinson's disease. Science 302: 819–822

Di Lisa F, Bernardi P (1998) Mitochondrial function as a determinant of recovery or death in cell response to injury. Mol Cell Biochem 184: 379–391

Di Lisa F, Blank PS, Colonna R, Gambassi G, Silverman HS, Stern MD, Hansford RG (1995) Mitochondrial membrane potential in single living adult rat cardiac myocytes exposed to anoxia or metabolic inhibition. J Physiol 486 (pt 1): 1–13

Farber JL, Gerson RJ (1984) Mechanisms of cell injury with hepatotoxic chemicals. Pharmacol Rev 36: 71S–75S

Garlid KD, Paucek P, Yarov-Yarovoy V, Murray HN, Darbenzio RB, D'Alonzo AJ, Lodge NJ, Smith MA, Grover GJ (1997) Cardioprotective effect of diazoxide and its interaction with mitochondrial ATP-sensitive K+ channels. Possible mechanism of cardioprotection. Circ Res 81: 1072–1082

Gottlieb RA, Burleson KO, Kloner RA, Babior BM, Engler RL (1994) Reperfusion injury induces apoptosis in rabbit cardiomyocytes. J Clin Invest 94: 1621–1628

Griffiths EJ, Halestrap AP (1993) Protection by Cyclosporin A of ischemia/reperfusion-induced damage in isolated rat hearts. J Mol Cell Cardiol 25: 1461–1469

Halestrap A (2005) Biochemistry: a pore way to die. Nature 434: 578–579

Halestrap AP, Clarke SJ, Javadov SA (2004) Mitochondrial permeability transition pore opening during myocardial reperfusion–a target for cardioprotection. Cardiovasc Res 61: 372–385

Halestrap AP, Connern CP, Griffiths EJ, Kerr PM (1997) Cyclosporin A binding to mitochondrial cyclophilin inhibits the permeability transition pore and protects hearts from ischaemia/reperfusion injury. Mol Cell Biochem 174: 167–172

Hausenloy D, Wynne A, Duchen M, Yellon D (2004) Transient mitochondrial permeability transition pore opening mediates preconditioning-induced protection. Circulation 109: 1714–1717

Imahashi K, Schneider MD, Steenbergen C, Murphy E (2004) Transgenic expression of Bcl-2 modulates energy metabolism, prevents cytosolic acidification during ischemia, and reduces ischemia/reperfusion injury. Circ Res 95: 734–741

Javadov SA, Clarke S, Das M, Griffiths EJ, Lim KH, Halestrap AP (2003) Ischaemic preconditioning inhibits opening of mitochondrial permeability transition pores in the reperfused rat heart. J Physiol 549: 513–524

Jennings RB, Ganote CE (1976) Mitochondrial structure and function in acute myocardial ischemic injury. Circ Res 38: I80–91

Jennings RB, Murry CE, Steenbergen C Jr, Reimer KA (1990) Development of cell injury in sustained acute ischemia. Circulation 82: II2–12

Jennings RB, Reimer KA (1981) Lethal myocardial ischemic injury. Am J Pathol 102: 241–255

Jennings RB, Steenbergen C Jr (1985) Nucleotide metabolism and cellular damage in myocardial ischemia. Annu Rev Physiol 47: 727–749

Juhaszova M, Zorov DB, Kim SH, Pepe S, Fu Q, Fishbein KW, Ziman BD, Wang S, Ytrehus K, Antos CL, Olson EN, Sollott SJ (2004) Glycogen synthase kinase-3beta mediates convergence of protection signaling to inhibit the mitochondrial permeability transition pore. J Clin Invest 113: 1535–1549

Karmazyn M (1988) Amiloride enhances postischemic ventricular recovery: possible role of Na^+-H^+ exchange. Am J Physiol 255: H608–615

Kroemer G, Dallaporta B, Resche-Rigon M (1998) The mitochondrial death/life regulator in apoptosis and necrosis. Annu Rev Physiol 60: 619–642

Kuznetsov AV, Schneeberger S, Seiler R, Brandacher G, Mark W, Steurer W, Saks V, Usson Y, Margreiter R, Gnaiger E (2004) Mitochondrial defects and heterogeneous cytochrome c release after cardiac cold ischemia and reperfusion. Am J Physiol Heart Circ Physiol 286: H1633–1641

Levin S, Bucci TJ, Cohen SM, Fix AS, Hardisty JF, LeGrand EK, Maronpot RR, Trump BF (1999) The nomenclature of cell death: recommendations of an ad hoc Committee of the Society of Toxicologic Pathologists. Toxicol Pathol 27: 484–490

Leyssens A, Nowicky AV, Patterson L, Crompton M, Duchen MR (1996) The relationship between mitochondrial state, ATP hydrolysis, $[Mg^{2+}]i$ and $[Ca^{2+}]i$ studied in isolated rat cardiomyocytes. J Physiol 496 (Pt 1): 111–128

Maloyan A, Sanbe A, Osinska H, Westfall M, Robinson D, Imahashi K, Murphy E, Robbins J (2005) Mitochondrial dysfunction and apoptosis underlie the pathogenic process in alpha-B-crystallin desmin-related cardiomyopathy. Circulation 112: 3451–3461

Murphy E, Perlman M, London RE, Steenbergen C (1991) Amiloride delays the ischemia-induced rise in cytosolic free calcium. Circ Res 68: 1250–1258

Nakagawa T, Shimizu S, Watanabe T, Yamaguchi O, Otsu K, Yamagata H, Inohara H, Kubo T, Tsujimoto Y (2005) Cyclophilin D-dependent mitochondrial permeability transition regulates some necrotic but not apoptotic cell death. Nature 434: 652–658

O'Rourke B (2004) Evidence for mitochondrial K+ channels and their role in cardioprotection. Circ Res 94: 420–432

Paradies G, Petrosillo G, Pistolese M, Di Venosa N, Federici A, Ruggiero FM (2004) Decrease in mitochondrial complex I activity in ischemic/reperfused rat heart: involvement of reactive oxygen species and cardiolipin. Circ Res 94: 53–59

Rodriguez M, Schaper J (2005) Apoptosis: measurement and technical issues. J Mol Cell Cardiol 38: 15–20

Rouslin W, Broge CW (1996) IF1 function in situ in uncoupler-challenged ischemic rabbit, rat, and pigeon hearts. J Biol Chem 271: 23638–23641

Schneider MD (2005) Cyclophilin D: knocking on death's door. Sci STKE 2005 pe26

Shiraishi J, Tatsumi T, Keira N, Akashi K, Mano A, Yamanaka S, Matoba S, Asayama J, Yaoi T, Fushiki S, Fliss H, Nakagawa M (2001) Important role of energy-dependent mitochondrial pathways in cultured rat cardiac myocyte apoptosis. Am J Physiol Heart Circ Physiol 281: H1637–1647

Steenbergen C, Murphy E, Levy L, London RE (1987) Elevation in cytosolic free calcium concentration early in myocardial ischemia in perfused rat heart. Circ Res 60: 700–707

Steenbergen C, Murphy E, Watts JA, London RE (1990) Correlation between cytosolic free calcium, contracture, ATP, and irreversible ischemic injury in perfused rat heart. Circ Res 66: 135–146

Zamzami N, Larochette N, Kroemer G (2005) Mitochondrial permeability transition in apoptosis and necrosis. Cell Death Differ 12 Suppl 2: 1478–1480

Zhu L, Yu Y, Chua BH, Ho YS, Kuo TH (2001) Regulation of sodium-calcium exchange and mitochondrial energetics by Bcl-2 in the heart of transgenic mice. J Mol Cell Cardiol 33: 2135–2144

Zong WX, Thompson CB (2006) Necrotic death as a cell fate. Genes Dev 20: 1–15

Part 5
Mitochondria as Modulators of Cell Death

14
Mitochondria and Their Role in Ischemia/Reperfusion Injury

Sebastian Phillip, James M. Downey and Michael V. Cohen

14.1. Myocardial Ischemia

Acute coronary occlusion is the leading cause of morbidity and mortality in the western world and according to the World Health Organization will be the major cause of death in the world as a whole by the year 2020 (Murray and Lopez 1997). The major complication of the sudden occlusion of a coronary artery is the loss of contractile myocardium served by that artery. The contractile dysfunction resulting from infarction of the ventricle is essentially permanent as the lost heart muscle cannot regenerate. The size of the resultant infarct is the decisive determinant of the extent and severity of remodeling (Pfeffer et al. 1991) and of the prognosis of patients after myocardial infarction (St. John Sutton et al. 1997). Fast revascularization will result in less myocardial necrosis; nevertheless congestive heart failure secondary to myocardial infarction is still common.

14.2. Preconditioning the Heart

In 1986, Murry et al. first described the phenomenon of ischemic preconditioning (IPC). After dog hearts were pretreated with 4 episodes of 5 minutes of ischemia, each followed by brief reperfusion, the resulting infarct s after 40 minutes of ischemia was about one-quarter of that in animals experiencing only the 40-minute coronary occlusion. Thus, the short ischemic periods caused a rapid adaptation of hearts against infarction. This process is the most powerful protection known against myocardial infarction. Since the initial report there has been intensive research to uncover the underlying mechanisms. The protective effect of IPC is receptor-mediated. It was first discovered that activation of the adenosine A_1 receptor can elicit it (Liu et al. 1991; Tsuchida et al. 1993). Subsequently other G protein-coupled receptors were found to trigger IPC like bradykinin (Wall et al. 1994; Goto et al. 1995; Cohen et al. 2001), opioid (Schultz et al. 1995; Miki et al. 1998; Cohen et al. 2001), angiotensin AT_1 (Liu et al. 1995), endothelin ET_1 (Wang et al. 1996), α_1 adrenergic (Tsuchida et al. 1994) and acetylcholine (Thornton et al. 1993; Qin et al. 2003; Krieg et al. 2004b) receptors.

Reactive oxygen species act as second messengers in the mechanism by which bradykinin and opioid receptors trigger the preconditioned state. G_i-coupled receptors cause the production of ROS by a complex pathway in which phosphatidylinositol 3 (PI3)-kinase activates Akt through phosphoinositide-dependent kinases 1 and 2 which in turn activate endothelial nitric oxide synthase (NOS) (Krieg et al. 2002; Krieg et al. 2004b) (Figure 14.1).

The resulting NO stimulates guanylyl cyclase which then activates protein kinase G (PKG). PKG opens mK_{ATP} by a process that involves an intra-mitochondrial protein kinase C (PKC) isoform (Oldenburg et al. 2004; Costa et al. 2005a). mK_{ATP} channel opening then leads to the production of reactive oxygen species (ROS) by the mitochondria. Those ROS are thought to then trigger kinase cascades which ultimately lead to cardioprotection (Pain et al. 2000; Oldenburg et al. 2004; Krieg et al. 2004b). In this scheme mitochondria play an important role by being part of a signal transduction pathway and by generating ROS which serve as critical second messengers.

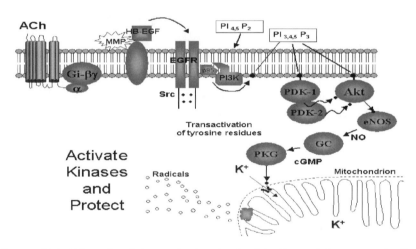

FIGURE 14.1. Schematic of the mechanism involved in G_i-coupled receptors leading to ROS generation. After ligand activation of the G_i-coupled receptor, the epidermal growth factor receptor (EGFR) is transactivated at its tyrosine groups through metalloproteinase (MMP)-dependent cleavage of heparin-binding EGF-like growth factor (HB-EGF) from proHB-EGF in the membrane. This in turn results in the formation of a complex with Src tyrosine kinase and phosphatidylinositol 3-kinase (PI3K) which can activate both 3' phosphoinositide-dependent kinase-1 (PDK 1) and -2 (PDK 2). This results in phosphorylation of Akt at Thr 308 and Ser 473, respectively. Akt activates endothelial nitric oxide synthase (eNOS), which in turn activates guanylyl cyclase (GC) to produce cGMP. The latter activates protein kinase G (PKG) which opens the mK_{ATP} channel via PKC (not shown). Radical production by the electron transport chain is increased and these radicals are released to activate downstream kinases and protect the heart.

14.3. Trigger, Ischemic, and Reperfusion Phases

When considering preconditioning it is useful to think in terms of triggers, mediators, memory, end-effectors, and reperfusion (Figure 14.2). The preconditioning ischemia triggers a change in the physiology of the heart, rendering it very resistant to infarction. A sequential series of snal transduction steps carry the signal for protection resulting in ROS production by mitochondria during the brief reperfusion period (Figure 14.1). Even if events in this phase are removed the heart remains protected hence the term trigger. The ROS then activate a kinase cascade beginning with cytosolic PKC. The kinase activity of PKC is not needed until the index ischemia begins and, therefore, it is classified as a *mediator*. Because the heart stays in a preconditioned state for up to 2 hours after the trigger has been removed (Van Winkle et al. 1991), there must be a *memory* somewhere. When PKC is activated, several isoforms, specifically PKC-ε and -η in rabbit hearts, translocate from the cytosol to membranes (Ping et al. 1997). This translocation may act as a memory step, perhaps by keeping PKC bound near its substrate. In one study, however, Nozawa et al. (2003) found that PKC isoforms remained translocated even after protection had waned or been blocked with 5-hydroxydecanoate (5-HD), a specific antagonist of mitochondrial ATP-sensitive potassium channels (mK_{ATP}), thus arguing against them being the memory. Other mediator kinases possibly include p38MAPK

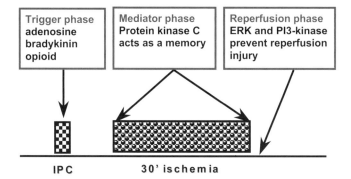

FIGURE 14.2. Preconditioning can be divided into three phases: trigger, mediator, and reperfusion. The trigger phase occurs before the index ischemia, while the mediator phase encompasses the index ischemia. The reperfusion phase begins after the onset of reperfusion or reoxygenation. The trigger phase is initiated by binding of released agonists from ischemic myocardial cells to G protein-coupled receptors which initiates a signaling cascade involving transactivation of growth factors, activation of Akt and nitric oxide synthase, production of nitric oxide with ultimate activation of protein kinase G and opening of mitochondrial K_{ATP} channels. The latter produce reactive oxygen species during the brief reperfusion interval before the long ischemia and these act as second messengers to activate protein kinase C. The mediator phase involves a kinase cascade including the activation of survival kinases, Akt and ERK. These kinases are effective during early reperfusion to suppress opening of the mitochondrial permeability transition pore.

(Nakano et al. 2000; Mocanu et al. 2000). These kinases eventually activate one or more *end-effectors*. The *end-effector(s)* actually cause the protection. For a long time it was thought that the protection was exerted during the ischemic period, but we now believe that it occurs during the subsequent *reperfusion* period. Activation of Akt and ERK 1/2, termed survival kinases, during early reperfusion was found to be critical (Hausenloy et al. 2005; Solenkova et al. 2005), and this activation appears to be in response to adenosine A_1 and/or A_{2B} occupancy during reperfusion (Solenkova et al. 2005). The identity of the end-effector(s) has not yet been proven, but there is compelling evidence that the mitochondrial permeability transition pore (see below) may be the elusive target. IPC may act to suppress its opening and thus prevent death of myocytes. Hence, mitochondria appear to serve dual roles in preconditioning. They are critical signaling elements and may also be end-effectors.

14.4. The Role of Mitochondria During Ischemia

14.4.1. Cell Death

The viability of the ischemic myocardium is dependent on ATP availability and changes in intracellular Ca^{2+} which are related to derangements in mitochondrial function. Mitochondria play a key function in cell survival not only because of their bioenergetic role as a supplier of ATP for vital intracellular processes but also because of their critical involvement in programmed cell death or apoptosis. The organelle plays an integral part in the regulatory and signaling events that occur in response to physiological stresses, including myocardial ischemia and reperfusion, hypoxia, oxidative stress, and hormonal and cytokine stimulation. It is known that cells die from necrosis when they can not maintain adequate ATP levels. Necrotic cells swell because of a failure to maintain the Donnan equilibrium causing rapid membrane failure. But they can also die from apoptosis in which ATP is not depleted. Rather mitochondria actively participate in molecular signaling events leading to apoptosis by releasing cytochrome c and initiating activation of a caspase cascade that attacks nuclear DNA (Newmeyer and Ferguson-Miller 2003; Borutaite and Brown 2003). Although apoptosis is initiated during ischemia/reperfusion, the number of otherwise viable cells killed by this process is unknown.

14.4.2. Ischemia/Reperfusion

Acute myocardial ischemia/reperfusion (I/R) injury involves a variable mix of necrosis and apoptosis. The role of the mitochondria during I/R is particularly critical since both apoptosis by the mitochondrial pathway and necrosis by irreversible opening of the mitochondrial permeability transition pore are involved. Due to leaks in the respiratory chain, mitochondria continuously produce oxygen free radicals at a rate of 1–2% of consumed oxygen. Opening mK$_{ATP}$ leads to K$^+$ entry, matrix alkalinization, a small degree of depolarization of the mitochondrial membrane potential ($\Delta\psi_m$), mitochondrial swelling, and finally increased ROS formation. Free radicals can trigger cardioprotection (Baines et al. 1997) and do so during pathophysiological processes such as I/R (see below). Changes in the mitochondrial membrane potential affect other mitochondrial membrane proteins, such as the mitochondrial permeability transition pore (mPTP). Opening of the mPTP can be triggered by a collapse of the mitochondrial membrane potential, increased formation of ROS, and increased levels of intracellular Ca^{2+}. All of these are present when the heart is reperfused following a lethal ischemic insult.

Apoptosis is defined by a pivotal event – mitochondrial outer membrane permeabilization. This occurs suddenly during apoptosis, leading to the release of proteins normally found in the space between the inner and outer mitochondrial membranes (Figure 14.3). Before, during or after membrane permeabilization, there is frequently a dissipitation of the mitochondrial inner transmembrane potential. The mechanisms responsible for this breakdown in the outer membrane barrier during apoptosis remain controversial, although it is clear that many proteins influence membrane permeabilization via local effects on mitochondrial membranes.

Swelling and rupture of membranes of ischemic mitochondria signal loss of the ability to maintain critical volume homeostasis. To avoid rupture of the mitochondrial outer membrane due to swelling, as well as limit unnecessary energy wastage, mitochondrial cation flux is typically tightly regulated. The

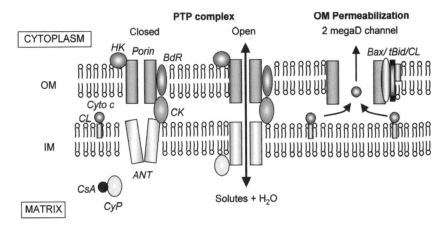

FIGURE 14.3. Proteins involved in the formation and/or regulation of mPTP in the inner membrane (IM) and the permeabilization of the outer membrane (OM) to cytochrome C and larger molecules. BdR = benzodiazepine receptor; CL = cardiolipin; CK = creatine kinase; Cyto c = cytochrome c; CyP = cyclophilin D; HK = hexokinase; porin = VDAC or voltage-dependent anion channel; ANT = adenine nucleotide translocator. CsA inhibits mPTP opening by binding to CyP and preventing its interaction with ANT. BdR, HK and CK are regulatory components of mPTP (reproduced with permission from Weiss et al. 2003).

mitochondrial K^+ cycle (Garlid and Paucek 2003) with K^+/H^+ exchange counterbalances mitochondrial uniporter activity and is an important component of mitochondrial volume homeostasis which is overwhelmed when the mitochondrial permeability transition pore opens. Opening of the mK_{ATP} channel also leads to K^+ influx accompanied by the movement of permeable anions and H_2O. This influx can transiently exceed the extrusion capacity of the K^+/H^+ exchanger resulting in an increase in steady-state matrix volume. However, this is a highly controlled process.

14.4.3. The Role of Mitochondrial ATP-Sensitive Potassium Channels in Cardioprotection

In 1991 the mK_{ATP} channels were discovered in the liver mitochondrial inner membrane (Inoue et al. 1991). This finding was bolstered by evidence that a variety of K^+ channel openers and inhibitors influenced mitochondrial function (Paucek et al. 1992; Garlid 1996). A link between mK_{ATP} channels and cardioprotection was first suggested by Gross and Auchampach (Gross and Auchampach 1992) who observed that cardioprotection caused by IPC was blocked by glibenclamide. Thus the opening of K_{ATP} channels exerted a powerful effect against infarction, whereas channel blockers abrogated IPC's protection (Gross and Auchampach 1992; Auchampach and Gross 1993; Critz et al. 1997; Pain et al. 2000; Forbes et al. 2001). Originally it was thought that sarcolemmal K_{ATP} channels were the end-effectors of PC's protection, and this protection was originally ascribed to shortening of the action potential (Gross and Fryer, 1999). At that time it was not appreciated that cardiomyocytes contained not only sarcolemmal but also mK_{ATP} channels and each had a distinct pharmacological profile. In subsequent investigations by Garlid et al. (1997), Liu et al. (1998) and Sato et al. (2000) evidence was provided that indeed mK_{ATP} and not sarcolemmal channels were the ones involved in cardioprotection. Garlid et al. (1997) observed that diazoxide selectively opened mK_{ATP} (EC_{50} ~30 μM) and that 30 μM diazoxide improved post-ischemic ventricular function by an amount similar to that of IPC. Furthermore, diazoxide mimicked IPC in an infarct model and the mK_{ATP} closer 5-HD blocked IPC's protection (Baines et al. 1999). These critical observations led to the proposal that opening of mK_{ATP} rather than the sarcolemmal channel evoked cardioprotection.

If the mK_{ATP} channel were the *end-effector* of IPC as proposed by some and its opening the cause of cardioprotection, then it should behave as a mediator. Pain et al. (2000) and also Wang et al. (2001) used the mK_{ATP} channel blocker 5-HD in isolated rabbit hearts and applied it either during the preconditioning ischemia (trigger phase) or during the index ischemia (mediator phase). While both groups could block protection with 5-HD when it was limited to the trigger phase, Pain could not block the protective effect when 5-HD was applied during only the mediator phase. On the other hand, Wang did document blockade of protection with 5-HD added during the index ischemia, but only when the dose of 5-HD was increased 4-fold. In a recent study by Schulz et al. (2003) the K_{ATP}

blocker glibenclamide was infused in a porcine model of ischemia/reperfusion. When the blocker was administered 5 minutes after the short preconditioning phase until the end of the index ischemia (mediator phase) the protection could not be blocked. Conversely, early application of glibenclamide during the trigger phase abolished the infarct-sparing effect of IPC.

While a mediator role of mK_{ATP} is uncertain, all now agree the mK_{ATP} channel acts as a trigger. Therefore these data suggest mK_{ATP} cannot be the end-effector of IPC. But how does channel opening protect? Interestingly, antioxidants abolish the anti-infarct effect of diazoxide (Forbes et al. 2001) as well as of IPC (Baines et al. 1997; Tritto et al. 1997; Pain et al. 2000; Forbes et al. 2001). These observations led to the proposal that opening of mK_{ATP} led to the generation of reactive oxygen species (ROS) which act as second messengers and activate kinases further downstream (Pain et al., 2000). It is still not known exactly how opening of the mK_{ATP} channel causes ROS generation. Most likely ROS are generated within the mitochondrial electron transport chain due to the potassium influx. The potassium ionophore valinomycin also induces ROS generation (Oldenburg et al. 2002), while myxothiazol, a mitochondrial electron transport blocker, abolishes it (Yue et al. 2001; Krenz et al. 2002; Oldenburg et al. 2003). Costa et al. (2005b) propose that ROS production is a direct result of matrix alkalinization which follows potassium entry.

The triggering sequence depicted in Figure 14.1 has been established in isolated rabbit cardiomyocytes in which mitochondrial production of ROS was measured by the radical-induced oxidation of reduced MitoTracker red to its fluorescing oxidized form (Krenz et al. 2002; Oldenburg et al. 2002; Oldenburg et al. 2003; Oldenburg et al. 2004; Krieg et al. 2004a). Thus the end-point for each of the steps was mitochondrial production of ROS. G-protein coupled receptor agonists, epidermal growth factor, nitric oxide donors, mK_{ATP} openers and activation of PKG all led to increased ROS generation. By using appropriate blockers and activators we could establish upstream/downstream relationships for each of these. For example the PI3-kinase inhibitor wortmannin did not block ROS production from a nitric oxide donor; therefore it was assumed that PI3-kinase was upstream of the nitric oxide step. At the same time 5-HD did block ROS production, implying that mK_{ATP} must be located downstream of nitric oxide involvement.

14.5. Function of mK_{ATP}

The structure of the mK_{ATP} channel is not known but it is assumed that it incorporates elements of the two known ATP-sensitive inward rectifier K^+ channel proteins, Kv6.1 (uK_{ATP}) and Kv6.2 (sK_{ATP}). These K_{ATP} channels contain 4 pore-forming channel proteins and 4 sulfonylurea-binding receptor (SUR) proteins. The latter have an ATP binding site and confer sensitivity to glibenclamide (and related sulfonylurea drugs) (Inagaki et al. 1995a; Inagaki et al. 1995b; Inagaki et al. 1996; Aguilar-Bryan et al. 1998; Lorenz and Terzic 1999; Grover and Garlid 2000). Mitochondria are negatively charged. Although the potassium concentration in mitochondria is similar to that in the cell, the electrical gradient favors potassium entry. That allows mitochondria to regulate their volume by controlling their content of potassium which is osmotically active at the inner membrane.

The mitochondria make ATP by causing positively charged hydrogen ions to be extruded by the electron transport apparatus; they then reenter along a strong electrochemical gradient through the F1 apparatus. Reentry causes phosphorylation of ADP to ATP. Potassium enters along the electrochemical gradient through the mK_{ATP} channel resulting in swelling of the matrix. The potassium-hydrogen exchanger on the inner mitochondrial membrane readily exchanges intramitochondrial potassium for extra- mitochondrial H^+ which causes the matrix to shrink. The mK_{ATP} has a relatively low conductance and the change in $\Delta\psi_m$ caused by the influx of K^+ when the channels are fully opened is rather small (~5mV). This meager change is most likely related to the low density of these channels in the inner membrane (Garlid 2000).

14.5.1. Do mK_{ATP} Really Exist?

As mentioned above the actual construct of the mK_{ATP} channels is yet to be identified. In fact some investigators even question whether they actually exist (Das et al., 2003). The pharmacology of mK_{ATP} is distinctive. 5-HD is a selective blocker of the mK_{ATP} channel, while diazoxide selectively opens it (Garlid et al. 1997). Some have proposed that these agents are producing effects by intracellular actions other than the postulated effects on mK_{ATP} (Hanley et al. 2002; Das et al. 2003). Electrophysiological recordings are perhaps the most concrete proof of the existence of mK_{ATP} channels. Indeed direct detection of single-channel activity in the mitochondrial inner membrane provided the foundation for subsequent interpretations. Inoue et al. (Inoue et al. 1991) directly patch-clamped mitoblasts (which are mitochondria stripped of the outer membrane) prepared from fused liver mitochondria. Single potassium selective channels were identified and inhibited by ATP applied to the mitochondrial matrix face of the channel. They were blocked by the potassium channel inhibitor 4-aminopyridine. Importantly, the channels were inhibited by glibenclamide, clear evidence that a K_{ATP}-like channel was present.

A second line of evidence supporting a specific mK_{ATP} channel is the measurement of potassium uptake into phospholiposomes reconstituted with purified mitochondrial protein fractions. This method has been primarily used by Garlid's group (Paucek et al. 1992; Paucek et al. 1996; Garlid et al. 1997; Yarov-Yarovoy et al. 1997; Bajgar et al. 2001) to characterize the channel's regulation by Mg^{2+}, ATP, GTP, and acyl-CoA esters, and its sensitivity to pharmacological openers and inhibitors.

Furthermore potassium influx into mitochondria is directly increased by diazoxide and inhibited by ATP and 5-HD (Costa et al. 2005b). It is also important to point out that the effects of diazoxide and 5-HD on mitochondrial volume changes are absent when tetraethyl ammonium is substituted for potassium in the medium (Garlid 2000). Finally, the effects of mK_{ATP} openers and blockers were compared to those seen with the potassium ionophore valinomycin, and four independent assays of mK_{ATP} activity yielded quantitatively identical results for mK_{ATP}-mediated potassium transport (Costa et al. 2005b). These observations strongly support the existence of mK_{ATP}.

14.5.2. Mechanisms of Protection

While mK_{ATP} are known to have a trigger function, there is also evidence that they may be involved as mediators of preconditioning as well. There is good evidence that mK_{ATP} activation decreases ROS production at reperfusion (Vanden Hoek et al. 2000; Ozcan et al. 2002), blunts mitochondrial Ca^{2+} accumulation during ischemia (Holmuhamedov et al. 1998; Holmuhamedov et al. 1999; Murata et al. 2001), and improves mitochondrial energy production after ischemia. Swelling of mitochondria occurs as a consequence of activation of the mitochondrial K^+ cycle (Garlid and Paucek 2003). Expansion of the mitochondrial matrix improves fatty acid oxidation, respiration, and ATP production (Halestrap 1994). It is apparent that although mitochondrial swelling is a likely consequence of mK_{ATP} channel opening and has been demonstrated in mitochondria, linking this effect directly to mitochondrial preservation has proven to be a challenging task.

14.5.3. ROS Production

We have already discussed the role mK_{ATP} play in the production of ROS which act as a trigger for the preconditioned state. In contrast, a flood of ROS is produced on reperfusion after a long period of ischemia (Zweier et al. 1987), and if enough are produced they can cause irreversible cell death by oxidizing critical intracellular enzymes and proteins and by attacking membranes. Interestingly this postischemic ROS burst can be suppressed by pretreatment with mK_{ATP} channel openers (Vanden Hoek et al. 2000; Ozcan et al. 2002). Whether mK_{ATP} are involved in this process directly or simply act to trigger signal transduction pathways is not clear.

14.5.4. Ca^{2+} Homeostasis

Another hypothesis is that mitochondrial Ca^{2+} accumulation during ischemia and reperfusion may be attenuated by mK_{ATP} channel opening (Liu et al. 1998). It has been shown that the selective mK_{ATP} channel opener diazoxide decreases the magnitude and rate of Ca^{2+} uptake into isolated mitochondria and that this effect can be inhibited by the selective mK_{ATP} channel blocker 5-HD. This effect was attributed to a partial depolarization of $\Delta\psi_m$ which was potassium-dependent and could be inhibited by a K^+channel blocker (Holmuhamedov et al. 1998; Holmuhamedov et al. 1999). Mitochondrial Ca^{2+} accumulation during simulated ischemia is attenuated by mK_{ATP} channel opening, resulting in reduced mPTP opening upon reperfusion (Murata et al., 2001).

14.5.5. PKG Interaction

Recent evidence suggests that PKG is the final cytosolic component in a signaling cascade that opens mK_{ATP} (Costa et al. 2005a). Activated PKG binds to the outer mitochondrial membrane, and presumably phosphorylates some membrane protein that ultimately causes mK_{ATP} opening on the inner membrane. It is unclear how the signal gets across the inter-membrane space, but we know that the activation is dependent on PKC resident within the mitochondria. Possibly PKC shuttles between the outer and inner membranes and stimulates mK_{ATP} to open. PKC is not a known substrate for PKG, so it is uncertain what lies between them. Nor is it known whether mK_{ATP} have PKC phosphorylation sites. These data amplify the interaction between PKC and mK_{ATP} required to elicit cardioprotection. Recently it was shown that mitochondria were protected from anoxic injury by opening mK_{ATP} which prevented opening of the mPTP upon reoxygenation (Korge et al. 2002). This effect was mimicked by phorbol 12-myristate 13-acetate again showing that mK_{ATP} are under the control of PKC.

14.6. Mitochondrial Permeability Transition Pore (mPTP)

When the mPTP opens, the permeability barrier of the inner membrane becomes disrupted with two major consequences: release of proapoptotic cytochrome c and the cessation of ATP production (Halestrap et al. 2004). Some investigators have proposed that mPTP is the lethal event in reperfusion injury (Halestrap 1999; Zorov et al. 2000). Halestrap et al. (2004) proposed that the mPTP opens during reperfusion and that prevention of mPTP opening is the primary target of cardioprotective interventions against reperfusion injury. Recent studies propose a significant role for the mitochondrial permeability transition pore (mPTP) in IPC (Javadov et al. 2000; Hausenloy et al. 2002). It is not known for certain whether the mPTP is the end-effector of IPC and whether prevention of mPTP opening is the final step in a long signaling cascade. There is evidence that opening of mPTP during early reperfusion leads to apoptotic and necrotic cell death (see above). The distinct relationship between ROS generation, PKC activation and opening of the mK_{ATP} channel (Korge et al. 2002), and the mechanisms that regulate mPTP opening are not fully understood.

mPTP are multiprotein complexes that are capable of forming large nonselective pores in the inner membrane of the mitochondria (Figure 14.3). Current evidence suggests there are key contributions from three key structural elements, the voltage-dependent anion channel (VDAC) located in the outer mitochondrial membrane, the adenine nucleotide translocase (ANT-1) located in the inner mitochondrial membrane, and cyclophilin D which seems to facilitate a calcium-triggered conformational change in ANT, converting it into an open pore (Baines et al. 2005). Under stress conditions VDAC and ANT-1 line up across the mitochondrial intermembrane space and form a high conductance pore, the mPTP, which, among other things, causes release of cytochrome c.

This activates caspase 9 which, in turn, activates caspase 3 and triggers apoptosis (Kluck et al. 1997; Green and Reed 1998; Haunstetter and Izumo 1998). In addition, the mPTP allows extruded hydrogen ions to reenter the matrix freely, thus uncoupling the mitochondria and halting ATP production. A critical issue for further investigation is to identify which kinases are capable of regulating mPTP. Reasonable candidates in the heart include PI3-kinase, Akt, p38 MAPK, ERK 1/2, PKG, and PKC-ε because each one contributes to IPC's protection and some have also been identified in mitochondria (Baines et al. 2002; Baines et al. 2003).

It has been proposed that mPTP can open either in a low or high conductance mode (Weiss et al. 2003). The existence of the reversible low-conductance mode allowing permeation of small solutes that depolarize $\Delta\psi_m$ transiently is still under discussion. In the high conductance mode, which causes matrix swelling and ion-driven water influx, mPTP opening is distinct and typically irreversible. Extensive matrix swelling of the inner membrane (IM) unfolds the cristae causing a rupture of the outer membrane (OM), which in turn leads to release of proapoptotic molecules residing in the intermembrane space. These molecules (cytochrome c and others) promote apoptotic cell death via caspase-dependent and -independent mechanisms (Weiss et al. 2003). This is one possible explanation for apoptotic cell death in I/R injury.

Typically ROS and high cellular Ca^{2+} are the most important inducers of mPTP opening. Oxidation of critical thiol groups on ANT-1 leads to formation of disulfide bonds which allows cyclophilin D normally present in the matrix to bind to ANT-1 thus promoting mPTP to open. It is possible that ROS may act at this site to oxidize thiol groups to promote mPTP opening.

14.6.1. The Role of the Mitochondrial Transition Pore During Myocardial Ischemia

It is less clear how IPC actually protects. A number of hypotheses have been proposed, including reduced free radical production at reperfusion, reduced osmotic swelling, reduced apoptosis, and reduced Ca^{2+} overload (Yellon and Downey 2003). There is increasing evidence that IPC's protection is mediated, at least in part, by preventing formation of mPTP. This high conductance pore forms during reperfusion and acts to depolarize the matrix, thus stopping energy production. The classic stimuli for mPTP opening during reperfusion are Ca^{2+} and ROS (Halestrap 1999), whereas during ischemia the fall in intracellular pH and the increase in Mg^{2+} and ADP may instead reduce the likelihood of mPTP opening. During ischemia latent susceptibility of mitochondria to mPTP opening increases probably because of the accumulation of long-chain fatty acids and ROS. This results in leakiness of the internal mitochondrial membrane and probably some loss of cytochrome c. However, although the mPTP may be primed, it is not necessarily destined to open at reperfusion. If it doesn't, the mitochondria and myocytes can recover. However, if the interplay between mPTP inducers and inhibitors present during reperfusion is unfavorable and the ability to increase electron transport capacity and regenerate $\Delta\psi_m$ is impaired, mPTP will open leading to permanent mitochondrial dysfunction and cell death. Because of the intracellular acidosis present at the onset of reperfusion, mPTP is actually inhibited from opening for a short period of time until the resumed coronary flow can wash out the accumulated H^+. Apparently ischemic preconditioning acts to prevent opening of the mPTP during the reperfusion period (Hausenloy et al. 2002). Cyclosporin A (CSA) is an mPTP blocker which binds to cyclophilin and interferes with the binding of the latter to ANT-1. Infusion of CSA at reperfusion following 35 min of ischemia mimicked the protective effect of IPC and diminished infarction in isolated rat hearts (Hausenloy et al. 2002). By contrast, infusion of atractyloside, an mPTP promoter, at reperfusion in IPC-treated or pharmacologically preconditioned hearts blocked protection (Hausenloy et al. 2002). Furthermore, blockade of mPTP with CSA or N-methyl-4-valine-CSA increased the time to contracture in rat cardiomyocytes, an effect that was mimicked by hypoxic and pharmacological preconditioning (Hausenloy et al. 2004). Further support for this hypothesis was obtained when mitochondria isolated from the ischemic zone of preconditioned hearts were found to be much more resistant to calcium-triggered mPTP formation than those from non-preconditioned hearts (Argaud et al. 2004). Together these data suggest that IPC protects against infarction by keeping mPTP closed during reperfusion.

Recent observations have provided convincing evidence that preconditioning prevents mPTP opening by activating the survival kinases ERK and Akt (Juhaszova et al. 2004). A number of seemingly disparate cardioprotective interventions all seem to revolve around a mechanism involving activation of the survival kinases which then act to suppress mPTP opening during reperfusion.

Interestingly, the newly described postconditioning, like preconditioning, acts to prevent mPTP opening as well (Argaud et al. 2005). In postconditioning several very brief ischemic periods are interposed within the first minutes of reperfusion after a prolonged coronary occlusion (Zhao et al., 2003; Yang et al. 2004). Therefore, in contradistinction to IPC, the brief ischemia/reperfusion cycles follow rather than precede the index ischemia. Postconditioning is also dependent on activation of Akt and ERK (Yang et al., 2004). Its interference with mPTP opening was demonstrated in a rather simple model in which calcium was added to isolated mitochondria (Argaud et al. 2005). Mitochondria sequester this calcium until the mPTP opens at which point they suddenly release the calcium back into the medium. The amount of calcium added until mPTP opening was measured. While the mPTP of mitochondria from the ischemic zone of untreated hearts opened when the added calcium load reached 16 µM/mg mitochondrial protein, mPTP in mitochondria from postconditioned hearts didn't open until the calcium load reached 41 µM/mg protein. And 47 µM/mg protein was required to trigger mPTP opening in mitochondria from preconditioned hearts.

Thus there seems to be a pattern emerging here. Drugs and interventions that activate survival kinase pathways during the reperfusion period protect the heart, and they do so by preventing mPTP opening. Mitochondria are increasingly involved in the pathways leading to cardioprotection. They are both signaling elements and potentially critical end-effectors. Their dysfunction contributes to both apoptosis and necrosis. A rational goal, therefore, in the development of cardioprotective strategies is to preserve mitochondrial function and integrity.

References

Aguilar-Bryan L, Clement JP, IV, Gonzalez G, Kunjilwar K, Babenko A, Bryan J (1998) Toward understanding the assembly and structure of K_{ATP} channels. Physiol Rev 78:227–245

Argaud L, Gateau-Roesch O, Chalabreysse L, Gomez L, Loufouat J, Thivolet-Béjui F, Robert D, Ovize M (2004) Preconditioning delays Ca^{2+}-induced mitochondrial permeability transition. Cardiovasc Res 61:115–12

Argaud L, Gateau-Roesch O, Raisky O, Loufouat J, Robert D, Ovize M (2005) Postconditioning inhibits mitochondrial permeability transition. Circulation 111:194–197

Auchampach JA, Gross GJ (1993) Adenosine A_1 receptors, K_{ATP} channels, and ischemic preconditioning in dogs. Am J Physiol 264:H1327–H1336

Baines CP, Goto M, Downey JM (1997) Oxygen radicals released during ischemic preconditioning contribute to cardioprotection in the rabbit myocardium. J Mol Cell Cardiol 29:207–216

Baines CP, Kaiser RA, Purcell NH, Blair NS, Osinska H, Hambleton MA, Brunskill EW, Sayen MR, Gottlieb RA, Dorn GW II, Robbins J, Molkentin JD (2005) Loss of cyclophilin D reveals a critical role for mitochondrial permeability transition in cell death. Nature 434:658–662

Baines CP, Liu GS, Birincioglu M, Critz SD, Cohen MV, Downey JM (1999) Ischemic preconditioning depends on interaction between mitochondrial K_{ATP} channels and actin cytoskeleton. Am J Physiol 276:H1361–H1368

Baines CP, Song C-X, Zheng Y-T, Wang G-W, Zhang J, Wang O-L, Guo Y, Bolli R, Cardwell EM, Ping P (2003) Protein kinase Cepsilon interacts with and inhibits the permeability transition pore in cardiac mitochondria. Circ Res 92:873–880

Baines CP, Zhang J, Wang G-W, Zheng Y-T, Xiu JX, Cardwell EM, Bolli R, Ping P (2002) Mitochondrial PKCepsilon and MAPK form signaling modules in the murine heart: enhanced mitochondrial PKCepsilon-MAPK interactions and differential MAPK activation in PKCepsilon-induced cardioprotection. Circ Res 90:390–397

Bajgar R, Seetharaman S, Kowaltowski AJ, Garlid KD, Paucek P (2001) Identification and properties of a novel intracellular (mitochondrial) ATP-sensitive potassium channel in brain. J Biol Chem 276:33369–33374.

Borutaite V, Brown GC (2003) Mitochondria in apoptosis of ischemic heart. FEBS Lett 541:1–5

Cohen MV, Yang X-M, Liu GS, Heusch G, Downey JM (2001) Acetylcholine, bradykinin, opioids, and phenylephrine, but not adenosine, trigger preconditioning by generating free radicals and opening mitochondrial K_{ATP} channels. Circ Res 89:273–278

Costa ADT, Garlid KD, West IC, Lincoln TM, Downey JM, Cohen MV, Critz SD (2005a) Protein kinase G transmits the cardioprotective signal from cytosol to mitochondria. Circ Res 97:329–336

Costa ADT, Quinlan CL, Andrukhiv A, West IC, Jaburek M, Garlid KD (2005b) The direct physiological effects of MitoK_{ATP} opening on heart mitochondria. Am J Physiol (in press)

Critz SD, Liu G-S, Chujo M, Downey JM (1997) Pinacidil but not nicorandil opens ATP-sensitive K^+ channels and protects against simulated ischemia in rabbit myocytes. J Mol Cell Cardiol 29:1123–1130

Das M, Parker JE, Halestrap AP (2003) Matrix volume measurements challenge the existence of diazoxide/glibenclamide-sensitive K_{ATP} channels in rat mitochondria. J Physiol 547:893–902

Forbes RA, Steenbergen C, Murphy E (2001) Diazoxide-induced cardioprotection requires signaling through a redox-sensitive mechanism. Circ Res 88:802–809

Garlid KD (1996) Cation transport in mitochondria - the potassium cycle. Biochim Biophys Acta 1275:123–126

Garlid KD (2000) Opening mitochondrial K_{ATP} in the heart - what happens, and what does not happen. Basic Res Cardiol 95:275–279

Garlid KD, Paucek P (2003) Mitochondrial potassium transport: the K^+ cycle. Biochim Biophys Acta 1606:23–41

Garlid KD, Paucek P, Yarov-Yarovoy V, Murray HN, Darbenzio RB, D'Alonzo AJ, Lodge NJ, Smith MA, Grover GJ (1997) Cardioprotective effect of diazoxide and its interaction with mitochondrial ATP-sensitive K^+ channels: possible mechanism of cardioprotection. Circ Res 81:1072–1082

Goto M, Liu Y, Yang X-M, Ardell JL, Cohen MV, Downey JM (1995) Role of bradykinin in protection of ischemic preconditioning in rabbit hearts. Circ Res 77:611–621

Green DR, Reed JC (1998) Mitochondria and apoptosis. Science 281:1309–1312

Gross GJ, Auchampach JA (1992) Blockade of ATP-sensitive potassium channels prevents myocardial preconditioning in dogs. Circ Res 70:223—233

Gross GJ, Fryer RM (1999) Sarcolemmal versus mitochondrial ATP-sensitive K^+ channels and myocardial preconditioning. Circ Res 84:973–979

Grover GJ, Garlid KD (2000) ATP-sensitive potassium channels: a review of their cardioprotective pharmacology. J Mol Cell Cardiol 32:677–695

Halestrap AP (1994) Regulation of mitochondrial metabolism through changes in matrix volume. Biochem Soc Trans 22:522–529

Halestrap AP (1999) The mitochondrial permeability transition: its molecular mechanism and role in reperfusion injury. Biochem Soc Symp 66:181–203

Halestrap AP, Clarke SJ, Javadov SA (2004) Mitochondrial permeability transition pore opening during myocardial reperfusion–a target for cardioprotection. Cardiovasc Res 61:372–385

Hanley PJ, Mickel M, Löffler M, Brandt U, Daut J (2002) K_{ATP} channel-independent targets of diazoxide and 5-hydroxydecanoate in the heart. J Physiol 542:735–741

Haunstetter A, Izumo S (1998) Apoptosis: basic mechanisms and implications for cardiovascular disease. Circ Res 82:1111–1129

Hausenloy DJ, Maddock HL, Baxter GF, Yellon DM (2002) Inhibiting mitochondrial permeability transition pore opening: a new paradigm for myocardial preconditioning? Cardiovasc Res 55:534–543

Hausenloy DJ, Tsang A, Mocanu MM, Yellon DM (2005) Ischemic preconditioning protects by activating prosurvival kinases at reperfusion. Am J Physiol 288:H971–H976

Hausenloy DJ, Yellon DM, Mani-Babu S, Duchen MR (2004) Preconditioning protects by inhibiting the mitochondrial permeability transition. Am J Physiol 287:H841–H849

Holmuhamedov EL, Jovanovic S, Dzeja PP, Jovanovic A, Terzic A (1998) Mitochondrial ATP-sensitive K^+ channels modulate cardiac mitochondrial function. Am J Physiol 275:H1567–H1576

Holmuhamedov EL, Wang L, Terzic A (1999) ATP-sensitive K^+ channel openers prevent Ca^{2+} overload in rat cardiac mitochondria. J Physiol 519:347–360

Inagaki N, Gonoi T, Clement JP IV, Namba N, Inazawa J, Gonzalez G, Aguilar-Bryan L, Seino S, Bryan J (1995a) Reconstitution of I_{KATP}: an inward rectifier subunit plus the sulfonylurea receptor. Science 270:1166–1170

Inagaki N, Gonoi T, Clement JP IV, Wang C-Z, Aguilar-Bryan L, Bryan J, Seino S (1996) A family of sulfonylurea receptors determines the pharmacological properties of ATP-sensitive K^+ channels. Neuron 16:1011–1017

Inagaki N, Tsuura Y, Namba N, Masuda K, Gonoi T, Horie M, Seino Y, Mizuta M, Seino S (1995b) Cloning and functional characterization of a novel ATP-sensitive potassium channel ubiquitously expressed in rat tissues, including pancreatic islets, pituitary, skeletal muscle, and heart. J Biol Chem 270:5691–5694

Inoue I, Nagase H, Kishi K, Higuti T (1991) ATP-sensitive K^+ channel in the mitochondrial inner membrane. Nature 352:244–247

Javadov SA, Lim KHH, Kerr PM, Suleiman M-S, Angelini GD, Halestrap AP (2000) Protection of hearts from reperfusion injury by propofol is associated with inhibition of the mitochondrial permeability transition. Cardiovasc Res 45:360–369

Juhaszova M, Zorov DB, Kim S-H, Pepe S, Fu Q, Fishbein KW, Ziman BD, Wang S, Ytrehus K, Antos CL, Olson EN, Sollott SJ (2004) Glycogen synthase kinase-3b mediates convergence of protection signaling to inhibit the mitochondrial permeability transition pore. J Clin Invest 113:1535–1549

Kluck RM, Bossy-Wetzel E, Green DR, Newmeyer DD (1997) The release of cytochrome c from mitochondria: a primary site for Bcl-2 regulation of apoptosis. Science 275:1132–1136

Korge P, Honda HM, Weiss JN (2002) Protection of cardiac mitochondria by diazoxide and protein kinase C: implications for ischemic preconditioning. Proc Natl Acad Sci 99:3312–3317

Krenz M, Oldenburg O, Wimpee H, Cohen MV, Garlid KD, Critz SD, Downey JM, Benoit JN (2002) Opening of ATP-sensitive potassium channels causes generation of free radicals in vascular smooth muscle cells. Basic Res Cardiol 97:365–373

Krieg T, Cui L, Qin Q, Cohen MV, Downey JM (2004a) Mitochondrial ROS generation following acetylcholine-induced EGF receptor transactivation requires metalloproteinase cleavage of proHB-EGF. J Mol Cell Cardiol 36:435–443

Krieg T, Qin Q, McIntosh EC, Cohen MV, Downey JM (2002) ACh and adenosine activate PI3-kinase in rabbit hearts through transactivation of receptor tyrosine kinases. Am J Physiol 283:H2322–H2330

Krieg T, Qin Q, Philipp S, Alexeyev MF, Cohen MV, Downey JM (2004b) Acetylcholine and bradykinin trigger preconditioning in the heart through a pathway that includes Akt and NOS. Am J Physiol 287:H2606–H2611

Liu GS, Thornton J, Van Winkle DM, Stanley AWH, Olsson RA, Downey JM (1991) Protection against infarction afforded by preconditioning is mediated by A_1 adenosine receptors in rabbit heart. Circulation 84:350–356

Liu Y, Sato T, O'Rourke B, Marban E (1998) Mitochondrial ATP-dependent potassium channels: novel effectors of cardioprotection? Circulation 97:2463–2469

Liu Y, Tsuchida A, Cohen MV, Downey JM (1995) Pretreatment with angiotensin II activates protein kinase C and limits myocardial infarction in isolated rabbit hearts. J Mol Cell Cardiol 27:883–892

Lorenz E, Terzic A (1999) Physical association between recombinant cardiac ATP-sensitive K^+ channel subunits Kir6.2 and SUR2A. J Mol Cell Cardiol 31:425–434

Miki T, Cohen MV, Downey JM (1998) Opioid receptor contributes to ischemic preconditioning through protein kinase C activation in rabbits. Mol Cell Biochem 186:3–12

Mocanu MM, Baxter GF, Yue Y, Critz SD, Yellon DM (2000) The p38 MAPK inhibitor, SB203580, abrogates ischaemic preconditioning in rat heart but timing of administration is critical. Basic Res Cardiol 95:472–478

Murata M, Akao M, O'Rourke B, Marbán E (2001) Mitochondrial ATP-sensitive potassium channels attenuate matrix Ca^{2+} overload during simulated ischemia and reperfusion: possible mechanism of cardioprotection. Circ Res 89:891–898

Murray CJL, Lopez AD (1997) Alternative projections of mortality and disability by cause 1990-2020: Global Burden of Disease Study. Lancet 349:1498–1504

Murry CE, Jennings RB, Reimer KA (1986) Preconditioning with ischemia: a delay of lethal cell injury in ischemic myocardium. Circulation 74:1124–1136

Nakano A, Baines CP, Kim SO, Pelech SL, Downey JM, Cohen MV, Critz SD (2000) Ischemic preconditioning activates MAPKAPK2 in the isolated rabbit heart: evidence for involvement of p38 MAPK. Circ Res 86:144–151

Newmeyer DD, Ferguson-Miller S (2003) Mitochondria: releasing power for life and unleashing the machineries of death. Cell 112:481–490

Nozawa Y, Miura T, Miki T, Ohnuma Y, Yano T, Shimamoto K (2003) Mitochondrial K_{ATP} channel-dependent and -independent phases of ischemic preconditioning against myocardial infarction in the rat. Basic Res Cardiol 98:50–58

Oldenburg O, Critz SD, Cohen MV, Downey JM (2003) Acetylcholine-induced production of reactive oxygen species in adult rabbit ventricular myocytes is dependent on phosphatidylinositol 3- and Src-kinase activation and mitochondrial K_{ATP} channel opening. J Mol Cell Cardiol 35:653–660

Oldenburg O, Qin Q, Krieg T, Yang X-M, Philipp S, Critz SD, Cohen MV, Downey JM (2004) Bradykinin induces mitochondrial ROS generation via NO, cGMP, PKG, and mitoK_{ATP} channel opening and leads to cardioprotection. Am J Physiol 286: H468–H476

Oldenburg O, Qin Q, Sharma AR, Cohen MV, Downey JM, Benoit JN (2002) Acetylcholine leads to free radical production dependent on K_{ATP} channels, Gi proteins, phosphatidylinositol 3-kinase and tyrosine kinase. Cardiovasc Res 55:544–552

Ozcan C, Bienengraeber M, Dzeja PP, Terzic A (2002) Potassium channel openers protect cardiac mitochondria by attenuating oxidant stress at reoxygenation. Am J Physiol 282:H531-H539

Pain T, Yang X-M, Critz SD, Yue Y, Nakano A, Liu GS, Heusch G, Cohen MV, Downey JM (2000) Opening of mitochondrial K_{ATP} channels triggers the preconditioned state by generating free radicals. Circ Res 87:460–466

Paucek P, Mironova G, Mahdi F, Beavis AD, Woldegiorgis G, Garlid KD (1992) Reconstitution and partial purification of the glibenclamide-sensitive, ATP-dependent K^+ channel from rat liver and beef heart mitochondria. J Biol Chem 267:26062–26069

Paucek P, Yarov-Yarovoy V, Sun X, Garlid KD (1996) Inhibition of the mitochondrial K_{ATP} channel by long-chain acyl-CoA esters and activation by guanine nucleotides. J Biol Chem 271:32084–32088.

Pfeffer JM, Pfeffer MA, Fletcher PJ, Braunwald E (1991) Progressive ventricular remodeling in rat with myocardial infarction. Am J Physiol 260:H1406–H1414

Ping P, Zhang J, Qiu Y, Tang X-L, Manchikalapudi S, Cao X, Bolli R (1997) Ischemic preconditioning induces selective translocation of protein kinase C isoforms e and h in the heart of conscious rabbits without subcellular redistribution of total protein kinase C activity. Circ Res 81:404–414

Qin Q, Downey JM, Cohen MV (2003) Acetylcholine but not adenosine triggers preconditioning through PI3-kinase and a tyrosine kinase. Am J Physiol 284:H727–H734

Sato T, Sasaki N, Seharaseyon J, O'Rourke B, Marbán E (2000) Selective pharmacological agents implicate mitochondrial but not sarcolemmal K_{ATP} channels in ischemic cardioprotection. Circulation 101:2418–2423

Schultz JEJ, Rose E, Yao Z, Gross GJ (1995) Evidence for involvement of opioid receptors in ischemic preconditioning in rat hearts. Am J Physiol 268:H2157–H2161

Schulz R, Gres P, Heusch G (2003) Activation of ATP-dependent potassium channels is a trigger but not a mediator of ischaemic preconditioning in pigs. Br J Pharmacol 139:65–72

Solenkova N, Cohen MV, Downey JM (2006) Endogenous adenosine protects the preconditioned heart during the early minutes of reperfusion by activating Akt. Am J Physiol 290:H441–449

St..John Sutton M, Pfeffer MA, Moye L, Plappert T, Rouleau JL, Lamas G, Rouleau J, Parker JO, Arnold MO, Sussex B, Braunwald E, for the SAVE Investigators (1997) Cardiovascular death and left ventricular remodeling two years after myocardial infarction. Baseline predictors and impact of long-term use of captopril: information from the Survival and Ventricular Enlargement (SAVE) trial. Circulation 96: 3294–3299

Thornton JD, Liu GS, Downey JM (1993) Pretreatment with pertussis toxin blocks the protective effects of preconditioning: evidence for a G-protein mechanism. J Mol Cell Cardiol 25:311–320

Tritto I, D'Andrea D, Eramo N, Scognamiglio A, De Simone C, Violante A, Esposito A, Chiariello M, Ambrosio G (1997) Oxygen radicals can induce preconditioning in rabbit hearts. Circ Res 80:743–748

Tsuchida A, Liu GS, Wilborn WH, Downey JM (1993) Pretreatment with the adenosine A_1 selective agonist, 2-chloro-N6-cyclopentyladenosine (CCPA), causes a sustained limitation of infarct size in rabbits. Cardiovasc Res 27:652–656

Tsuchida A, Liu Y, Liu GS, Cohen MV, Downey JM (1994) a1-Adrenergic agonists precondition rabbit ischemic myocardium independent of adenosine by direct activation of protein kinase C. Circ Res 75:576–585

Van Winkle DM, Thornton JD, Downey DM, Downey JM (1991) The natural history of preconditioning: cardioprotection depends on duration of transient ischemia and time to subsequent ischemia. Coron Artery Dis 2:613–619

Vanden Hoek TL, Becker LB, Shao Z-H, Li C-Q, Schumacker PT (2000) Preconditioning in cardiomyocytes protects by attenuating oxidant stress at reperfusion. Circ Res 86:541–548

Wall TM, Sheehy R, Hartman JC (1994) Role of bradykinin in myocardial preconditioning. J Pharmacol Exp Ther 270:681–689

Wang P, Gallagher KP, Downey JM, Cohen MV (1996) Pretreatment with endothelin-1 mimics ischemic preconditioning against infarction in isolated rabbit heart. J Mol Cell Cardiol 28:579–588

Wang S, Cone J, Liu Y (2001) Dual roles of mitochondrial K_{ATP} channels in diazoxide-mediated protection in isolated rabbit hearts. Am J Physiol 280:H246–H255

Weiss JN, Korge P, Honda HM, Ping P (2003) Role of the mitochondrial permeability transition in myocardial disease. Circ Res 93:292–301

Yang X-M, Proctor JB, Cui L, Krieg T, Downey JM, Cohen MV (2004) Multiple, brief coronary occlusions during early reperfusion protect rabbit hearts by targeting cell signaling pathways. J Am Coll Cardiol 44:1103–1110

Yarov-Yarovoy V, Paucek P, Jaburek,M, Garlid KD (1997) The nucleotide regulatory sites on the mitochondrial K_{ATP} channel face the cytosol. Biochim Biophys Acta 1321:128–136

Yellon DM, Downey JM (2003) Preconditioning the myocardium: from cellular physiology to clinical cardiology. Physiol Rev 83:1113–1151

Yue Y, Krenz M, Cohen MV, Downey JM, Critz SD (2001) Menadione mimics the infarct-limiting effect of preconditioning in isolated rat hearts. Am J Physiol 281: H590–H595

Zhao Z-Q, Corvera JS, Halkos ME, Kerendi F, Wang N-P, Guyton RA, Vinten-Johansen J (2003) Inhibition of myocardial injury by ischemic postconditioning during reperfusion: comparison with ischemic preconditioning. Am J Physiol 285:H579–H588

Zorov DB, Filburn CR, Klotz L-O, Zweier JL, Sollott SJ (2000) Reactive oxygen species (ROS)-induced ROS release: a new phenomenon accompanying induction of the mitochondrial permeability transition in cardiac myocytes. J Exp Med 192: 1001–1014

Zweier JL, Flaherty JT, Weisfeldt ML (1987) Direct measurement of free radical generation following reperfusion of ischemic myocardium. Proc Natl Acad Sci 84: 1404–1407

15
Mitochondrial DNA Damage and Repair

Inna N. Shokolenko, Susan P. Ledoux and Glenn L. Wilson

15.1. Mitochondrial Genome Overview

One of the unique features of mitochondria is that these organelles possess their own DNA (mtDNA). The mitochondrial genome, like any DNA is subject to continuous attack on its integrity from both endogenous and exogenous sources. In order to understand the consequences of such an attack, one must consider key aspects of mtDNA organization and maintenance. Mammalian cells contain one to several thousand copies of mtDNA per cell, which are characterized as being enclosed in multiple mitochondria at 1 to 11 copies per organelle (Cavelier et al. 2000). Human mtDNA is a circular negatively supercoiled double-stranded molecule that is 16,569 bp long (Figure 15.1). It encodes 13 polypeptides, 22 tRNAs and 2 rRNAs. All the proteins encoded by mtDNA and synthesized within this organelle on mitochondrial ribosomes are key subunits of the electron transport chain (ETC) and ATP synthase. These include 7 of the 46 subunits of complex I (NADH dehydrogenase), one (cytochrome b) of the 11 subunits of complex III (cytochrome bc1n), three of the 13 subunits of complex IV (cytochrome c oxidase), and two of the 16 subunits of complex V (F_0F_1 ATPase) (Attardi and Schatz 1988). The rest of the polypeptides of the ETC complexes, including all the subunits of complex II (succinate dehydrogenase), as well as approximately 1500 other proteins which function in mitochondria are encoded by nuclear genes, synthesized in the cytosol and imported into mitochondria through various protein import systems.

The organization of mammalian mtDNA is very different from the nuclear genome in that mitochondrial genes have no 5' or 3' non-coding sequences, no introns, and intergenic sequences are absent or limited to a few bases. mtDNA is totally dependent on nuclear-encoded proteins for its maintenance. Furthermore, this DNA replicates throughout the lifespan of an organism in both proliferating and post-mitotic cells in order to maintain a constant supply of genetic material so that the organelles can undergo continuous turnover. The mean lifetime of

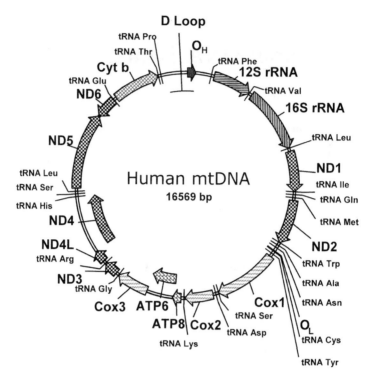

FIGURE 15.1. Map of human mtDNA. OH and OL, origins of heavy and light strand replication respectively; NDi-ND6, NADH dehydrogenase (ETC complex I) subunits 1-6; CoxI-Cox3, cytochrome oxidase subunits 1-3 (ETC complex IV); ATP6 and ATP8, subunits 6 and 8 of mitochondrial ATPase (complex V); Cyt b, cytochrome b (complex III).

a mtDNA molecule has been estimated to be 2 and 4 weeks in rat liver and brain cells respectively (Cross et al. 1969). According to the strand-asymmetric model, mtDNA replication occurs bi-directionally, initiated at two spatially and temporally distinct origins of replication O_H and O_L, for the heavy and light strands, respectively [for review, see (Shadel and Clayton 1997; Taanman 1999)]. However, this paradigm recently has been challenged, and evidence has been offered suggesting the presence of conventional duplex mtDNA replication intermediates, indicative of coupled leading and lagging-strand DNA synthesis (Yang et al. 2002). The replication of mtDNA is conducted by DNA polymerase gamma (DNA pol γ). The minimal mitochondrial "replisome" has been reconstituted *in vitro* and consists of several proteins which include DNA pol γ, a 5'-3'DNA helicase Twinkle, and a mitochondrial single-strand DNA-binding protein (mtSSB)(Korhonen et al. 2004). Initiation of mtDNA replication is coupled to transcription. Transcription of each mtDNA strand in humans starts from two promoters, the light- and heavy-strand promoters (LSP and HSP) and requires mitochondrial transcription factor TFAM and RNA polymerase. Large, near-genomic length polycistronic transcripts are produced from each strand and processed to generate mature mitochondrial tRNAs, rRNAs, and mRNAs. The primer needed to initiate DNA replication at O_H is generated by the transcription from LSP and subsequent RNA processing [reviewed in (Shadel and Clayton, 1997)]. This mechanism of transcription-primed replication in mitochondria is conserved among vertebrates. Spatial organization of mtDNA within organelles also is different from that of nuclear DNA. Mitochondria do not have histones and mtDNA is not organized in nucleosomes, as is nuclear DNA. However, it is by no means "naked" DNA as was originally believed. In fact, mammalian mtDNA is organized in nucleoids, which can be seen under the microscope as punctate structures containing mtDNA and proteins which localize to the matrix surface of the mitochondrial inner membrane. These large mtDNA-protein complexes are similar to the nucleoids described in lower eukaryotes. Human mtDNA is organized into several hundred nucleoids which contain a mean of 2-8 mtDNA molecules each and are distributed throughout the mitochondrial compartment (Legros et al. 2004). Nucleoids are dynamic structures able to divide and redistribute in the mitochondrial network and are recognized as heritable units during mitochondrial division (Garrido et al. 2003). mtDNA binding proteins such as TFAM, mtSSB and Twinkle have been shown to co-localize with mtDNA in intramitochondrial foci in live cells (Garrido et al. 2003). Another study showed that nucleoids purified from *Xenopus* oocytes, contained additional proteins persistently associated with mtDNA in a relatively detergent-resistant complex, including adenine nucleotide translocator 1 (ANT1), which may be responsible for the attachment of mtDNA to the inner mitochondrial membrane, E2 subunits of pyruvate and α-ketoacid dehydrogenases, and mitochondrial chaperone prohibitin2 (Bogenhagen et al. 2003). The molecular characterization of the protein component of nucleoids remains a challenging task. Active research, currently underway in this area, will undoubtedly produce a better understanding of these structures in the near future.

15.2. Drugs, Toxins, Oxidative Stress and Other Hazards in the Life of mtDNA

15.2.1. Vulnerability of mtDNA to Environmental Toxins and Drugs

Mitochondrial DNA, just like its counterpart in the nucleus, is constantly exposed to damaging agents from both environmental toxins, as well as many drugs used for therapeutic purposes. Some of these chemicals may pose a greater danger to mtDNA as a consequence of its unique location. A number of bulky carcinogens have been shown to cause adducts to form in mtDNA at a higher frequency than nuclear DNA. Because the mitochondrial membrane potential generates a negative charge on the matrix-side of the inner membrane, lipophilic cations tend to accumulate in mitochondria, specifically in mitochondrial membranes. Therefore, mitochondria can take up lipophilic cations from the cytosol and concentrate them up to 1000-fold (Singer et al,. 1990). Unfortunately, many chemical carcinogens as well as medically beneficial drugs have a lipophilic hydrocarbon structure with positive charges or may require activation by mixed-function oxidases in the mitochondria. Thus, activated carcinogens may have easy access to mtDNA owing to its association with the mitochondrial inner membrane. For example, polycyclic aromatic hydrocarbons and nitrosoamines accumulate in mitochondria (Allen and Coombs 1980; Takayama and Muramatsu 1969). Also, the mutagenic and carcinogenic benzo[a]pyrene derivatives have been shown to cause bulky covalent modifications in mtDNA to a much greater extent than in nDNA (Backer and Weinstein 1980). These bulky adducts lead to strong distortion of the DNA double helix and block transcription and DNA replication.

Other compounds that preferentially damage mtDNA are: methyl methane-sulphonate (MMS) (Pirsel and Bohr 1993), N-nitroso-N-nitrosoguanidine (MNNG) (Miyaki et al. 1977), N-methyl-N-nitrosourea (MNU) (Wunderlich et al. 1970), all causing alkylation of DNA; 4-nitroquinoline 1-oxide (4NQO) which induces a wide range of DNA lesions including single-strand breaks, pyrimidine-dimer formation, abasic sites, and oxidized bases (Mambo et al., 2003); cyclophosphamide (CPA) (Neubert et al. 1981) – an alkylating and DNA crosslinking agent; chromium (Rossi et al. 1988), carbon tetrachloride (Tomasi et al. 1987), and aflatoxin B1 (Niranjan et al. 1982) – which can cause oxidative damage, depurinations, and open ring structures.

Antitumor drugs such as Adriamycin and related compounds undergo redox cycling within the electron transport chains in mitochondria and thereby generate oxygen radicals (Doroshow and Davies 1986). Also mtDNA is as vulnerable as nDNA to ionizing radiation and radiomimetic drugs such as bleomycin and neocarzinostatin which are known to lead to strand breaks.

15.3. Endogenous Sources of ROS Pose a Risk of Oxidative Damage to mtDNA

Many normal cellular metabolic processes are well established sources for the endogenous production of reactive oxygen species (ROS) (for a comprehensive review see (Droge 2002) and references therein). ROS include the oxygen free radicals superoxide anion (O_2^-) and the very active hydroxyl radical (.OH), as well as non-radical oxidants such as hydrogen peroxide (H_2O_2) and singlet oxygen. Living organisms use ROS for a variety of physiological functions including phagocytosis, reactions of microsomal P_{450} oxygenases and peroxisomal oxidative metabolism, regulation of vascular tone, oxygen tension control, prostaglandin metabolism. ROS are typically generated in these cases by tightly regulated enzymes. Excessive amounts of ROS also may arise from less well-regulated sources such as the mitochondrial electron-transport chain (ETC) during oxidative phosphorylation. Although the ETC is a very efficient system, a small portion of electrons are passed not to the next electron carrier in the chain but directly to oxygen, generating O_2^-. Superoxide is converted to H_2O_2 by the mitochondrial matrix enzyme Mn superoxide dismutase (MnSOD) or by Cu/Zn superoxide dismutase (Cu/ZnSOD) which is located in both the mitochondrial intermembrane space and in the cytosol. H_2O_2 is more stable than superoxide and can readily diffuse across membranes. It can be converted further to water by mitochondrial and cytosolic glutathione peroxidase or to O_2 and H_2O by peroxisomal catalase as part of the cellular antioxidant defense mechanism. However, H_2O_2 also can be converted to hydroxyl radical in the presence of reduced transition metals such as iron and copper, which are abundant in the inner mitochondrial membrane. Traditionally, it was considered that mitochondrial generation of ROS represents quantitatively the major intracellular source of oxygen radicals under physiological conditions. In addition to the ETC toxic by-products produced in the inner mitochondrial membrane, there is another source of mitochondria-generated ROS that may have been underestimated. The mitochondrial outer membrane enzyme monoamine oxidase catalyzes the oxidative deamination of biogenic amines and is a quantitatively large source of the H_2O_2 which contributes to an increase in the steady state concentrations of ROS within both the mitochondrial matrix and cytosol (Cadenas and Davies 2000).

Additionally, recent studies from two independent groups (Starkov et al. 2004; Tretter and Adam-Vizi 2004) reveal that a key enzyme in the Krebs' cycle may rival the ETC complexes as the major site for ROS generation. Alpha-ketoglutarate dehydrogenase is able to generate hydrogen peroxide (and probably superoxide) at a high rate in normally functioning mitochondria. The rate of H_2O_2 production is strongly dependent upon an elevated intramitochondrial NADH/NAD$^+$ ratio (e.g. condition occurring during ETC inhibition).

Because of the presence of such strong ROS producing systems in mitochondria, these organelles also are one of the main intracellular targets of the ROS - induced damage. Oxidized proteins, lipids, nucleic acids and nucleotides can be detected under normal metabolic conditions in mitochondria (Raha and Robinson 2000), although under normal circumstances, the levels of ROS in mitochondria is rather low because it is efficiently removed through anti-oxidant defense systems which include superoxide dismutases, glutathione peroxidase and catalase. However, there are conditions when the rate of ROS production by mitochondria may dramatically increase, such as during exposure to toxic chemicals or chemotherapeutic drugs with redox-cycling ability or when local tissue anoxia during ischemia is followed by reperfusion with oxygen. Elevated ROS levels pose a greater threat of oxidative damage to mtDNA, simply because mtDNA is located in close proximity to mitochondrial sources of ROS production (Cadenas and Davies 2000).

15.4. Oxidative Damage to mtDNA

Reactive oxygen radicals can cause a wide variety of lesions in DNA. Reactions with purines and pyrimidines result in multiple products in DNA. Currently, more than 50 oxidative DNA base lesions have been characterized [reviewed in (Cooke et al. 2003; Bjelland and Seeberg 2003; Martinez et al. 2003)], and many of these were identified *in vitro* or *in vivo* following exposure of mammalian cells to ROS (Dizdaroglu 1998). Other lesions include DNA –protein cross-links, sugar modifications, and single- and double-strand breaks [for more detailed reviews see (Cooke et al. 2003; Dizdaroglu 1998; Wallace 1998)]. Additionally, ROS generated during prolonged oxidative stress in a fibroblast cell line has been shown to induce extensive fragmentation and deletions in mtDNA (Yoneda et al. 1995). Of the four bases in DNA guanine is the most easily oxidized, as it has the lowest redox potential (Steenken 1997). Presently, about 15 oxidized forms of guanine have been identified in DNA (Bjelland and Seeberg 2003). Of these the most studied lesion is 8-oxoguanine (8-oxoG), which is formed by a variety of oxidative treatments. 8-oxoG has been found to be a mutagenic lesion. It does not pose a significant block to DNA polymerases, instead, it is recognized as a substrate, with polymerases inserting either C or A opposite 8-oxoG. Mispairing of 8-oxoG with adenine results in G-C to T-A transversions during the subsequent round of replication. This has been well documented in both bacterial systems (Cheng et al. 1992; Moriya et al. 1991; Wood et al. 1990) and mammalian cells (Klein et al. 1992; Le Page et al. 1995; Moriya 1993). All replicative and repair DNA polymerases studied demonstrate different degrees of A and C incorporation opposite 8-oxoG. DNA polymerase gamma (Pol γ), which is both the repair and replicative polymerase in mitochondria, has been shown in *in vitro* studies to insert A opposite 8-oxoG 27% of the time and thus potentially lead to mutations during mtDNA replication (Pinz et al. 1995). Free dGTP is also a target of oxygen radicals, which represents another potentially mutagenic event. 8-oxodGTP can be used by viral, bacterial and mammalian polymerases to incorporate 8-oxo-GMP into DNA- both opposite cytosine and adenine and thereby induce A to C transversions (Cheng et al. 1992; Moriya et al. 1991; Wood et al. 1990; Klein et al. 1992; Le Page et al. 1995; Moriya 1993; Pinz et al. 1995; Pavlov et al. 1994).

In the previous decade a significant amount of effort was put into the evaluation of the *in vivo* levels of 8-oxoG in nuclear and mtDNA. Since the initial report in 1988 by Richter et al. (1988) describing 16-fold higher levels of 8-oxoG in mtDNA than in nDNA, a number of groups have carried out comparative studies which revealed greater oxidative damage in mtDNA (Hamilton et al. 2001; Zastawny et al. 1998). However, others reported that damage in mitochondria was overestimated, and it's level is comparable with that of nuclear DNA (Anson et al. 2000; Lim et al. 2005). A comparison of the reported levels of oxidation in mtDNA revealed that the range of measurements spanned over four orders of magnitude in different studies (Beckman and Ames 1999). Variation also existed in the measurement of nuclear DNA oxidation (Beckman and Ames 1999). More recently it has become apparent that such variations between published values for 8-oxoG in DNA are due to the methods employed, rather than to endogenous levels (Helbock et al. 1998 and Senturker et al. 1999). Thus, there has been an argument in the literature over the validity of the various methods used. At issue here is whether spurious oxidation of DNA bases occurred during sample preparation for analysis by chromatographic methods (GS-MS, HPLC) with electrochemical detection (HPLC-ECD) or HPLC with tandem mass spectrometry (HPLC-MS/MS) and, if so, ways to minimize it. Recently, to address this issue, a consortium of 28 laboratories – the European Standards Committee on Oxidative DNA damage (ESCODD) – was created. ESCODD has been testing the ability of different laboratories, using a variety of methods, to measure 8-oxoG in standard samples (ESCODD, 2002 and ESCODD, 2003). In addition to chromatographic analysis, the testing was performed using an alternative approach based on the use of the DNA repair enzyme formamidopyrimidine DNA N-glycosylase (FPG). This enzyme, in conjunction with alkali treatment, makes breaks in DNA at sites of oxidized bases. The techniques used were the Comet assay, alkaline elution and alkaline unwinding. The enzymatic approach, which is less prone to artifactual oxidation, gave much lower values than chromatography based methods, and demonstrated greater sensitivity. All methods had their strengths and weaknesses and required a great amount of experience on the part of the operator to be reliable. It was proposed that the background level of DNA oxidation in normal human cells is likely to be around 0.3-4.2 8-oxoG per 10^6 guanines (Collins et al. 2004).

The findings discussed above are pertinent to total cellular DNA. Since mtDNA comprises about only 1% of the DNA content in the cell, even if the levels of oxidation in mtDNA were higher, the actual amount would be masked by the relatively low values in nDNA and, as a result, underestimated. On the other hand, the accurate measurement of oxidative damage in mtDNA is particularly challenging because the process of organellar purification with subsequent mtDNA isolation may induce an additional oxidative damage (Anson et al., 2000).

Although, a consensus still has not been reached on whether the level of endogenous oxidative damage is greater in mtDNA, some indirect evidence supports this notion. The detection of DNA damage by the method of quantitative extended length PCR (QXL-PCR) developed by Van Houten et al. (1997), circumvents some of the problems associated with the analytical methods. This gene-specific PCR assay does not need mtDNA isolation, and it only requires nanogram quantities of total DNA. The method is based on the premise that damage in DNA, such as single-strand breaks and some sugar and base lesions, blocks thermostable polymerases and reduces the efficiency of amplification. Therefore, the product yield is inversely correlated with the level of damage in the starting template. It has been demonstrated that treatment of cells with exogenous hydrogen peroxide induces more lesions in mtDNA than in nuclear DNA in the cells studied, and they take a longer time to repair (Yakes and Van Houten 1997). Thus, it seems likely that in addition to the well established vulnerability to exogenous chemical mutagens, mtDNA also is vulnerable to exogenous oxidative mutagens as well.

15.5. Mitochondrial DNA Damage, Mutations and Disease

The combination of two factors, a great vulnerability to chemical and oxidative damage and the continuous turnover of the mitochondrial genome in post mitotic tissues, can cause an accumulation of mutations. Therefore, higher mutation rates in mtDNA than in nuclear would be expected. In actuality, mutation rates of human mtDNA have been reported to be several hundredfold greater than nuclear gene mutation rates (Khrapko et al. 1997). Consistent with this high mutation rate is the fact that genes encoded by mtDNA evolve 10-fold faster than nuclear genes (Brown et al. 1979).

In the 1980s it was discovered that mutations in mtDNA could be pathogenic. Several groups reported that both mtDNA point mutations (Wallace et al. 1988) and deletions (Holt et al., 1988; Lestienne et al. 1988) could be the underlying cause of a variety of defined human pathologies. Subsequently there has been a rapid growth in the number of mtDNA mutations implicated in human disease, and new mutations are still being identified. Base substitution and rearrangement mutations can occur throughout the mitochondrial genome and include both maternally inherited and sporadic cases. The recently released Mitomap (Human Mitochondrial Genome Database) lists almost 200 pathogenic point mutations, single nucleotide deletions and insertions (http://mitomap.org). Mitochondrial diseases reveal not only a causative link between mtDNA mutations and pathology, but also an increased oxidative burden in patients suffering from these disorders (Geromel et al. 2001; Kunishige et al. 2003; Lu et al. 2003). As the number of mtDNA – associated diseases identified has increased, it has become clear that many mitochondrial diseases typically have a delayed onset and progressive course and that they result in the same clinical manifestations as are observed in age-related diseases. Accumulation of mutations and deletions in mtDNA with their associated defects in energy metabolism have been implicated in ischemic heart disease, Parkinson's disease, Alzheimer's disease, amyotrophic lateral sclerosis (ALS), diabetes mellitus and aging. Specific symptoms include blindness, deafness, movement disorders, dementias, renal dysfunction, cardiovascular disease and endocrine disorders including hyperglycemia [reviewed in (DiMaura and Schon 2003; Kang and Hamasaki 2005; Wallace 2005)].

Direct involvement of mtDNA mutations in aging has been recently demonstrated in homozygous knock-in mice expressing a proof-reading deficient version of DNA pol γ, the mtDNA polymerase (Trifunovic et al. 2004). The knock-in mice develop a mtDNA mutator phenotype with an increase in the levels of point mutations and deletions in somatic mtDNA. This increase was associated with reduced life span and the premature onset of aging (Trifunovic et al. 2004). In other studies, transgenic mice expressing proof-reading deficient mtDNA polymerase under a cardiac-specific promoter accumulate mutations specifically in heart mtDNA and subsequently develop a severe dilated cardiomyopathy (Zhang et al. 2003). Interestingly, in both models, subsequent investigations failed to detect signs of increased oxidative stress, but found that tissue dysfunction could be attributed to the loss of a critical number of cells through the induction of apoptosis (Zhang et al. 2003; Kujpoth et al. 2005).

Somatic mtDNA mutations also have been found in a variety of human malignancies including breast, colorectal, liver, lung, prostate, skin, bladder, head and neck cancers (Copeland et al. 2002; Durham et al. 2003; Petros et al. 2005). mtDNA mutations often are found in primary tumors but not in the surrounding tissue. The mutations are found over the entire region of the mitochondrial genome, including point, deletion and insertion mutations which are homoplasmic in nature. This indicates that the mutant mtDNA becomes dominant in these cells (Fliss et al. 2000; Polyak et al. 1998). A demonstration of the physiological significance of these mtDNA mutations was provided recently by studies with transmitochondrial cybrids ("chimeric" cells in which their own mtDNA is destroyed and replaced by the mtDNA from another cell). The known pathogenic mtDNA mutation in ATP6 was introduced into the PC3 prostate cancer cell line via cybrid transfer. Injection of the resulting ATP6 mutant cybrids in nude mice resulted in the generation of tumors that were 7 times larger than those generated from wild type cybrids (Petros et al. 2005). In addition, the ATP6 mutant cybrid tumors generated much more reactive oxygen species (ROS) than the wild type cybrid tumors (Petros et al. 2005). These findings are a part of a growing body of evidence which indicates that cancer cells are under increased intrinsic oxidative stress [reviewed in (Pelicano et al. 2004)]. Since mitochondria are the major source for ROS production in cells, the vulnerability of mtDNA to ROS-mediated damage appears to be a mechanism for amplifying ROS stress in cancer cells. Also, mitochondrial ROS may play a role in carcinogenesis, which is supported by the finding that mice heterozygous for MnSOD (MnSOD +/- mice) have a 100% increase in tumor incidence which is accompanied by an increase in oxidative damage in nuclear and mitochondrial DNA (Van Remmen et al. 2003).

15.6. Repair of DNA Damage in Mammalian Mitochondria

The ability to maintain genomic integrity through DNA repair is a fundamental feature of all living organisms. Unless repaired, DNA damage may cause detrimental biological consequences in living cells such as cell death or mutations, which may increase genomic instability and thus enhance the risk for the development of cancer and other diseases. Characterization of DNA repair mechanisms has generally focused on these processes in nuclear DNA. Several major repair pathways processing a wide variety of DNA lesions are operative in the nucleus [reviewed in (Christmann et al. 2003; Sancar et al. 2004; Scharer 2003)]. They include nucleotide excision repair (NER), base excision repair (BER), direct repair/reversal (DR), mismatch repair (MMR), homologous recombination (HR), nonhomologous end joining (NHEJ) and translesion synthesis

(TLS). These incredibly complex systems, involving dozens of proteins and multi-protein complexes, continue to be actively investigated.

Until recently it was commonly believed that DNA repair mechanisms were either non-existent or very inefficient in mitochondria and that damaged DNA molecules were simply degraded, while multiple undamaged copies served as templates for DNA re-synthesis. Such views were based on early experiments showing that UV-induced pyrimidine dimers were not repaired in mtDNA (Clayton et al. 1974). In later studies, it was observed that mitochondria also did not repair complex alkyl damage caused by nitrogen mustard and intrastrand cross-links induced by cisplatin (LeDoux et al. 1992). Psoralen - induced cross-links also appeared to persist in mitochondria (Magana-Schwencke et al. 1982). However, evidence indicating that mitochondria do possess some repair systems started accumulating at this time. The first reports revealed that simple alkyl damage from streptozotocin (Pettepher et al. 1991), methylnitrosourea and dimethylsulfoxide (LeDoux et al. 1992) was repaired in mtDNA. Subsequent studies on repair of alloxan induced oxidative damage indicated that mitochondria were able to efficiently repair injury to their DNA caused by ROS (Drigger et al. 1993). The consistent findings of numerous studies that the repair capacity in mitochondria depends upon the type of lesion produced has led to the conclusion that mechanisms for repair of complex strand distorting lesions (bulky adducts, cross-links and double-strand breaks), namely nucleotide excision repair and homologous recombination, are absent in mitochondria. However, mtDNA is well protected against damage to DNA bases resulting from deamination, simple alkylation and oxidation [reviewed in (Croteau et al. 1999; LeDoux et al. 1999; Sawyer and Van Houten 1999)] by a robust base excision repair pathway.

15.7. Base Excision Repair (BER) Pathway in Mitochondria

BER is highly conserved from bacteria to humans. It is a multiprotein pathway which is different from NER in that the substrate specificity depends upon diverse DNA glycosylases initiating the first step in the pathway, rather than a multi-protein complex. The overall pathway for mitochondrial BER is similar to that observed for the short-patch pathway of BER in the eukaryotic nucleus (Bohr et al. 2002). This pathway consists of several enzymes working in succession (Figure 15.2). The BER pathway is initiated by DNA glycosylases, a class of enzymes that recognize a modified base, such as 8-oxoG or thymine glycol, and remove it from DNA by hydrolyzing the N-glycosidic bond between the modified base and the sugar. Simple glycosylases only cleave the N-glycosidic bond leaving an abasic (AP) site, while glycosylases with an associated AP lyase activity cleave both the N-glycosidic bond and the DNA phosphate backbone 3' of the lesion. Generally, the lyase action is associated with DNA glycosylases specific for oxidatively modified DNA bases, and not with those that remove alkylated bases (Dizdaroglu 2005). Following the glycosylase step, an

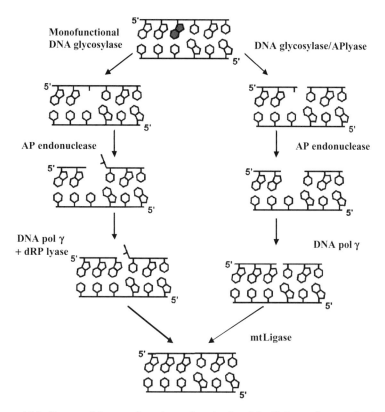

FIGURE 15.2. Base excision repair pathway in mitochondria. Refer to the text for details.

AP endonuclease is required to incise the DNA backbone 5' of the AP site, if it is following the action of a simple glycosylase. If it follows the action of a glycosylase/AP lyase, an AP endonuclease must remove the 3'- deoxyribosephosphate moiety. In either case, the role of AP endonuclease is to generate a 3'-hydroxyl group, which can be extended by a polymerase γ. The 5'- deoxyribosephosphate, left after AP endonuclease incision of an AP site, is removed either by DNA polymerase γ or DNA ligase III. After DNA polymerase γ inserts a new nucleotide into the lesion site, the process is completed by DNA ligase III, which seals the nick. Although, each step in BER is mediated by an individual enzyme, the process is highly coordinated in order to ensure un-interrupted repair of a basic sites and strand-break intermediates, which are themselves highly toxic lesions.

At the time when the evidence for mammalian mitochondrial DNA repair started accumulating, little was known about the particular DNA repair enzymes involved in those processes. Over the past several years an active investigation into this area has started to clarify the nature of many of these proteins and mechanisms of their actions.

15.8. DNA-Glycosylases in Mitochondria

The first DNA glycosylase activity identified in mitochondria was uracil DNA glycosylase (UNG) (Anderson and Friedberg 1980). This enzyme removes uracil from DNA which occurs as a result of deamination of cytosine or through misincorporation of dUMP. In eukaryotic cells there are two isoforms of UNG, mitochondrial (UNG1) and nuclear (UNG2), which are encoded by the same nuclear gene. UNG1 and UNG2 utilize different promoters and are processed further by alternative splicing which gives them either a strong mitochondrial or nuclear localization signal [reviewed in (Krokan et al. 2001)]. UNG1 is expressed in most tissues with the highest expression being in heart, skeletal muscles and testis. UNG1 is the only mitochondrial DNA glycosylase that can act on both single-strand and double-strand DNA, whereas all others require double-strand DNA for activity. UNG1 may have a slightly different mechanism of action in mitochondria, because it must work through short-patch BER, which is the only one known in mitochondria. In the nucleus uracils generally are repaired by long-patch BER (Stierum et al. 1999). Recently, it has been shown that $Ung^{-/-}$ mice develop normally and have no obvious phenotype after 18 months (Nilsen et al. 2000). These Ung knockout mice are deficient in UNG activity, but display low uracil-DNA-glycosylase activities in extracts from different organs. These "backup" activities differ in biochemical properties from UNG and may indicate a previously unrecognized uracil-DNA –glycosylase (Krokan et al. 2001).

8-oxo-guanine DNA glycosylase (OGG1) is the predominant enzyme identified in mammalian cells for the repair of 8-oxoG [reviewed in (Boiteux and Radicella 2000)]. This is a bi-functional DNA glycosylase with an associated AP lyase activity. It preferentially removes 8-oxoG opposite C, as well as several other substrates (Dizdaroglu 2005). Two major isoforms of OGG1, αOGG1 and βOGG1 are generated by alternative splicing of the nuclear encoded transcript (Nishioska et al. 1999). All isoforms have a mitochondrial targeting signal (MTS) on their N-termini. αOGG1 has a nuclear localization signal (NLS) which directs it to the nucleus, while βOGG1 has been found to localize exclusively in mitochondria (Nishioka et al. 1999). Human $Ogg1$ gene is localized to chromosome 3p26.2, which is a frequent site of loss of heterozygosity in many types of human tumors (Lu et al. 1997). The importance of OGG1 for mitochondrial repair was demonstrated in several studies. One study showed that mitochondria isolated from $Ogg1^{-/-}$ mice (Klungland et al. 1999) were completely deficient in incision at 8-oxoG, while other lesions were removed (de Souza-Pinto et al. 2001). Additionally, it has been estimated that mtDNA extracted from $Ogg1^{-/-}$ mice contained a 20-fold increase in 8-oxoG compared to wt mice, while nuclear DNA from $Ogg1$ null mice showed only a 2 fold increase. This indicates that OGG1 is much more important for removal of 8-oxoG from mtDNA than nuclear DNA (de Souza-Pinto et al. 2001). A study conducted in our laboratory, revealed that overexpression of OGG1 in mitochondria of HeLa cells, which are not proficient in mtDNA repair, improves repair and enhances cell survival following oxidative stress (Dobsan et al. 2000). The levels of 8-oxoG in mtDNA have been shown to increase in mammalian mitochondria in association with aging as well as during pathological conditions contributing to development of neurodegenerative diseases (Kang and Hamasaki 2005). This increase in 8-oxoG levels can be attributed to inadequate or dysfunctional OGG1 activity in mitochondria. This hypothesis is strengthened by studies on overexpression of mitochondrially targeted OGG1 in oligodendrocytes. Cells expressing OGG1 showed a significant enhancement in repair of oxidative lesions in mtDNA and were better protected against apoptosis following oxidative stress (Druzhyna et al. 2003). Other studies have proposed that there is an age-dependent deficiency in the mitochondrial import of βOGG1 which is responsible for the age-dependent increase in 8-oxoG levels in mtDNA (Szczesny et al. 2003).

8-oxoG within the mtDNA may not be repaired prior to replication. In this case DNA polymerase γ may insert adenine opposite 8-oxoG in the template strand during replication (Pinz et al. 1995). When this occurs, adenine is removed from A/8-oxoG as well as from A/G mispairs by a monofunctional adenine DNA glycosylase MYH, which is a homologue of MutY in *E.coli* (Parker et al. 2000). Human MYH (hMYH), in contrast to *E. coli* also can remove 2-OH-A opposite any normal base in the template (Ohtsubo et al. 2000). Mammalian hMYH is localized in both nuclei and mitochondria. A single human *MYH* gene produces multiple transcripts through different transcription initiation sites and alternative splicing, which results in the generation of an N-terminal mitochondrial targeting signal in a number of products (Nakabeppu 2001). Electron microscopic immunocytochemistry demonstrated that hMYH in mitochondria associates with the inner mitochondrial membrane (Nakabeppu, 2001). However, the actual size of the active mitochondrial form of hMYH requires further clarification. Mice deficient in MYH activity are viable but show an increased occurrence of cancers in various tissues 1.5 years after birth (Nakabeppu 2004). In humans, germline defects in MYH cause a multiple colorectal adenoma and carcinoma phenotype (Al-Tassan et al. 2002; Jones et al., 2002; Sieber et al. 2003).

NTH1 glycosylase is the main homologue of *E.coli* Endonuclease III (product of *nth* gene) cloned and characterized in humans (Aspinwall et al. 1997; Hilbert et al. 1997; Ikeda et al. 1998). It is a bifunctional glycosylase/AP lyase with broad substrate specificity. NTH1 recognizes a wide range of oxidized pyrimidine derivatives, including ring-saturated and ring-fragmented derivatives such as thymine glycol (Tg), 5-hydroxycytosine,5, 6-dihydrouracil (DHU), and urea. It also can remove formamidopyrimidines such as Fapy-G (Asagoshi et al. 2000 and Luna et al. 2000). Several studies have reported that human NTH1 can localize both to the nucleus and mitochondria (Ikeda et al. 1998; Asagoshi et al. 2000 Luna et al. 2000; Takao et al. 1998). The presence of NTH1 in mitochondria also is supported by a study, which showed that any incision activity against Tg was absent in liver mitochondrial extracts from mice deficient in NTH1 (Karahalil et al. 2003). Also, this indicates that the mitochondrial form of NTH1 is a product of a single nuclear gene. The importance of NTH1 activity in the protection of mitochondria against oxidized pyrimidines is supported by a study where HeLa cells expressing Endonuclease III from *E.coli* showed enhanced repair of mtDNA as well as increased cell survival, after treatment with menadione, an oxidizing agent (Rachek et al. 2004). Several groups have generated $nth1^{-/-}$ mice (Ocampo et al. 2002; Takao et al. 2002). Homozygous *nth1* mutant mice showed no phenotypical abnormalities. Tissue extracts from these mice contained enzymatic activities against Tg, although biochemically different from NTH1, which suggests that there are back-up mechanisms in mammalian cells which compensate for the loss of NTH1.

In addition to the glycosylases mentioned above, mitochondria also are equipped with mechanisms which protect its genome from misincorporation of oxidized nucleotides. MTH1, an *E.coli* MutT homologue, is an oxidized purine nucleoside triphosphatase. Mammalian MTH1 hydrolyzes 8-oxo-dGTP to a monophosphate form thus removing it from the nucleotide pool. In contrast to its *E.coli* homologue, MTH1 also has the ability to efficiently hydrolyze oxidized dATP and ATP, such as 2-OH-dATP and 2-OH-ATP [reviewed in (Nakabeppu et al. 2001)], which makes it an important defense mechanism, because it is believed that most of the 2-hydroxyadenine in DNA is derived from misincorporation of 2-OH-dATP and not from a direct oxidation of adenine in DNA (Ames et al. 1993). Multiple transcription products and polypeptides originate from a single *MTH1* gene by alternative splicing and translation initiation mechanisms (Nakabeppu et al. 2004). It has been shown by electron microscopy that in HeLa cells the MTH1 proteins mainly localize in the cytosol and in mitochondria (Kang et al. 1995). In MTH1 deficient mice (Tsuzuki et al. 2001), the incidence of spontaneous carcinogenesis in the liver, and to a lesser extent in the lung and stomach, increased severalfold compared to that observed in wild-type mice.

The combination of three enzymes present in mitochondria – OGG1, MYH and MTH1 – indicates that mitochondria contain all of the components of the so-called GO system of *E.coli*, which responds to oxidative damage to guanosine either as a free nucleotide, or within DNA (Grollman and Moriya 1993).

15.9. DNA Polymerase γ

DNA pol γ, the only DNA polymerase characterized in mitochondria, is involved in replication of mtDNA as well as its repair. In higher eukaryotes DNA pol γ contains a 120-140 kDa catalytic subunit related to the family-A DNA polymerases, such as E.coli DNA pol I, and a small subunit of 35-50 kDa which serves as a processivity factor [reviewed in (Copeland and Longley 2003; Kaguni 2004)]. Recently it has been shown that human DNA pol γ is actually a heterotrimer consisting of a catalytic subunit and two identical subunits serving as processivity factors (Yabubovskaya et al. 2005). In the BER pathway operative in mitochondria, AP endonuclease incision at an AP site is followed by the next repair step which requires both dRP lyase activity to remove the 5'-terminal dRP sugar moiety and DNA polymerase activity to insert new nucleotide, thus preparing the DNA for ligation. It has been shown that DNA polγ possesses an intrinsic dRP lyase activity which resides in its catalytic core (Longley et al. 1998; Pinz and Bogenhagen 2000), although this reaction is slow compared to that of DNA pol β the nuclear DNA repair polymerase. The high affinity of DNA pol γ for incised AP sites and the slow resolution of the enzyme-dRP intermediate may indicate that DNA pol γ is not capable of dealing with high levels of base damage. Thus, although it is clear that DNA pol γ is capable of acting as a dRP lyase in BER *in vitro*, it is uncertain whether this is the enzyme which performs this reaction *in vivo* (Bogenhagen et al. 2001). DNA pol γ itself is a target of oxidative damage and has been shown to be one of the major oxidized mitochondrial matrix proteins. This damage can lead to a detectable decline in polymerase activity, a condition which can cause a reduction in mitochondrial DNA replication and repair capacities (Graziewicz et al. 2002).

15.10. Mitochondrial DNA Ligase

DNA ligases perform the final step in BER, which is the resealing of the DNA nick between the newly inserted nucleotide and the rest of the DNA strand. All DNA ligases characterized in eukaryotes are ATP-dependent enzymes which act through an enzyme-adenylate intermediate to reseal single-strand breaks. mtDNA ligase was purified for the first time as ~100 kDa protein from it Xenopus oocyte mitochondria (Pinz and Bogenhagen 1998). Subsequent studies revealed that the only DNA ligase in mitochondria is the product of DNA ligase III gene. The mitochondrial localization signal is generated on the N-terminus of mtDNA ligase during translation through an alternative upstream translation initiation site (Lakshmipathy and Campbell 1999). The participation of DNA ligase III in mtDNA maintenance and repair was confirmed using an antisense strategy to decrease levels of ligase III. DNA ligase III antisense mRNA – expressing cells had reduced mtDNA content, accumulated single-strand nicks and did not repair mtDNA damaged by gamma-irradiation (Lakshmipathy and Campbell 2001). Unlike nDNA ligase III, which requires an interaction with an accessory nuclear protein XRCC1 for stability, mtDNA ligase III has been shown to act independently of this protein (Lakshmipathy and Campbell 2000).

15.11. Modulation of mtDNA Repair In Vivo

It is now well accepted that there are mechanisms for repair of endogenous damage in mtDNA. However, what is not as clear is how important mtDNA repair is for the cellular defenses of normal cells. Considering the differential sensitivities which exist among cell types and tissues to DNA damaging agents, an important question that can be asked is whether there are cell-specific differences in mtDNA repair. In research performed in our laboratory, we used a well-characterized culture system of pure primary cell cultures of oligodendrocytes, microglia, and astrocytes derived from the same region of rat brain, to evaluate mtDNA repair following alkylation and oxidative damage. The results showed a consistent and strong correlation between high capacity to repair mtDNA and resistance to apoptosis. Astrocytes which repair mtDNA proficiently are resistant to apoptosis, while oligodendrocytes and microglia which are less able to repair mtDNA damage are more vulnerable to programmed cell death (Ledoux et al. 1998). An evaluation of the antioxidant defenses in these cell types showed that there were no significant differences in catalase, glutathione peroxidase, or CuZnSOD activity between any of the cell types. Moreover, astrocytes had significantly lower levels of glutathione than oligodendrocytes or microglia. The same also was true for MnSOD (Hollensworth et al. 2000). Thus, while antioxidant and repair capacity are both involved in protecting cells from oxidant insults, it appears that cells with an efficient DNA repair capacity may be spared, even in the presence of very low antioxidant levels, and that cells with less efficient repair are susceptible to apoptosis even in the presence of higher levels of antioxidants. From these experiments, it can be concluded that mtDNA repair plays a pivotal role in cellular defense mechanisms. This leads to the notion that modulation of mtDNA repair could have a profound effect on general cell survival under conditions of genotoxic stress. Subsequent work performed by our laboratory and others supports this notion. In these studies, targeting DNA repair enzymes to mitochondria modulated mitochondrial DNA repair and cell survival. Targeting hOGG1 8-oxo-G glycosylase to mitochondria in HeLa cells protects them against oxidative damage (Dobson et al. 2000; Rachek et al. 2002). A similar protection is afforded by targeting hOGG1 into mitochondria of rat pulmonary artery endothelial cells (Dobson et al., 2002), and primary rat oligodendrocytes (Druzhyna et al. 2003). On the other hand, targeting of specific repair enzymes into mitochondria also can lead to an imbalance in base excision repair thus making otherwise resistant cells more vulnerable to the effects of cytotoxic drugs. Expressing the bacterial Exonuclease III in mitochondria of breast cancer cells renders them more susceptible to oxidative stress (Shokolenko et al. 2003). Also, expression of the mitochondrially targeted N-methylpurine DNA glycosylase dramatically increases the sensitivity of breast cancer cells' to alkylation damage (Fishel et al. 2003). mtDNA repair plays an important role in normal cell physiology, making it a potential target for therapeutic intervention. Recent successes in application of a new delivery system through protein transduction (Wadia and Dowdy 2005) have prompted us to adapt this methodology for the purpose of mitochondrial targeting of repair enzymes (Shokolenko et al. 2005), thus circumventing the drawbacks associated with the approaches using DNA delivery methods.

15.12. Conclusions

Interest in DNA damage and repair in mitochondria reflects an increasing awareness of the central role that these organelles play in cellular physiology in both normal and diseased states. Research in the past several years has shown that mitochondrial DNA is more susceptible to various carcinogens and ROS than nDNA. Damage in mtDNA is implicated in a wide variety of common pathological states, including metabolic and neurodegenerative diseases, cancer and aging. Studies into the mechanisms involved in mtDNA repair have only just begun and already it has become apparent how efficient and well-coordinated these processes can be. However, many questions remain to be answered. Undoubtedly, future research will address such topics as: a more precise definition of the components involved in mtDNA repair, a better comprehension of how they are regulated, and a more thorough understanding of how they can malfunction to precipitate disease states. This knowledge will be essential for the development of future therapeutic strategies to combat a variety of human diseases.

References

Al-Tassan N, Chmiel NH, Maynard J, Fleming N, Livingston AL, Williams GT, Hodges AK, Davies DR, David SS, Sampson JR, Cheadle JP (2002) Inherited variants of MYH associated with somatic G:C—T:A mutations in colorectal tumors. Nat Genet 30: 227–232

Allen JA, Coombs MM Covalent binding of polycyclic aromatic compounds to mitochondrial and nuclear DNA (1980) Nature 287: 244–245

Ames BN, Shigenaga MK, Hagen TM (1993) Oxidants, antioxidants, and the degenerative diseases of aging. Proc Natl Acad Sci U S A 90: 7915–7922

Anderson CT, Friedberg EC (1980) The presence of nuclear and mitochondrial uracil-DNA glycosylase in extracts of human KB cells. Nucleic Acids Res 8: 875–888

Anson RM, Hudson E, Bohr VA (2000) Mitochondrial endogenous oxidative damage has been overestimated. FASEB J 14: 355–360

Asagoshi K, Yamada T, Okada Y, Terato H, Ohyama Y, Seki S, Ide H (2000) Recognition of formamidopyrimidine by Escherichia coli and mammalian thymine glycol glycosylases. Distinctive paired base effects and biological and mechanistic implications. J Biol Chem 275: 24781–24786

Aspinwall R, Rothwell DG, Roldan-Arjona T, Anselmino C, Ward CJ, Cheadle JP, Sampson JR, Lindahl T, Harris PC, Hickson ID (1997) Cloning and characterization of a functional human homolog of Escherichia coli endonuclease III. Proc Natl Acad Sci USA 94: 109–114

Attardi G, Schatz G (1988) Biogenesis of mitochondria. Annu Rev Cell Biol 4: 289–333

Backer JM, Weinstein IB (1980) Mitochondrial DNA is a major cellular target for a dihydrodiol-epoxide derivative of benzo[a]pyrene. Science 209: 297–299

Beckman KB, Ames BN (1999) Endogenous oxidative damage of mtDNA. Mutat Res 424: 51–58

Bjelland S, Seeberg E (2003) Mutagenicity, toxicity and repair of DNA base damage induced by oxidation. Mutat Res 531: 37–80

Bogenhagen DF, Pinz KG, Perez-Jannotti RM (2001) Enzymology of mitochondrial base excision repair. Prog Nucleic Acid Res Mol Biol 68: 257–271

Bogenhagen DF, Wang Y, Shen EL, Kobayashi R (2003) Protein components of mitochondrial DNA nucleoids in higher eukaryotes. Mol Cell Proteomics 2: 1205–1216

Bohr VA, Stevnsner T, de Souza-Pinto NC (2002) Mitochondrial DNA repair of oxidative damage in mammalian cells. Gene 286: 127–134

Boiteux S, Radicella JP (2000) The human OGG1 gene: structure, functions, and its implication in the process of carcinogenesis. Arch Biochem Biophys 377: 1–8

Brown WM, George M Jr, Wilson AC (1979) Rapid evolution of animal mitochondrial DNA. Proc Natl Acad Sci USA 76: 1967–1971

Cadenas E, Davies KJ (2000) Mitochondrial free radical generation, oxidative stress, and aging. Free Radic Biol Med 29: 222–230

Cadet J, Delatour T, Douki T, Gasparutto D, Pouget JP, Ravanat JL, Sauvaigo S (1999) Hydroxyl radicals and DNA base damage. Mutat Res 424: 9–21

Cavelier L, Johannisson A, Gyllensten U (2000) Analysis of mtDNA copy number and composition of single mitochondrial particles using flow cytometry and PCR. Exp Cell Res 259: 79–85

Cheng KC, Cahill DS, Kasai H, Nishimura S, Loeb LA (1992) 8-Hydroxyguanine, an abundant form of oxidative DNA damage, causes G—T and A—C substitutions. J Biol Chem 267: 166–172

Christmann M, Tomicic MT, Roos WP, Kaina B (2003) Mechanisms of human DNA repair: an update. Toxicology 193: 3–34

Clayton DA, Doda JN, Friedberg EC (1974) The absence of a pyrimidine dimer repair mechanism in mammalian mitochondria. Proc Natl Acad Sci USA 71: 2777–2781

Collins AR, Cadet J, Moller L, Poulsen HE, Vina J (2004) Are we sure we know how to measure 8-oxo-7,8-dihydroguanine in DNA from human cells? Arch Biochem Biophys 423: 57–65

Cooke MS, Evans MD, Dizdaroglu M, Lunec J (2003) Oxidative DNA damage: mechanisms, mutation, and disease. FASEB J 17: 1195–1214

Copeland WC, Longley MJ (2003) DNA polymerase gamma in mitochondrial DNA replication and repair. Scientific World Journal 3: 34–44

Copeland WC, Wachsman JT, Johnson FM, Penta JS (2002) Mitochondrial DNA alterations in cancer. Cancer Invest 20: 557–569

Croteau DL, Stierum RH, Bohr VA (1999) Mitochondrial DNA repair pathways. Mutat Res 434: 137–148

de Souza-Pinto NC, Eide L, Hogue BA, Thybo T, Stevnsner T, Seeberg E, Klungland A, Bohr VA (2001) Repair of 8-oxodeoxyguanosine lesions in mitochondrial dna depends on the oxoguanine DNA glycosylase (OGG1) gene and 8-oxoguanine accumulates in the mitochondrial dna of OGG1-defective mice. Cancer Res 61: 5378–5381

DiMauro S, Schon EA (2003) Mitochondrial respiratory-chain diseases. N Engl J Med 348: 2656–2668

Dizdaroglu M (2005) Base-excision repair of oxidative DNA damage by DNA glycosylases. Mutat Res 591:45–59

Dizdaroglu M (1998) Mechanisms of free radical damage to DNA. In: O. I. H. Aruoma (ed.), DNA & Free Radicals: Techniques, Mechanisms & Applications, pp. 3–26. Saint Lucia: OICA International

Dobson AW, Grishko V, LeDoux SP, Kelley MR, Wilson GL, Gillespie MN (2002) Enhanced mtDNA repair capacity protects pulmonary artery endothelial cells from oxidant-mediated death. Am J Physiol Lung Cell Mol Physiol 283: L205–210

Dobson AW, Xu Y, Kelley MR, LeDoux SP, Wilson GL (2000) Enhanced mitochondrial DNA repair and cellular survival after oxidative stress by targeting the human 8-oxoguanine glycosylase repair enzyme to mitochondria. J Biol Chem 275: 37518–37523

Doroshow JH, Davies KJ (1986) Redox cycling of anthracyclines by cardiac mitochondria. II. Formation of superoxide anion, hydrogen peroxide, and hydroxyl radical. J Biol Chem 261: 3068–3074

Driggers WJ, LeDoux SP, Wilson GL (1993) Repair of oxidative damage within the mitochondrial DNA of RINr 38 cells. J Biol Chem 268: 22042–22045

Droge W (2002) Free radicals in the physiological control of cell function. Physiol Rev 82: 47–95

Druzhyna NM, Hollensworth SB, Kelley MR, Wilson GL, Ledoux SP (2003) Targeting human 8-oxoguanine glycosylase to mitochondria of oligodendrocytes protects against menadione-induced oxidative stress. Glia 42: 370–378

Durham SE, Krishnan KJ, Betts J, Birch-Machin MA (2003) Mitochondrial DNA damage in non-melanoma skin cancer. Br J Cancer 88: 90–95

ESCODD (2002) Comparative analysis of baseline 8-oxo-7,8-dihydroguanine in mammalian cell DNA, by different methods in different laboratories: an approach to consensus. Carcinogenesis 23: 2129–2133

ESCODD (2003) Measurement of DNA oxidation in human cells by chromatographic and enzymic methods. Free Radic Biol Med 34: 1089–1099

Fishel ML, Seo YR, Smith ML, Kelley MR (2003) Imbalancing the DNA base excision repair pathway in the mitochondria; targeting and overexpressing N-methylpurine DNA glycosylase in mitochondria leads to enhanced cell killing. Cancer Res 63: 608–615

Fliss MS, Usadel H, Caballero OL, Wu L, Buta MR, Eleff SM, Jen J, Sidransky D (2000) Facile detection of mitochondrial DNA mutations in tumors and bodily fluids. Science 287: 2017–2019

Garrido N, Griparic L, Jokitalo E, Wartiovaara J, van der Bliek AM, Spelbrink JN (2003) Composition and dynamics of human mitochondrial nucleoids. Mol Biol Cell 14: 1583–1596

Geromel V, Kadhom N, Cebalos-Picot I, Ouari O, Polidori A, Munnich A, Rotig A, Rustin P (2001) Superoxide-induced massive apoptosis in cultured skin fibroblasts harboring the neurogenic ataxia retinitis pigmentosa (NARP) mutation in the ATPase-6 gene of the mitochondrial DNA. Hum Mol Genet 10: 1221–1228

Graziewicz MA, Day BJ, Copeland WC (2002) The mitochondrial DNA polymerase as a target of oxidative damage. Nucleic Acids Res 30: 2817–2824

Grollman AP, Moriya M (1993) Mutagenesis by 8-oxoguanine: an enemy within. Trends Genet 9: 246–249

Gross NJ, Getz GS, Rabinowitz M (1969) Apparent turnover of mitochondrial deoxyribonucleic acid and mitochondrial phospholipids in the tissues of the rat. J Biol Chem 244: 1552–1562

Hamilton ML, Guo Z, Fuller CD, Van Remmen H, Ward WF, Austad SN, Troyer DA, Thompson I, Richardson AA (2001) Reliable assessment of 8-oxo-2-deoxyguanosine levels in nuclear and mitochondrial DNA using the sodium iodide method to isolate DNA. Nucleic Acids Res 29: 2117–2126

Helbock HJ, Beckman KB, Shigenaga MK, Walter PB, Woodall AA, Yeo, HC, Ames BN (1998) DNA oxidation matters: the HPLC-electrochemical detection assay of 8-oxo-deoxyguanosine and 8-oxo-guanine. Proc Natl Acad Sci USA 95: 288–293

Hilbert TP, Chaung W, Boorstein RJ, Cunningham RP, Teebor GW (1997) Cloning and expression of the cDNA encoding the human homologue of the DNA repair enzyme, Escherichia coli endonuclease III. J Biol Chem 272: 6733–6740

Hollensworth SB, Shen C, Sim JE, Spitz DR, Wilson GL, LeDoux SP (2000) Glial cell type-specific responses to menadione-induced oxidative stress. Free Radic Biol Med 28: 1161–1174

Holt IJ, Harding AE, Morgan-Hughes JA (1998) Deletions of muscle mitochondrial DNA in patients with mitochondrial myopathies. Nature 331: 717–719

Ikeda S, Biswas T, Roy R, Izumi T, Boldogh I, Kurosky A, Sarker AH, Seki S, Mitra S (1998) Purification and characterization of human NTH1, a homolog of Escherichia coli endonuclease III. Direct identification of Lys-212 as the active nucleophilic residue. J Biol Chem 273: 21585–21593

Jones S, Emmerson P, Maynard J, Best JM, Jordan S, Williams GT, Sampson JR, Cheadle JP (2002) Biallelic germline mutations in MYH predispose to multiple colorectal adenoma and somatic G:C–>T:A mutations. Hum Mol Genet 11: 2961–2967

Kaguni LS (2004) DNA polymerase gamma, the mitochondrial replicase. Annu Rev Biochem 73: 293–320

Kang D, Hamasaki N (2005) Alterations of mitochondrial DNA in common diseases and disease states: aging, neurodegeneration, heart failure, diabetes, and cancer. Curr Med Chem 12: 429–441

Kang D, Nishida J, Iyama A, Nakabeppu Y, Furuichi M, Fujiwara T, Sekiguchi M, Takeshige K (1995) Intracellular localization of 8-oxo-dGTPase in human cells, with special reference to the role of the enzyme in mitochondria. J Biol Chem 270: 14659–14665

Karahalil B, de Souza-Pinto NC, Parsons JL, Elder RH, Bohr VA (2003) Compromised incision of oxidized pyrimidines in liver mitochondria of mice deficient in NTH1 and OGG1 glycosylases. J Biol Chem 278: 33701–33707

Khrapko K, Coller HA, Andre PC, Li XC, Hanekamp JS, Thilly WG (1997) Mitochondrial mutational spectra in human cells and tissues. Proc Natl Acad Sci USA 94: 13798–13803

Klein JC, Bleeker MJ, Saris CP, Roelen HC, Brugghe HF, van den Elst H, van der Marel GA, van Boom JH, Westra JG, Kriek E, Berens AJM (1992) Repair and replication of plasmids with site-specific 8-oxodG and 8-AAFdG residues in normal and repair-deficient human cells. Nucleic Acids Res 20: 4437–4443

Klungland A, Rosewell I, Hollenbach S, Larsen E, Daly G, Epe B, Seeberg E, Lindahl T, Barnes DE (1999) Accumulation of premutagenic DNA lesions in mice defective in removal of oxidative base damage. Proc Natl Acad Sci USA 96: 13300–13305

Korhonen JA, Pham XH, Pellegrini M, Falkenberg M (2004) Reconstitution of a minimal mtDNA replisome in vitro. Embo J 23: 2423–2429

Krokan HE, Otterlei M, Nilsen H, Kavli B, Skorpen F, Andersen S, Skjelbred C, Akbari M, Aas PA, Slupphaug G (2001) Properties and functions of human uracil-DNA glycosylase from the UNG gene. Prog Nucleic Acid Res Mol Biol 68: 365–386

Kujoth GC, Hiona A, Pugh TD, Someya S, Panzer K, Wohlgemuth SE, Hofer T, Seo AY, Sullivan R, Jobling WA, Morrow JD, Van Remmen H, Sedivy JM, Yamasoba T, Tanokura M, Weindruch R, Leeuwenburgh C, Prolla TA (2005) Mitochondrial DNA mutations, oxidative stress, and apoptosis in mammalian aging. Science 309: 481–484

Kunishige M, Mitsui T, Akaike M, Kawajiri M, Shono M, Kawai H, Matsumoto T (2003) Overexpressions of myoglobin and antioxidant enzymes in ragged-red fibers of skeletal muscle from patients with mitochondrial encephalomyopathy. Muscle Nerve 28: 484–492

Lakshmipathy U, Campbell C (2001) Antisense-mediated decrease in DNA ligase III expression results in reduced mitochondrial DNA integrity. Nucleic Acids Res 29: 668–676

Lakshmipathy U, Campbell C (2000) Mitochondrial DNA ligase III function is independent of Xrcc1. Nucleic Acids Res 28: 3880–3886

Lakshmipathy U, Campbell C (1999) The human DNA ligase III gene encodes nuclear and mitochondrial proteins. Mol Cell Biol 19: 3869–3876

Le Page F, Margot A, Grollman AP, Sarasin A, Gentil A (1995) Mutagenicity of a unique 8-oxoguanine in a human Ha-ras sequence in mammalian cells. Carcinogenesis 16: 2779–2784

LeDoux SP, Driggers WJ, Hollensworth BS, Wilson GL (1999) Repair of alkylation and oxidative damage in mitochondrial DNA. Mutat Res 434: 149–159

Ledoux SP, Shen CC, Grishko VI, Fields PA, Gard AL, Wilson GL (1998) Glial cell-specific differences in response to alkylation damage. Glia 24: 304–312

LeDoux SP, Wilson GL, Beecham EJ, Stevnsner T, Wassermann K, Bohr VA (1992) Repair of mitochondrial DNA after various types of DNA damage in Chinese hamster ovary cells. Carcinogenesis 13: 1967–1973

Legros F, Malka F, Frachon P, Lombes A, Rojo M (2004) Organization and dynamics of human mitochondrial DNA. J Cell Sci 117: 2653–2662

Lestienne P, Ponsot G (1988) Kearns-Sayre syndrome with muscle mitochondrial DNA deletion. Lancet 1: 885

Lim KS, Jeyaseelan K, Whiteman M, Jenner A, Halliwell B (2005) Oxidative damage in mitochondrial DNA is not extensive. Ann N Y Acad Sci 1042: 210–220

Longley MJ, Prasad R, Srivastava DK, Wilson SH, Copeland WC (1998) Identification of 5'-deoxyribose phosphate lyase activity in human DNA polymerase gamma and its role in mitochondrial base excision repair in vitro. Proc Natl Acad Sci USA 95: 12244–12248

Lu CY, Wang EK, Lee HC, Tsay HJ, Wei YH (2003) Increased expression of manganese-superoxide dismutase in fibroblasts of patients with CPEO syndrome. Mol Genet Metab 80: 321–329

Lu R, Nash HM, Verdine GLA (1997) Mammalian DNA repair enzyme that excises oxidatively damaged guanines maps to a locus frequently lost in lung cancer. Curr Biol 7: 397–407

Luna L, Bjoras M, Hoff E, Rognes T, Seeberg E (2000) Cell-cycle regulation, intracellular sorting and induced overexpression of the human NTH1 DNA glycosylase involved in removal of formamidopyrimidine residues from DNA. Mutat Res 460: 95–104

Magana-Schwencke N, Henriques J A, Chanet R, Moustacchi E The fate of 8-methoxypsoralen photoinduced crosslinks in nuclear and mitochondrial yeast DNA: comparison of wild-type and repair-deficient strains. Proc Natl Acad Sci U S A 79: 1722–1726, 1982

Mambo E, Gao X, Cohen Y, Guo Z, Talalay P, Sidransky D (2003) Electrophile and oxidant damage of mitochondrial DNA leading to rapid evolution of homoplasmic mutations. Proc Natl Acad Sci USA 100: 1838–1843

Martinez GR, Loureiro AP, Marques SA, Miyamoto S, Yamaguchi LF, Onuki J, Almeida EA, Garcia CC, Barbosa LF, Medeiros MH, Di Mascio P (2003) Oxidative and alkylating damage in DNA. Mutat Res 544: 115–127

Miyaki M, Yatagai K, Ono T (1977) Strand breaks of mammalian mitochondrial DNA induced by carcinogens. Chem Biol Interact 17: 321–329

Moriya M (1993) Single-stranded shuttle phagemid for mutagenesis studies in mammalian cells: 8-oxoguanine in DNA induces targeted G.C->T.A transversions in simian kidney cells. Proc Natl Acad Sci USA 90: 1122–1126

Moriya M, Ou C, Bodepudi V, Johnson F, Takeshita M, Grollman AP (1991) Site-specific mutagenesis using a gapped duplex vector: a study of translesion synthesis past 8-oxodeoxyguanosine in E. coli. Mutat Res 254: 281–288

Nakabeppu Y (2001) Molecular genetics and structural biology of human MutT homolog, MTH1. Mutat Res 477: 59–70

Nakabeppu Y (2001) Regulation of intracellular localization of human MTH1, OGG1, and MYH proteins for repair of oxidative DNA damage. Prog Nucleic Acid Res Mol Biol 68: 75–94

Nakabeppu Y, Tsuchimoto D, Ichinoe A, Ohno M, Ide Y, Hirano S, Yoshimura D, Tominaga Y, Furuichi M, Sakumi K (2004) Biological significance of the defense mechanisms against oxidative damage in nucleic acids caused by reactive oxygen species: from mitochondria to nuclei. Ann N Y Acad Sci 1011: 101–111

Neubert D, Hopfenmuller W, Fuchs G (1981) Manifestation of carcinogenesis as a stochastic process on the basis of an altered mitochondrial genome. Arch Toxicol 48: 89–125

Nilsen H, Rosewell I, Robins P, Skjelbred CF, Andersen S, Slupphaug G, Daly G, Krokan HE, Lindahl T, Barnes DE (2000) Uracil-DNA glycosylase (UNG)-deficient mice reveal a primary role of the enzyme during DNA replication. Mol Cell 5: 1059–1065

Niranjan BG, Bhat NK, Avadhani NG (1982) Preferential attack of mitochondrial DNA by aflatoxin B1 during hepatocarcinogenesis. Science 215: 73–75

Nishioka K, Ohtsubo T, Oda H, Fujiwara T, Kang D, Sugimachi K, Nakabeppu Y (1999) Expression and differential intracellular localization of two major forms of human 8-oxoguanine DNA glycosylase encoded by alternatively spliced OGG1 mRNAs. Mol Biol Cell 10: 1637–1652

Ocampo MT, Chaung W, Marenstein DR, Chan MK, Altamirano A, Basu AK, Boorstein RJ, Cunningham RP, Teebor GW (2002) Targeted deletion of mNth1 reveals a novel DNA repair enzyme activity. Mol Cell Biol 22: 6111–6121

Ohtsubo T, Nishioka K, Imaiso Y, Iwai S, Shimokawa H, Oda H, Fujiwara T, Nakabeppu Y (2000) Identification of human MutY homolog (hMYH) as a repair enzyme for 2-hydroxyadenine in DNA and detection of multiple forms of hMYH located in nuclei and mitochondria. Nucleic Acids Res 28: 1355–1364

Parker A, Gu Y, Lu AL (2000) Purification and characterization of a mammalian homolog of Escherichia coli MutY mismatch repair protein from calf liver mitochondria. Nucleic Acids Res 28: 3206–3215

Pavlov YI, Minnick DT, Izuta S, Kunkel TA (1994) DNA replication fidelity with 8-oxodeoxyguanosine triphosphate. Biochemistry 33: 4695–4701

Pelicano H, Carney D, Huang P (2004) ROS stress in cancer cells and therapeutic implications. Drug Resist Updat 7: 97–110

Petros JA, Baumann AK, Ruiz-Pesini E, Amin MB, Sun CQ, Hall J, Lim S, Issa MM, Flanders WD, Hosseini SH, Marshall FF, Wallace DC (2005) mtDNA mutations increase tumorigenicity in prostate cancer. Proc Natl Acad Sci USA 102: 719–724

Pettepher CC, LeDoux SP, Bohr VA, Wilson GL (1991) Repair of alkali-labile sites within the mitochondrial DNA of RINr 38 cells after exposure to the nitrosourea streptozotocin. J Biol Chem 266: 3113–3117

Pinz KG, Bogenhagen DF (2000) Characterization of a catalytically slow AP lyase activity in DNA polymerase gamma and other family A DNA polymerases. J Biol Chem 275: 12509–12514

Pinz KG, Bogenhagen DF (1998) Efficient repair of abasic sites in DNA by mitochondrial enzymes. Mol Cell Biol 18: 1257–1265

Pinz KG, Shibutani S, Bogenhagen DF (1995) Action of mitochondrial DNA polymerase gamma at sites of base loss or oxidative damage. J Biol Chem 270: 9202–9206

Pirsel M, Bohr VA (1993) Methyl methanesulfonate adduct formation and repair in the DHFR gene and in mitochondrial DNA in hamster cells. Carcinogenesis 14: 2105–2108

Polyak K, Li Y, Zhu H, Lengauer C, Willson JK, Markowitz SD, Trush MA, Kinzler KW, Vogelstein B (1998) Somatic mutations of the mitochondrial genome in human colorectal tumours. Nat Genet 20: 291–293

Rachek LI, Grishko VI, Alexeyev MF, Pastukh VV, LeDoux SP, Wilson, GL (2004) Endonuclease III and endonuclease VIII conditionally targeted into mitochondria enhance mitochondrial DNA repair and cell survival following oxidative stress. Nucleic Acids Res 32: 3240–3247

Rachek LI, Grishko VI, Musiyenko SI, Kelley MR, LeDoux SP, Wilson GL (2002) Conditional targeting of the DNA repair enzyme hOGG1 into mitochondria. J Biol Chem 277: 44932–44937

Raha S, Robinson BH (2000) Mitochondria, oxygen free radicals, disease and ageing. Trends Biochem Sci 25: 502–508

Richter C, Park JW, Ames BN (1998) Normal oxidative damage to mitochondrial and nuclear DNA is extensive. Proc Natl Acad Sci USA 85: 6465–6467

Rossi SC, Gorman N, Wetterhahn KE (1988) Mitochondrial reduction of the carcinogen chromate: formation of chromium(V). Chem Res Toxicol 1: 101–107

Sancar A, Lindsey-Boltz LA, Unsal-Kacmaz K, Linn S (2004) Molecular mechanisms of mammalian DNA repair and the DNA damage checkpoints. Annu Rev Biochem 73: 39–85

Sawyer DE, Van Houten B (1999) Repair of DNA damage in mitochondria. Mutat Res 434: 161–176

Scharer OD (2003) Chemistry and biology of DNA repair. Angew Chem Int Ed Engl 42: 2946–2974

Senturker S, Dizdaroglu M (1999) The effect of experimental conditions on the levels of oxidatively modified bases in DNA as measured by gas chromatography-mass spectrometry: how many modified bases are involved? Prepurification or not? Free Radic Biol Med 27: 370–380

Shadel GS, Clayton DA (1997) Mitochondrial DNA maintenance in vertebrates. Annu Rev Biochem 66: 409–435

Shokolenko IN, Alexeyev MF, LeDoux SP, Wilson GL (2005) TAT-mediated protein transduction and targeted delivery of fusion proteins into mitochondria of breast cancer cells. DNA Repair (Amst) 4: 511–518

Shokolenko IN, Alexeyev MF, Robertson FM, LeDoux SP, Wilson GL (2003) The expression of Exonuclease III from E. coli in mitochondria of breast cancer cells diminishes mitochondrial DNA repair capacity and cell survival after oxidative stress. DNA Repair (Amst) 2: 471–482

Sieber OM, Lipton L, Crabtree M, Heinimann K, Fidalgo P, Phillips RK, Bisgaard ML, Orntoft TF, Aaltonen LA, Hodgson SV, Thomas HJ, Tomlinson IP (2003) Multiple colorectal adenomas, classic adenomatous polyposis, and germ-line mutations in MYH. N Engl J Med 348: 791–799

Singer TP, Ramsay RR (1990) Mechanism of the neurotoxicity of MPTP. An update. FEBS Lett 274: 1–8

Starkov AA, Fiskum G, Chinopoulos C, Lorenzo BJ, Browne SE, Patel MS, Beal MF (2004) Mitochondrial alpha-ketoglutarate dehydrogenase complex generates reactive oxygen species. J Neurosci 24: 7779–7788

Steenken S (1997) Electron transfer in DNA? Competition by ultra-fast proton transfer? Biol Chem 378: 1293–1297

Stierum RH, Dianov GL, Bohr VA (1999) Single-nucleotide patch base excision repair of uracil in DNA by mitochondrial protein extracts. Nucleic Acids Res 27: 3712–3719

Szczesny B, Hazra TK, Papaconstantinou J, Mitra S, Boldogh I (2003) Age-dependent deficiency in import of mitochondrial DNA glycosylases required for repair of oxidatively damaged bases. Proc Natl Acad Sci USA 100: 10670–10675

Taanman JW (1999) The mitochondrial genome: structure, transcription, translation and replication. Biochim Biophys Acta 1410: 103–123

Takao M, Aburatani H, Kobayashi K, Yasui A (1998) Mitochondrial targeting of human DNA glycosylases for repair of oxidative DNA damage. Nucleic Acids Res 26: 2917–2922

Takao M, Kanno S, Shiromoto T, Hasegawa R, Ide H, Ikeda S, Sarker AH, Seki S, Xing JZ, Le XC, Weinfeld M, Kobayashi K, Miyazaki J, Muijtjens M, Hoeijmakers JH, van der Horst G, Yasui A (2002) Novel nuclear and mitochondrial glycosylases revealed by disruption of the mouse Nth1 gene encoding an endonuclease III homolog for repair of thymine glycols. Embo J 21: 3486–3493

Takayama S, Muramatsu M (1969) Incorporation of tritiated dimethylnitrosamine into subcellular fractions of mouse liver after long term administration of dimethylnitrosamine. Z Krebsforsch 73: 172–179

Tomasi A, Albano E, Banni S, Botti B, Corongiu F, Dessi MA, Iannone A, Vannini V, Dianzani MU (1987) Free-radical metabolism of carbon tetrachloride in rat liver mitochondria. A study of the mechanism of activation. Biochem J 246: 313–317

Tretter L, Adam-Vizi V (2004) Generation of reactive oxygen species in the reaction catalyzed by alpha-ketoglutarate dehydrogenase. J Neurosci 24: 7771–7778

Trifunovic A, Wredenberg A, Falkenberg M, Spelbrink JN, Rovio AT, Bruder CE, Bohlooly YM, Gidlof S, Oldfors A, Wibom, R, Tornell J, Jacobs HT, Larsson NG (2004) Premature ageing in mice expressing defective mitochondrial DNA polymerase. Nature 429: 417–423

Tsuzuki T, Egashira A, Igarashi H, Iwakuma T, Nakatsuru Y, Tominaga Y, Kawate H, Nakao K, Nakamura K, Ide F, Kura S, Nakabeppu Y, Katsuki M, Ishikawa T, Sekiguchi M (2001) Spontaneous tumorigenesis in mice defective in the MTH1 gene encoding 8-oxo-dGTPase. Proc Natl Acad Sci USA 98: 11456–11461

Van Remmen H, Ikeno Y, Hamilton M, Pahlavani M, Wolf N, Thorpe SR, Alderson NL, Baynes JW, Epstein CJ, Huang TT, Nelson J, Strong R, Richardson A (2003) Life-long reduction in MnSOD activity results in increased DNA damage and higher incidence of cancer but does not accelerate aging. Physiol Genomics 16: 29–37

Wadia JS, Dowdy SF (2005) Transmembrane delivery of protein and peptide drugs by TAT-mediated transduction in the treatment of cancer. Adv Drug Deliv Rev 57: 579–596

Wallace DCA (2005) Mitochondrial Paradigm of Metabolic and Degenerative Diseases, Aging, and Cancer: A Dawn for Evolutionary Medicine. Annu Rev Genet 39:359–407

Wallace DC, Singh G, Lott MT, Hodge JA, Schurr TG, Lezza AM, Elsas LJ (1988) 2nd, Nikoskelainen E K Mitochondrial DNA mutation associated with Leber's hereditary optic neuropathy. Science 242: 1427–1430

Wallace SS (1998) Enzymatic processing of radiation-induced free radical damage in DNA. Radiat Res 150: S60–79

Wood ML, Dizdaroglu M, Gajewski E, Essigmann JM (1990) Mechanistic studies of ionizing radiation and oxidative mutagenesis: genetic effects of a single 8-hydroxyguanine (7-hydro-8-oxoguanine) residue inserted at a unique site in a viral genome. Biochemistry 29: 7024–7032

Wunderlich V, Schutt M, Bottger M, Graffi A (1970) Preferential alkylation of mitochondrial deoxyribonucleic acid by N-methyl-N-nitrosourea. Biochem J 118: 99–109

Yakes FM, Van Houten B (1997) Mitochondrial DNA damage is more extensive and persists longer than nuclear DNA damage in human cells following oxidative stress. Proc Natl Acad Sci USA 94: 514–519

Yakubovskaya E, Chen Z, Carrodeguas JA, Kisker C, Bogenhagen DF (2006) Functional human mitochondrial DNA polymerase gamma forms a heterotrimer. J Biol Chem 281: 374–382

Yang MY, Bowmaker M, Reyes A, Vergani L, Angeli P, Gringeri E, Jacobs H T, Holt IJ (2002) Biased incorporation of ribonucleotides on the mitochondrial L-strand accounts for apparent strand-asymmetric DNA replication. Cell 111: 495–505

Yoneda M, Katsumata K, Hayakawa M, Tanaka M, Ozawa T (1995) Oxygen stress induces an apoptotic cell death associated with fragmentation of mitochondrial genome. Biochem Biophys Res Commun 209: 723–729

Zastawny TH, Dabrowska M, Jaskolski T, Klimarczyk M, Kulinski L, Koszela A, Szczesniewicz M, Sliwinska M, Witkowski P, Olinski R (1998) Comparison of oxidative base damage in mitochondrial and nuclear DNA. Free Radic Biol Med 24: 722–725

Zhang D, Mott JL, Farrar P, Ryerse JS, Chang SW, Stevens M, Denniger G, Zassenhaus HP (2003) Mitochondrial DNA mutations activate the mitochondrial apoptotic pathway and cause dilated cardiomyopathy. Cardiovasc Res 57: 147–157

Index

Acetyl CoA carboxylase (ACC)
   Effect of diabetes on, 45
   Effect of ischemia on, 47
   Isoforms of, 37
   Regulation of, 37, 38
   As regulator of fatty acid oxidation, 37, 38
Acidosis
   Activation of branched chain keto-acid dehydrogenase during, 142, 143
   Effect in ischemia-reperfusion injury, 46, 158, 211, 226, 227, 251, 255, 292, 296
   Enhanced aminogenesis and ammoniogenesis during, 123, 125
   Enhanced turnover of glutamine during, 124, 125
   Protein turnover during, 142
Acyl CoA binding protein, 34
Acyl CoA dehydrogenase, 33, 44, 45, 71
Adenine nucleotide translocase
   As component of mitochondrial permeability transition pore, 229, 232, 244, 246–249, 314
   Conformations of, 246, 247, 250, 251
   Dependence on mitochondrial membrane potential, 14–16
   Effect of calcium on, 247, 250
   Effect of cyclophilin D on, 244, 248, 250, 251
   Effect of knocking out, 248, 249
   Localization of, 3, 8, 250
   Oxidative damage to, 247
   Transport of ADP and ATP by, 9, 14
Adipocytes
   As source of fatty acid and glycerol, 73, 75–77
   Effect of PPARs on, 73–79
Adiponectin, 76
β-adrenergic agonists
   Effect on triacylglycerol, 35
   Involvement in heart failure, 279
   Stimulation of apoptosis by, 279

Aging
   Effect of reactive oxygen species on, 192, 193
   Effect on adipocytes, 82
   Formation of reactive oxygen species during, 190, 192, 193
   Oxidation of mitochondrial DNA in, 192, 330, 335
   Role of apoptosis in, 272, 279
A-kinase anchoring protein, 17
Akt and PI-3 kinase
   As components of preconditioning pathway, 297, 308, 311, 315
Alanine
   Biosynthesis of, 121
   Release by skeletal muscle, 131–133
Alzheimer's disease, 330
Amino acids
   As gluconeogenic substrates, 117, 120
   Effect on hypertrophic and ischemic heart, 158–160
   Release into blood, 118–120
   Transport of, 151–154, 158
   Use in energy metabolism, 118
Ammonia
   Detoxification of, 123, 124, 156, 157
   Effect of acidosis on, 123–125
   Formation of, 122, 123, 142
AMP kinase
   As regulator of Acetyl CoA carboxylase, 37
   Effect of adiponectin on, 76
   Effect of diabetes on, 45
   Effect of ischemia on, 47, 48
   Effect on energy metabolism, 38
   Regulation of, 11, 38
Anaplerosis, 12, 13
Angiotensin II
   Activation of, 278, 279
   Effect of DNase I, 279
   Generation of reactive oxygen species by, 279

Initiation of apoptosis by, 279, 282
Modulation of heart failure by antagonists of, 282
Antioxidant vitamins, 189, 193
Apoptosis
Cytochrome c release during, 189, 191, 274, 275
Extrinsic pathway of, 272
In cancer, 161, 162
In diabetes, 191, 272
In failing, infarcted and hibernating heart, 272, 276–278, 280, 295, 308
Intrinsic pathway of, 272, 273
Nuclear DNA fragmentation in, 275
Role in cardiac pathology and aging, 271, 272, 330, 331
Stimulators of, 273, 276–280
Therapeutic interventions against, 280–283
Triggered by β-adrenergic agonists, 279
Triggered by calcium, 279
Triggered by mtDNA damage, 282, 283, 335, 338
Triggered by mitochondrial membrane permeabilization, 245, 274, 294, 296, 309, 314, 315
Apoptosis inducing factor (AIF)
Induction of apoptosis by, 274
Involvement in chromatin condensation, 276
Release from mitochondria, 274
Apoptosis repressor with caspase recruitment domain (ARC), 281
Apoptosome
Activation of caspase 9 by, 275
Components of, 275
Formation during ischemia, 275
Apoptotic protease activating factor 1 (APAF-1), 275
Aquaporin, 76
Aralar 1 & 2
Distribution of, 152, 153
Exchange of aspartate and glutamate by, 156
Involvement in adult type II citrullinemia, 153
Involvement in urea cycle, 156
Arginase, 156
ATP
Consumption during ischemia-reperfusion, 291–296, 308
Depletion by mitochondrial permeability transition, 243, 294, 295, 315
Regulation by F_1F_0 ATPase, 7, 9, 223, 296
Synthesis of (see oxidative phosphorylation)

Atractyloside and carboxyatractyloside, 246, 316

Bad, Bak and Bax
Activation by Bid, 273, 274
Cellular localization of, 274
Effect of cell stretching on, 274, 278
Effect of training on, 281
Elevated expression by pro-apoptotic stimuli, 274
Formation of heterodimers with Bcl-2, 274, 277
Interaction with heat shock protein, 281
Interaction with mitochondrial permeability transition pore, 249, 296
Mitochondrial effects of, 229, 274
Overexpression of, 294
Release of cytochrome c by, 273, 274, 277
Translocation to the mitochondria, 273, 277
Base excision repair
Involvement of AP endonuclease in, 332, 333
Involvement of AP lyase in, 332
Involvement of DNA ligase III in, 333, 337
Involvement of DNA polymerase γ in, 333, 336, 337
Bcl-2 and Bcl-$_{XL}$
Association with mitochondria, 274, 296
Attenuation of ischemic injury by, 281, 295, 296
Cellular localization of, 274
Changes during development, 274
Effect on mitochondrial calcium uptake, 296
Effect on mitochondrial permeability transition, 296
Interaction with Bax and Bak, 274
Interaction with voltage dependent anion channel, 296, 298
Modulation of ATP consumption by, 296
Reduction of apoptosis by, 281, 291, 292, 296
Reduction of necrotic oncosis by, 295, 296
Bid
Activated by caspase 8, 273
Activation of Bax and Bak by, 273, 274
Bongkrekic acid, 153, 229, 246, 247
Branched chain keto-acid dehydrogenase complex
As rate limit step in leucine oxidation, 135, 138–141
Effect of acidosis on, 142, 143
Effect of dietary proteins on, 136
Effect of deficiency, 121
Effect of liver cirrhosis on, 141–143
Phosphorylation of, 136, 137

Reaction of, 121, 135
Regulation of, 135–137

Calcium
  As regulator of citric acid cycle flux, 12, 14, 197, 200, 224
  As regulator of pyruvate dehydrogenase, 130, 209
  Involvement in cell signaling, 204, 205
  Mitochondrial uptake of, 198
  Transients of, 207, 213
Cancer, 134, 135, 161, 162, 331, 335, 336, 338
Cariporide, 255
Carnitine, 35, 44–46
Carnitine palmitoyl transferases
  Effect of diabetes on, 45
  In hypertrophied heart, 44
  Isoforms of, 35, 36
  Localization of, 35
  Reactions of, 37, 39
  Regulation of, 35, 41, 44, 45, 71, 78
Caspase 9
  Activation by apoptosome, 275, 276
  Inhibitor of, 281
  Involvement in caspase 3 activation, 275
Caspases
  As initiators and effectors of apoptosis, 275
  Inhibitors of, 281
Catalase
  Effect on lifespan, 193
  Localization of, 190, 327
  Reaction of, 190
CD36/fatty acid translocase
  Function of, 31, 33
  Levels in hypertrophied heart, 44
  Regulation of, 32, 43
Ceramide, 280
CG137157, 198, 206, 207, 210–214
Chylomicrons, 29, 68, 69
Citrate synthase, 78, 134
Citrate transport protein
  As a homodimer, 110–113
  Role in lipid metabolism, 97
  Substrate binding to, 101–110
  Substrate translocation pathway of, 108–110
  Three dimensional model of, 105–113
  Transport kinetics of, 111
  Water accessible regions of, 99–105
Citric acid cycle (tricarboxylic acid cycle)
  Localization of, 3
  Regulation of, 11–14
Clonazepam, 198, 206, 207, 210, 211, 213
Complex 1
  Deficiency of, 213, 214
  Formation of superoxide by, 186, 191
  Phosphorylation of, 17
  Reactions of, 4, 5, 9
  Reduction of Doxorubicin by, 187
  Role in mitochondrial permeability transition, 251
  Role in Parkinson's disease, 191, 293
  Structure of, 6, 323
Complex II, 8, 186
Complex III
  Control coefficient of, 16
  Formation of superoxide by, 158, 186
  Reactions of, 5, 6, 186
  Structure of, 5, 6, 323
Complex IV
  Adenine nucleotide regulatory sites of, 17
  Phosphorylation of, 18
  Reactions of, 6, 186
  Structure of, 6, 323
Complex V
  Effect of ischemic preconditioning on, 7
  Flux control coefficient of, 16
  Reaction of, 7
  Structure of, 7
  Synthesis of ATP by, 7
Creatine phosphate cycle, 8
Cyclophilin D
  Encoded by, 244
  Interaction with adenine nucleotide translocase, 246, 247
  Peptidyl prolyl cis trans isomerase activity of, 244
  Role in mitochondrial permeability transition, 244–246, 248, 251, 296, 297
Cyclosporin A, 211, 229, 242, 244, 253, 254, 294, 316
Cysteine sulfinate, 152
Cytochrome b, 5, 6, 186, 323
Cytochrome c
  As component of apoptosome, 275
  Involvement in caspase activation, 275
  Reactions of, 186
  Release from permeabilized mitochondria, 229, 273, 274, 277, 280, 283
  Reoxidation by superoxide, 186
  Role in electron transport, 5, 6, 186
  Role in necrotic oncosis, 294
Cytochrome c oxidase, 6, 129, 185, 191, 192

Diabetes
  Effect on gluconeogenesis, 28, 29
  Effect on PPAR-α and PGC-1α, 45, 81
  Formation of reactive oxygen species in, 191, 277

Mitochondrial mutations in, 191, 330
PPAR-α and PPAR-γ overexpression as
model of, 79, 81, 82
Stimulation of fatty acid oxidation in, 44,
45, 70, 71, 126, 127
Dichloroacetate, 48, 123, 129, 130–133
Dihydroorotate dehydrogenase, 187
Diltiazem, 198, 206, 210, 213
DNA polymerase γ
dRP lyase activity of, 336, 337
Involvement in aging, 330
Involvement in DNA repair, 328, 336
Oxidative damage of, 337
Replication of mitochondrial DNA by,
324, 336
Doxorubicin (Adriamycin), 187, 252, 280, 326

Electron transfer flavoprotein (ETF), 8, 9, 11
Electron transport chain (respiratory chain)
Effect of reduced flux, 190, 293
Formation of superoxide by, 185, 186,
190–192, 283, 329
Localization of, 3
Reactions of, 3, 5–9, 13
Regulation of, 9, 10, 13–18, 292
Role in Friedreich's ataxia, 190
Endonuclease G
DNA fragmentation by, 276
Induction of apoptosis by, 274
Release from mitochondria, 274
Energy charge, 15
Ethyl pyruvate, 133, 134
Excitatory amino acid transporter (EAAT),
153, 160

Fatty acid binding proteins, 32–34, 44
Fatty acid oxidation
As activator of PPAR-α, 67, 71, 78
As primary energy source in the heart, 27,
70, 71
ATP synthesis by, 81
Effect of cardiac work on, 41
Effect of diabetes on, 44, 45
Effect of glucose metabolism on, 41, 48
Effect of ischemia-reperfusion on, 47, 71
Effect of sepsis on, 131
Effect of trimetazidine, ranolazine and
dichloroacetate on, 48
In hypertrophied and failing heart, 44
Modulation of pyruvate dehydrogenase by,
126, 129
Reactions of, 27, 28, 39, 40
Regulation of, 39–43
Fatty acid transport protein, 32, 33

Fatty acid uptake
By diffusion, 30
Regulation by PPAR-α, 45
Fatty acyl CoA synthetase (ACS)
Function of, 34
Isoforms of, 34
Localization of, 34
Regulation of, 42, 43
Fenton reaction, 188, 190
Fibrates
Effect on triglycerides, 74
Interaction with PPAR-α, 67–69
Flow adaptation
Cell signaling initiated by, 169, 170
Change in mitochondrial shape during, 176
Effect on adhesion, 170
Effect on microtubular structure, 173, 174
Hyperpolarization of endothelial cells during,
173, 178
Induction of connexin 43 and adhesion
proteins during, 170
Modulation of ATP during, 178
Reorganization of cell surface during, 170
Friedreich's ataxia, 190

GC1 & GC2
As a glutamate proton symporter, 153
Distribution of, 153
Modulation of amino acid deamination
by, 153
Glibenclamide
Effect on ischemic preconditioning, 224,
226, 310, 311
Effect on K_{ATP} channel, 226, 310–312
Gluconeogenesis
Carbon sources for, 129
Effect of sepsis on, 131
Effect of starvation and diabetes on, 129
Glucose metabolism
ATP synthesis by, 81
Effect of PPARs on, 64, 65, 71, 72
Effect of trimetazidine, ranolazine and
dichloroacetate on, 48
In normal and hypertrophic heart, 27, 160
Modulation by fatty acid oxidation, 41
Regulation of, 12, 41, 45
Glutamic dehydrogenase, 124, 161
Glutamate-oxaloacetate transaminase
Effect of cancer on, 161
Involvement in malate-aspartate shuttle,
155, 156
Glutaminase
Effect on hepatomas, 161
Reaction of, 123, 124

Index 353

Regulation of, 124, 125
Role in ammonia excretion, 124
Glutamine
　As source of ammonia shuttle, 122, 123
　As substrate in cancer, 161, 162
　Biosynthesis of, 122, 123, 135
　Involvement in ammonia detoxification, 123, 124
　Release by skeletal muscle, 123
　Transport of, 122–124, 154
Glutamine synthetase
　Effect of pyruvate dehydrogenase on, 123
　Reaction of, 122
　Regulation of, 122
Glutathione, 124, 190
Glutathione peroxidase, 124, 189, 327
Glutathione reductase, 189
Glycerol-3-phosphate dehydrogenase, 9, 186
Glycerophosphate shuttle, 9
Glycogen synthase kinase 3B
　Cardioprotective effect of, 227, 297

Heart
　Effect of afterload on, 282, 283
　Effect of elevated work on, 10, 14, 200, 201, 204
　Effect of angiotensin II antagonists on, 280
　Effect of anti-apoptotic agents on, 281–283
　Effect of fasting on, 80
　Effect of PPARs on, 65, 67, 70–72, 77, 81, 82
　Failure of, 81, 272, 277, 280–283
　Hypertrophy of, 44, 159, 160
　Initiation of apoptosis during ischemia of, 280, 295–297, 308–310
　Regulation of PGC-1α in, 79–81
Heat shock proteins, 281, 282
Htr/Omi
　Involvement in caspase activation, 274, 276
　Neutralization of inhibitors of apoptosis by, 276
　Release from permeabilized mitochondria, 274
3-Hydroxyacyl CoA dehydrogenase, 9
5-hydroxydecanoate
　Effect on ischemic preconditioning, 227
　Effect on mitochondrial K_{ATP} channel, 226, 227, 307, 310–312
Hydrogen peroxide, 185, 189, 192, 277, 327
Hyperglycemia
　Effect of antioxidants on, 278
　Elevation of reactive oxygen species by, 191, 278
　Initiation of apoptosis by, 191, 277

Inhibitors of apoptosis (IAPs), 276
Insulin
　Effect of PPARs on, 75, 81–83
　Effect of TNF-α on, 76
　Obesity-induced resistance of, 79–83
　Stimulation of triglyceride storage by, 67, 75
Ischemia-reperfusion (hypoxia-reoxygenation)
　Accumulation of inorganic phosphate during, 294
　Calcium overload and injury by, 51, 158, 197, 208, 209, 244, 251, 253–255, 258, 291, 292
　Cardioprotection of calcium antagonists during, 2, 10, 211, 213, 214, 255
　Complex 1 damage during, 293
　Depletion of adenine nucleotides by, 158, 208, 251, 292
　Dissociation of Bax-heat shock protein complex by, 281, 282
　Effect of amino acids on, 158, 159
　Effect of Bcl-2 on, 296
　Effect of caspase inhibition on, 281
　Effect of K_{ATP} channel agonists on, 225–227
　Effect of magnesium on, 255
　Effect of PPARs on, 71
　Effect of sodium-hydrogen exchange inhibitors on, 255, 292
　Effect on fatty acid metabolism, 28, 29, 47–49
　Effect on triglyceride lipase, 29
　Elevation of mitochondrial calcium by, 209, 211–213, 292, 309
　Elevation of reactive oxygen species during, 133, 191, 192, 251, 277, 294, 327
　Fatty acid metabolism during, 41, 46–48
　Initiation of apoptosis by, 295–298, 308
　Initiation of the mitochondrial permeability transition by, 252–257, 296, 308–310
　Mitochondrial membrane potential collapse during, 294, 309
　Mitochondrial swelling during, 309, 310
　Potential therapy for, 158, 159, 255–257, 280–283, 316
Ischemic preconditioning
　Effect of K^+ channel openers on, 227
　Effect on calcium overload, 213, 225, 258, 313
　Effect on free radical production, 227, 258, 313–315
　Effect on osmotic swelling, 313, 315
　Inhibition of F_1F_0 ATPase by, 7
　Initiation by chemical preconditioning agents, 310–313

Involvement of G-protein coupled receptors in, 305, 306
Mimicked by diazoxide, 310–313
Of the heart, 305
Role of Akt and PI-3 kinase in, 257, 258, 297, 298, 308
Role of ERK in, 297, 298, 308
Role of glycogen synthase kinase 3B in, 257, 297, 298
Role of K_{ATP} channel in, 227, 257, 297, 298, 306, 310–313
Role of mitochondrial permeability transition during, 253, 257, 258, 294, 297, 307, 308, 313–315
Role of nitric oxide synthase in, 306
Role of protein kinases C in, 257, 297, 306, 307
Role of protein kinase G in, 257, 298, 306, 314
Role of reactive oxygen species in, 257, 306, 307, 309, 311, 313
Stimulation of guanylyl cyclase by, 306
Translocation of connexin 43 during, 298
Isocitrate dehydrogenase, 12

Kearns-Sayre syndrome, 4
Ketogenesis, 72, 73
α-ketoglutarate dehydrogenase
Formation of superoxide by, 187, 327
Regulation of, 187
Kidney
As site of gluconeogenesis, 117, 129
Effect of acidosis on, 123, 124
Maintenance of acid-base balance by, 123

Lactate
As source of gluconeogenesis, 132
Link with pyruvate dehydrogenase, 132
Production by skeletal muscle, 121, 132
Leber hereditary ophthalmic neuropathy, 5
Leptin, 76
Leucine
Activation of mTOR by, 138–141
In dietary protein, 135
Oxidation of, 140, 141
Regulation of protein synthesis by, 136, 138–141
Leukotriene D_4, 67
Lipoic acid, 190
Lipoprotein lipase
Activity in cardiac muscle, 29, 82
Association with glycocalyx, 29
Link to PPARα activation, 69, 74, 82
Reaction of, 29, 30

Lipotoxicity
In ischemia, 48
Initiation of apoptosis, 277
Liver
Amino acid uptake by, 123, 124
As site of ammonia detoxification, 123
As site of gluconeogenesis, 117, 129, 131
Biosynthesis of glutamine by, 123
Effect of acidosis on, 124, 125
Effect of fibrate on, 74
Effect of PPAR-α on, 72–74, 80–83
Encephalopathy of, 141, 142
Oxidation of branched chain amino acids by, 136
Liver X receptor (LXR)
Activated by oxidized cholesterol, 63
Regulation of cholesterol homeostasis by, 64

Malate-aspartate shuttle
Reactions of, 153, 155, 156, 160
Regulation of, 160
Transport of NADH by, 8, 9, 155, 156
Malate dehydrogenase, 12, 155
Malonyl CoA
Effect of ischemia on, 47
Effect of PGC-1α on, 80
Inhibition of carnitine palmitoyl transferase by, 36
Regulation of fatty acid oxidation and ketogenesis by, 11, 36, 37
Synthesis and degradation of, 11, 37–40
Malonyl CoA decarboxylase (MCD)
Effect of diabetes on, 45
Localization of, 38
Regulation of, 37, 38
Role in fatty acid oxidation, 38, 39
Maple syrup urine disease, 137, 138
Mechanotransduction hypothesis, 169–179
Mitochondria
Antioxidant defenses of, 188–190
As energy source for apoptosis, 294, 295, 308
As mechanotransducers, 171–178
cAMP content of, 17
Content in flow adapted endothelial cells, 174–178
Effect of flow on, 173–179
Matrix volume of, 11, 221, 224, 310
Myocardial populations of, 4
Pro-apoptotic factors of, 277–280
Regulation by Bcl-2 family, 281
Role in heart failure, 280
Shape changes of, 176

Superoxide formation by, 185, 186, 190, 191, 313
Swelling of, 223, 226, 242, 309
Mitochondrial ATP-dependent potassium channel (K_{ATP})
 Activation of, 225–227, 298, 310–313
 As trigger of ischemic preconditioning, 311–313
 Cytoprotective activity of, 225, 226, 309–311
 Effect on K^+ and H^+, 311, 312
 Inhibition of, 225–227, 311–313
 Involvement in cell signaling, 305–317
 Involvement in ischemic preconditioning, 257, 258, 297, 298, 311–313
 Mitochondrial membrane swelling by, 309, 312
 Modulation of mitochondrial calcium by, 313
 Modulation of mitochondrial matrix pH by, 309
 Modulation of mitochondrial membrane potential by, 309
 Modulation of reactive oxygen species by, 309, 311, 313
 Regulation of, 312
Mitochondrial calcium
 As determinant of cell viability, 209–212
 Effect of acidosis on, 211
 Effect of cariporide on, 255, 256
 Effect of cell beating on, 206, 207
 Effect of cell stretch on, 278
 Effect of gender on, 209
 Effect of hypoxia-reoxygenation on, 208, 209, 211–213, 292, 309
 Effect of ischemic preconditioning on, 213, 258, 279, 313
 Effect of ruthenium red on, 255
 Effect on pyruvate dehydrogenase, 209
 Measurement of, 201–203
 Physiological role of, 197–201
 Regulation of apoptosis by, 279
 Release of, 223
 Role in apoptosis-linked heart failure, 280
 Role in ischemic-reperfusion injury, 241, 254, 256, 291, 292, 308–310
 Role in mitochondrial permeability transition, 254
 Transient of, 213
Mitochondrial calcium uniporter
 Effect of hypoxia on, 209, 211–213, 223
 Inhibition of, 198
 Reaction of, 198, 199, 222–224
 Role in contraction, 206, 207, 226
 Role in reduction of $[Ca^{2+}]_i$
Mitochondrial DNA (mtDNA)
 Damage, initiation of apoptosis by, 282, 283
 Effect of mutagens and carcinogens on, 325, 326
 Genes encoded by, 323
 Involvement in diabetes, 330
 Involvement in Kearns-Sayre syndrome, 4
 Involvement in malignancies, 331
 Mutation frequency of, 4, 330, 331
 Oxidative damage to, 279, 326–329
 Replication of, 324, 325
 Structure of, 323–325
 Turnover of, 323, 324
Mitochondrial DNA binding proteins, 325
Mitochondrial DNA glycosylases
 Abasic site lyase activity of, 332
 Cleavage of N-glycosidic bonds by, 332
 Isoforms of uracil DNA glycosylase, 334
 Role in DNA repair, 338
Mitochondrial DNA ligase, 337
Mitochondrial DNA repair
 By base excision repair, 331–333
 Protection from oxidative stress by, 281, 282, 332
 Relationship to apoptosis, 282, 283, 338
Mitochondrial inner membrane anion channel (IMAC)
 Effect of benzodiazepines on, 231
 Effect of oxidative stress on, 230, 231
 Effect on mitochondrial depolarization, 230, 231
Mitochondrial K_{Ca} channel
 Effect of openers, 227
 Inhibition of, 227
 Protection against ischemia-reperfusion injury, 227, 228
 Similarity to Big K^+ type K_{Ca} channel, 228
Mitochondrial K^+/H^+ exchanger, 310, 312
Mitochondrial membrane potential ($\Delta\psi_m$)
 Alteration by shear-stress, 173, 178
 Collapse of, 223, 229, 294, 309
 Effect of calcium on, 197, 199, 208, 209, 229, 249–251
 Effect of colchicine on, 173
 Effect of reactive oxygen species on, 229, 231, 244, 249–251
 Modulation by hypoxia, 211, 252, 294
 Modulation by mitochondrial permeability transition, 211, 229, 230, 309, 310, 314–317
Mitochondrial permeability transition
 Collapse of mitochondrial membrane potential by, 230, 244, 294, 296, 309, 315, 316

Cytochrome c release by, 191, 243, 275, 296, 314
Effect of acidosis on, 243, 315
Effect of adenine nucleotides on, 209, 229, 230, 244, 246, 294, 315
Effect of glycogen synthase kinase 3B on, 257
Effect of postconditioning on, 316
Effect of ryanodine on, 207
Inhibition by bongkrekic acid, 229, 255
Inhibition by Cyclosporin A and sanglifehrin, 229, 230, 246, 254
Initiation by calcium, 243, 244, 248, 250, 251, 279
Initiation by ischemia-reperfusion, 251–253, 294, 309, 315–317
Initiation by reactive oxygen species, 185, 186, 208, 244, 246, 247, 309, 315
Mitochondrial swelling during, 243, 315
Model of, 249–251
Regulation of, 229, 230, 247, 250, 251, 255–257, 279, 315
Reversal of F_1F_0 ATPase by, 243, 294, 296, 298
Role in apoptosis and necrotic oncosis, 296
Role in Doxorubicin toxicity, 252, 280
Role in ischemia-reperfusion injury, 294, 296, 309
Role in ischemic preconditioning, 257, 258, 314–317
Rupture of outer mitochondrial membrane during, 274, 315
Structural elements involved in, 230, 232, 246–251, 294, 314
Uncoupling of oxidative phosphorylation by, 243, 315
Mitochondrial K^+ channels, 224–228
Monoamine oxidase, 187, 191
MTH 1
Effect of deficiency, 336
Reaction of, 336
mTOR
Regulation of, 138–141
Role in protein synthesis, 138
MYH, 335
Myocardial hypertrophy
Effect on fatty acid oxidation, 44, 159
Effect on glucose metabolism, 160
Myocyte stretch, 278

Necrotic oncosis
ATP depletion during, 294, 296, 308
Effect of cell and mitochondrial swelling during, 292, 308
Initiation during ischemia-reperfusion, 291, 292, 295, 308
Nitric oxide as initiator of apoptosis, 278
Nitric oxide cGMP kinase, 227, 306
Nitric oxide synthase, 188, 192, 279

Osmotic stress, 278
Oxidative phosphorylation
Effect of calcium and ADP on, 14, 15, 199, 200, 224
Effect of elevated cardiac work on, 15
Effect of IF_1 on, 7
Effect of NADH on, 200
Proton gradient as driving force of, 6, 7, 221, 224, 312
Synthesis of ATP by, 3, 4, 6, 7
Oxidative stress
Cause of, 190
Damage to DNA by, 326–329
Detoxification by glutathione, 124
During hyperoxia, 192
Involvement in mitochondrial permeability transition, 208, 244, 246, 247
Link to apoptosis, 272
Production during ischemia-reperfusion, 158, 159
8-oxoguanine
Analysis of, 327, 329
As oxidative lesion of mitochondrial DNA, 328, 335
Cause of GC-to-TA transversions, 328
Effect of aging on, 334, 335
Repair of, 334, 335
8-oxoguanine glycosylase
Effect on survival, 334, 335, 338
Isoforms of, 334
Localization of, 334
Repair of 8-oxoguanine by, 332, 334

Parkinson's disease, 191, 213, 330
Peroxiredoxin, 189, 190
Peroxisome proliferators-activated receptors (PPARs)
As lipid sensors, 42, 64
Effect of specific response elements on, 67
Facilitation of fatty acid metabolism by, 41, 42, 63, 71, 75–77
Isoforms of, 42, 43
Recruitment of co-repressors by, 65–67
Structure of, 65,66
Peroxynitrite, 188
PGC-1α
Activation of PPARs by, 66, 79, 80
Activation by fatty acids, 68, 69

Distribution of, 79
Effect of starvation on, 80
Induction of, 79, 80
Regulation of fatty acid and glucose metabolism by, 67, 80, 81
Regulation of mitochondrial biogenesis by, 80, 83
Phosphate/OH$^-$ exchanger, 14
Phosphofructokinase, 12, 41
Postconditioning, 258–260, 316
PPAR-α
  As lipo-oxidative transcription factor, 64
  As target of fibrate and heptocarcinogens, 64, 67, 74
  Distribution of, 64
  Effect of downregulation, 42
  Effect on heart and liver, 69, 72–74, 77, 81, 82
  Modulation of pyruvate dehydrogenase and branched chain keto-acid
dehydrogenase by, 128, 129, 136
  Modulation of lipid metabolism by, 64, 69–71, 77, 78, 83, 129
  Modulation of pyruvate dehydrogenase kinase by, 71, 78, 129
  Regulation of, 77, 78
PPAR-δ (gPPARβ/γ)
  Activation of, 69
  Distribution of, 65
  Effect of deficiency of, 77
  Effect on adipocyte, 78,79
  Regulation of fatty acid oxidation by, 42, 65, 68, 77, 78
  Regulation of obesity and insulin resistance by, 76, 79
PPAR-γ
  As lipogenic transcription factor, 65
  As target of thiazolidinediones, 43, 67, 68
  Co-activators of, 75
  Distribution of, 42, 65, 75
  Effect on adipocyte, 75–77
  Effect on angiotensin II-induced hypertrophy, 42, 82
  Effect on insulin and leptin signaling, 68, 75, 76
  Phosphorylation of, 67
  Regulation of fatty acid oxidation by, 42, 43, 77
  Regulation of glucose metabolism by, 75–77
  Regulation of glyceroneogenesis and lipogenesis by, 65, 75–77
  Role in lipotoxicity, 82
  Variants of, 75

Propofol
  Anesthetic activity of, 256
  Free radical scavenger activity of, 256
  Inhibition of mitochondrial permeability transition by, 256
Protein kinase C
  Role in diabetes, 191
  Role in ischemic preconditioning, 227, 253, 257, 297, 306, 307, 314
Pyruvate
  Attenuation of mitochondrial permeability transition by, 256
  Effect of ATP/ADP and NADH, 200
  Effect on ischemia-reperfusion injury, 256
  Free radical scavenger activity of, 133, 256
  Oxidation to acetyl CoA, 121, 125
  Reduction to lactate, 121, 125, 132, 155
  Regulation of pyruvate dehydrogenase kinase by, 133
  Transamination to alanine, 121, 123
  Transport into the mitochondria, 121
Pyruvate dehydrogenase
  As determinant of glucose oxidation, 121, 125, 126
  As determinant of pyruvate metabolism, 121, 125–127
  Effect of ischemia on, 46–48
  Effect of sepsis on, 132
  Effect on gluconeogenesis, 129, 132, 133
  Effect on glutamine production, 123
  Reaction of, 40, 121, 123, 125
  Regulation of, 10, 41, 125–129, 209
Pyruvate dehydrogenase kinase
  As target of PPAR-α, 69, 71, 72, 78, 80, 128, 129
  Distribution of, 129
  Effect of diabetes and starvation on, 72, 126–129
  Effect of fat feeding on, 71, 72
  Inactivation of glucose metabolism by, 71
  Inhibition of pyruvate dehydrogenase by, 80, 125, 126
  Isoforms of, 72, 127, 128
  Modulation of fatty acid oxidation by, 128
  Regulation of, 78, 126–130
Pyruvate dehydrogenase phosphatase
  Distribution of, 130
  Isoforms of, 130
  Regulation of, 130,131

Ragaglitizar, 68
Ranolazine, 48, 209
Reactive oxygen species (ROS)
  Antioxidant therapy against, 278, 281

As product of ischemic preconditioning, 229, 309
Damage of macromolecules by, 189, 313, 327
Dependence on oxygen levels, 191
Effect of ischemic preconditioning on, 227, 257, 258, 313
Effect on mitochondrial membrane potential, 229, 231, 244, 249–251
Elevation by mitochondrial DNA damage, 323–329
Formation of, 185–187, 190, 309, 313
Generation by angiotensin II, 278
Generation by Doxorubicin, 280
Generation by electron transport chain, 185, 186, 190–192, 283, 293
Generation by α-ketoglutarate dehydrogenase, 187, 327
Generation by monoamine oxidase, 187, 327
Initiation of apoptosis by, 277
Initiation of the mitochondrial permeability transition by, 185, 186, 208, 244, 246, 247, 309, 315
Involvement in cell signaling, 305–317
Involvement in ischemia-reperfusion injury, 158, 159, 291
Involvement in metal-mediated pathology, 190, 191, 327
Role in aging, 192, 193
Role in carcinogenesis, 331
Role in diabetes, 191, 277
Role in heart failure, 277, 278
Role in hypoxia and hyperoxia, 191, 192, 251, 277
Role in Parkinson's disease and Wilson disease, 191
Translocation of Bax by, 277
Retinoid X receptor (RXR)
  Activation of, 67
  Distribution of, 67
  Interaction with PPARs, 65–67
  Isoforms of, 67
RU360, 198, 201,202, 205, 206, 210, 213
Ruthenium red, 198, 202, 204, 207, 210, 211, 213

Sanglefehrin, 230, 246, 254
Sarcolemmal K_{ATP} channel, 225
Sepsis, 131–133
Shear-stress
  Effect on mitochondrial content, 174–176
  Effect on mitochondrial membrane potential, 173, 178
  Effect on transcriptional activation, 170, 171
  Modulation of cytoskeleton during, 170, 172
Skeletal muscle
  As source of gluconeogenic substrates, 117, 119, 120
  Effect of acidosis on, 131, 142, 143
  Effect of PGC-1α on, 80
  Effect of PPARs on, 65, 78
  Effect of sepsis on, 131, 132
  Expression of PPARs in, 80
  Oxidation of amino acids by, 118
  Synthesis of alanine and glutamine by, 118, 119
  Utilization of ketone bodies by, 72
Smac/Diablo
  Involvement in caspase activation, 274–276
  Neutralization of inhibitors of apoptosis by, 276
  Release by permeabilized mitochondria, 274
Sodium-calcium exchange
  Effect of hypoxia on, 46, 210–213, 255, 292
  Efflux of calcium from the mitochondria via, 198, 212, 213, 223, 296
  Inhibition of, 210, 211
  Reaction of, 243, 255
Sodium-hydrogen exchange
  Effect of ischemia on, 46, 255, 256, 292
  Inhibition of, 255, 292
  Reaction of, 256
Sulpho-N-succinimidyl oleate, 31, 33
Superoxide
  As precursor of peroxynitrite, 188
  Cytochrome c oxidation by, 189
  Dismutation of, 187–189
  Formation of, 185–188, 283, 327
  Reduction of ferric iron by, 188
Superoxide dismutase
  Effect of aging on, 192
  Induction of, 189, 193
  Localization of, 186, 189, 327
  Reduction of superoxide by, 188, 189, 327

Taurine, 280
Thiazolidinediones
  Effect on hyperglycemia and hyperlipidemia, 68, 82
  Interaction with PPAR-γ, 43, 67, 68
Thioredoxin, 189
Thiolase, 11
Thyroid hormone, 160
TNF-α
  Effect of PPAR-γ on, 76
  Role in insulin resistance, 76
Trifluoperazine, 251
Triglyceride lipase

Effect of diabetes on, 35
Effect of ischemia on, 28, 29, 35
Regulation by catecholamines, 29, 35
Trimetazidine, 48

Ubiquinone
  As source of reactive oxygen species, 186, 190, 251
  Location of, 5
  Reduction of, 5, 190
  Scavenger of reactive oxygen species, 190
UCP3
  Effect on fatty acid oxidation, 40, 41
  Regulation by PPARs, 41, 42, 78
Urea cycle, 123, 125, 154, 156, 157

Very low density lipoproteins (VLDL)
  Effect of pyruvate dehydrogenase on, 129
  Receptor of, 33
  Source of fatty acids and lipids, 25, 35, 69
Voltage dependent anion channel (VDAC)
  As component of mitochondrial permeability transition pore, 249, 251, 271, 294, 296, 298

Wilson disease, 191

Printed in the United States of America